Waspan

Waspuk River

Mosquitia

Wawa River

Miskito Keys

● Bonanza

● Rosita

Cerro Saslaya
▲
Cerro El Toro ▲ ● Siuna

Mosquitia

Prinzapolka River

Puerto Cabezas

CARIBBEAN SEA

Cerro Musún ▲
● **Río Blanco**

Grande de Matagalpa River

Kurinwas River

Pearl
Lagoon

Pearl Keys

Corn Islands

El Rama ●

Escondido River
Bluefields ●

● Santo Tomás

Nueva Guinea ●

Punta Gorda River

San Miguelito
Wetlands

San Carlos ●

Sabalos ●
● El Castillo

San Juan del Norte ●

Guatuzo Plains

Costa Rica

San Juan River

Birds of Nicaragua

A FIELD GUIDE

LILIANA CHAVARRÍA-DURIAUX
DAVID C. HILLE
Illustrated by
ROBERT DEAN

A Zona Tropical Publication

FROM

Comstock Publishing Associates

an imprint of

Cornell University Press

Ithaca and London

Illustrations and Photo Credits

Front cover: male and female Elegant Euphonia (*Euphonia elegantissima*), Robert Dean
Back cover, top to bottom: female Three-wattled Bellbird (*Procnias tricarunculatus*),
Jabiru (*Jabiru mycteria*), male Three-wattled Bellbird, Robert Dean

p. ii: Rufous-vented Ground-Cuckoo (*Neomorphus geoffroyi*), Robert Dean
p. vi: King Vulture (*Sarcoramphus papa*), Jorge Chinchilla
p. 9: photos, Georges Duriaux
p. 11: photos, David C. Hille
p. 12: photo (left), David C. Hille; photo (right), Georges Duriaux
p. 13: photo (left), David C. Hille; photo (right), Georges Duriaux
p. 24: Great Green Macaw (*Ara ambiguus*), Glenn Bartley

First published 2018 by Cornell University Press

Printed in China

Library of Congress Cataloging-in-Publication Data

Names: Chavarría-Duriaux, Liliana, author. | Hille, David C., 1981– author. | Dean, Robert, 1955– illustrator.
Title: Birds of Nicaragua : a field guide / Liliana Chavarría-Duriaux, David C. Hille ; illustrated by Robert Dean.
Description: Ithaca : Comstock Publishing Associates, an imprint of Cornell University Press, 2018. | "A
    Zona Tropical publication." | Includes bibliographical references and indexes.
Identifiers: LCCN 2017045660 (print) | LCCN 2017046339 (ebook)
    ISBN 9781501709500 (pdf) | ISBN 9781501701580 | ISBN 9781501701580 (pbk. ; alk. paper)
Subjects: LCSH: Birds—Nicaragua—Identification.
Classification: LCC QL687.N5 (ebook) | LCC QL687.N5 C43 2018 (print) | DDC 598.097285—dc23
LC record available at https://lccn.loc.gov/2017045660

Zona Tropical Press ISBN 978-0-9894408-8-2

Book design: Gabriela Wattson

To my husband, Georges Duriaux, who was at my side on expeditions and as I did research and measurements work. And to my son, Jean-Yves.

Liliana Chavarría-Duriaux

To Sarah, Adele, and Henry, my constant sources of love and encouragement.

David C. Hille

To all the conservationists working tirelessly to protect birdlife in the Neotropics. They are needed now more than ever.

Robert Dean

Contents

Foreword

I lived, worked, and birded for more than seven years in Mesoamerica, but none of the places I traveled to holds as dear a place in my heart as Nicaragua. The reasons are many.

Nicaragua is the largest country in Central America and contains a wealth of intriguing, diverse habitats. Here, the montane and lowland pine forests of North America reach their southern limit, and you will find one of the largest freshwater lakes in the Americas, expanses of lowland rainforest, river systems that wend through tropical moist gallery and swamp forests, Pacific scrub and deciduous forest, semi-deciduous and evergreen montane forest, epiphyte-draped cloud forest, and large bays, beaches, and mangroves along both the Pacific and Caribbean coasts. These diverse habitats are home to a great variety of stunning resident and passage birds, a full 763 species by the latest count.

Yet, arguably, Nicaragua's avifauna remains the least explored in Mesoamerica. Despite a long history of influential naturalists and ornithologists like Thomas Belt, Ludlow Griscom, and Fr. Bernardo Ponsol, the first published, annotated country checklist didn't arrive until 2010, and no illustrated field guide existed until 2014! Now, this new field guide, comprehensive and richly illustrated, gives birders—new and returning—the means to add species to the evolving national list (I predict an easy dozen new birds within the next few years). There are also many species recorded in Nicaragua about which very little is known. For example, what *is* the actual distribution of the near-endemic Tawny-chested Flycatcher (*Aphanotriccus capitalis*) in Nicaragua? How many Rufous-vented Ground-Cuckoos remain in its forests, and where are they? What are the seasonal movements of Jabiru between the vast Mosquitia region and the lakeshore marshes of Lago Cocibolca (Lake Nicaragua)? With the help of this guide—and the thousands of eBird checklists that will result—many of these avian mysteries should be resolved.

Even today, there are entire sections of the country that remain little explored. These include the hilly region just northeast of the Gulf of Fonseca, the vast Bosawás UNESCO Biosphere Reserve and Cerro Saslaya in the north, and the biologically diverse Indio-Maíz Biological Reserve in the southeast, to name a few. I vividly remember the awe I felt canoeing just south of San Juan del Norte, slowly moving through the pristine freshwater lagoons, while surrounded by forested hills, each lagoon replete with wintering waterfowl and accessed via small connecting channels whose overarching tree branches were festooned with iguanas falling into the water, all to the raucous accompaniment of mysterious bird calls. It is without question one of the most beautiful places I've ever visited. While that was in 1991, and there has been much change since, adventurous naturalists may still find similar awe-inspiring landscapes.

Finally, the most important reason Nicaragua is a special place to enjoy birds is the Nicaraguan people themselves. Despite the fascinating cultural diversity of the country—Spanish speakers in the west, Caribbean English speakers in the east, and still many indigenous communities (Rama, Garifuna, Mayagna, and Miskitu) —there is nonetheless a recognizable national character that is distinctly Nicaraguan. Especially in the countryside, you will find people who are notably humble, generous, and down-to-earth. Good luck finding a campfire anywhere in the *campo* where Nicaraguans are not singing, with or (more often than not) without musical instruments. And if you ask a Nicaraguan whether he or she is a poet, the answer will inevitably be, "*¿Poeta? ¡Sí, por supuesto!*"

As you leaf through the pages of this guide, the colorful plumages depicted here are likely to excite the imagination, ignite the instinct to travel, and spur biological curiosity. I sincerely hope that this guide will not only aid in identifying birds but will also, through our ambassadors, the birds themselves, help connect international communities with those in Nicaragua and inspire folks to join forces with the growing number of Nicaraguan conservationists passionately dedicated to sustaining their natural legacy and their astounding birdlife.

—Tom Will
    Nongame Bird Coordinator, U.S. Fish and Wildlife Service

Acknowledgments

We are indebted to the many ornithologists and birders who, for nearly two centuries, have contributed to the knowledge about Nicaraguan avifauna. These include intrepid ornithologists of the early twentieth century, among them William B. Richardson, Dioclesanio Chaves, Waldron DeWitt Miller, Ludlow Griscom, Wharton Huber, and Bernardo Ponsol. Later in the century, Thomas R. Howell contributed invaluable specimen collections and writings.

From modern sources, we obtained a great deal of information from birders who have contributed sightings, photos, and audio recordings to eBird, the Macaulay Library, and xeno-canto, for which we are grateful. Keep sharing!

Museum data have fundamentally contributed to our work. We are grateful to Paul Sweet (Division of Vertebrate Zoology-Ornithology at the American Museum of Natural History), who created and shared a database of Nicaraguan specimen records that draws from sixteen museum collections. Some of these thousands of records have never been published elsewhere.

Terry Chesser and Tom Schulenberg graciously consulted with us on the latest taxonomical changes from the American Ornithological Society.

Needless to say, this book would not have been possible—at least in its current form—without the initial interest from John McCuen and Mark Roegiers at Zona Tropical Press, the spectacular design by Gabriela Wattson at Zona Creativa, and the guiding hand of Kitty Liu at Cornell University Press. A special thank you goes to John McCuen. His management skills, superb editing, and desire for perfection have made this a better book.

— The authors

The publication of *Birds of Nicaragua* was made possible thanks to many people who have helped in a variety of ways. Let's go back six years, in the early stages of this project. Thanks go to several people who believed in me and who encouraged me to work on this book: Tom Will, Oliver Komar, Rob Batchelder, and Stephen Paez. Special thanks go to John McCuen, our editor, who invited me to be part of the team. Thanks also to coauthor David Hille, whom it has been a great pleasure to work with. I was delighted to work with Robert Dean, whose magnificent illustrations grace these pages.

Special thanks go to Andrew Vallely (Field Associate, AMNH) for his generosity in sharing information, for answering my queries, and for accompanying me on two expeditions to Saslaya NP.

Expeditions to do research in remote areas of Nicaragua proved invaluable when it came time to write this book. Therefore I'm very grateful to Carlos Ramiro Mejía, from MARENA (*Ministerio del Ambiente y Recursos Naturales*), for granting the necessary permits to conduct research in all Nicaraguan protected areas. Thanks also to Atanasio Maldonado and Nachito Cruz, park rangers at Saslaya NP, who accompanied me on nine expeditions to the park. Moisés Siles and Wilmer Talavera, my field assistants, were vitally important on trips to collect data and do mist-netting.

Thanks also go to Juan Carlos Martínez-Sánchez, Osmar Arróliga, Manfred Bienert and Marvin Tórrez.

Finally, I'm grateful to my son, Jean-Yves, and my husband, Georges, for their patience as I spent so much time in front of the computer and sometimes left family matters to one side. Thanks to Georges for sharing my passion for birds, for helping me capture them in remote forests, and for organizing all the expeditions we made in Nicaragua. I also want to thank him for his constant support and encouragement, without which I wouldn't have been able to finish this marvelous project.

— Liliana Chavarría-Duriaux

I will never be able to express my full gratitude to my wife, Sarah, who believed in me from the very beginning, when this book was just a dreamer's notion, and who proceeded to be my mainstay for the long haul. Our children Adele and Henry played no small role in giving me encouragement and ample love, along with Monopoly Junior work breaks to clear my head. Their patience throughout my long work hours and my trips to Nicaragua still baffles me, and I am grateful for it. Thank you to all my family and friends, who provided love and support. Thank you also to the staff and faculty at the Oklahoma Biological Survey at the University of Oklahoma for supporting me in ways too numerous to list.

If we are lucky, we meet people who guide us toward the paths we end up taking or who walk with us along these paths. Dr. Leon Powers, of Northwest Nazarene University, enthusiastically introduced me to ornithology as a field of study. Once set down that path, I often found Scott Smithson and Matthew Strussis-Timmer at my side during some of my most formative birding years.

I am also grateful to Francisco (Chico) Muñoz, who joined me in the field for an incalculable number of hours while observing birds and exploring Nicaragua. It was a pleasure to bird with Georges Duriaux, who contributed in so many ways to our success. Bruce Mactavish, Orlando Jarquín, and Martín Vallecillo shared valuable records and other information. Jean Michel Maes (*Alianza para las Áreas Silvestres*) offered welcomed consultation in the beginning stages of this project.

Finally, I have talented and dedicated coauthors. Lili, thank you for your passion, energy, and love for birds and your desire to make this the best possible book. Robert, thank you for the insurmountable amount of work and care that is represented by your art.

— David C. Hille

I wish to thank the following people for being my support group during the fruition of this guide: Eduardo Amengual, David Rodriguez Arias, Patrick O'Donnell, Andrew Russell, Susan Blank, Lisa Erb, Lana Wedmore, Oscar Castillo, Cristian Chaves, Darren and Marie Mora, Simon Musselle, Angie Usher, Larry Landstrom, Ruth Rodriguez, Alex Villegas, Bitty Ramirez-Portilla, Adrian Forsyth, Alan and Karen Masters, and Daryl Loth.

My great appreciation goes to the two authors of this guide, David Hille and Liliana Chavarría-Duriaux, for their abundant knowledge of Nicaraguan avifauna and their assistance throughout the creative process.

— Robert Dean

Introduction

Wherever the roads, rivers, and trails of Nicaragua might take you, we trust that this book will enhance your ability to make correct bird id's. The number of birds in Nicaragua is impressively large—with 763 species confirmed to date. This wonderfully diverse group of birds resides in 77 avian families. This is the first field guide for Nicaragua that includes body measurements, country-specific range maps, key field marks for identification, comparisons with similar species, and vocalization descriptions. We hope that this book will contribute to an increase in birding activity in Nicaragua, which, in turn, would create more opportunities for ecotourism guides, increase the number of visitors to parks and reserves, and, ideally, spur conservation efforts. Nothing would delight us more than to see more people (of all ages and nationalities) begin to discover the avifauna of this country and, along the way, discover the country itself.

Family descriptions. For each family, we provide information on global distribution, general characteristics and anatomical features (as represented within Nicaragua) to help you identify birds to the family level, and relevant ecological and behavioral notes. And, we often give tips about how to distinguish among birds or groups of birds within the family.

Species name. Three names are given for each species, the English common name, Spanish common name, and scientific name. English common names and scientific names strictly follow the Checklist of North and Middle American Birds: 7th edition (updated through the 58th supplement), managed by the American Ornithological Society.

Body measurements. Measurements are listed first by inches (rounded to the nearest half inch) and followed in parenthesis by centimeters (rounded to the nearest centimeter). Total body length—the measurement from the tip of the bill to the tip of the tail—is listed for each species. When a bird has long, extended rectrices or tail coverts, the length of the extension is listed separately. If sexual dimorphism leads to noticeable differences in body size, male (M) and female (F) measurements are given separately. For birds that are most often seen soaring, the wingspan (WS) measurement is also listed.

Body measurements can be very helpful when making identifications. This is particularly the case when trying to distinguish between similar birds that are seen at the same time, or when using the size of a well-known species as a standard by which to judge a new bird. However, consider that the body length measurement can be misleading when a bird has a relatively long bill or tail. And, the perceived size of a bird is also influenced by leg length, body form, and weight.

Field marks. The opening sentences of each species account describe the most pertinent field marks needed to identify the bird; the most important characteristics are in bold font. These may include, but are not limited to, body size and form; bill size, form, and color; plumage patterns and color; eye color; and feet color. Definitions of anatomical terms used in these descriptions are found on page 14.

When relevant, we also describe the differences between male and female, adult and immature, breeding and nonbreeding, subspecies, and color morphs. We also give tips on how to distinguish between similar species.

Status. Each species is placed into one of 8 categories, each of which essentially refers to the time or times of year that the species is expected to occur in the country. Not all species are present year-round, of course, and some populations are very dynamic, migrating from North or South America. When a bird is not present year-round, we specify the normal range of months when it is found. Keep in mind that some birds may arrive earlier or leave later than the stated time.

The largest status group is the breeding resident (540 species). These are species that breed in Nicaragua and are found year-round. Some species perform short, seasonal movements within the country during the nonbreeding season. In some cases, a breeding resident population is joined by winter residents of the same species for a portion of the year. Unless otherwise specified in the text, you can assume that a given species is a breeding resident.

There are two status groups that migrate between the Neotropics and North America (NA). Winter residents (158 species) spend the entire boreal winter in Nicaragua, arriving in the boreal fall and departing in the boreal spring; they are generally present between August and May. Small numbers of some species will remain in Nicaragua during the boreal summer; these are typically immature individuals.

In the boreal fall, North American passage migrants (36 species) pass through Nicaragua on their way to wintering grounds that are farther south; in the boreal spring, they leave their southern wintering grounds and once again pass through Nicaragua on their return trip to North America. Note that in some cases a population of a given species may include both winter residents and NA passage migrants; in such cases, we only mention the movements of passage migrants if they add significantly to the size of the population of the winter residents.

There are two status groups that migrate between Nicaragua and South America (SA). Breeding migrants (10 species) arrive from the south to spend the breeding season in Nicaragua, February to August, before returning south during the austral summer. South American (SA) passage migrants only pass through Nicaragua, either traveling north from South America on their way to breeding grounds farther north or traveling south on the return trip. There is no species in Nicaragua whose population is composed solely of SA passage migrants; typically SA passage migrant birds simply add to the population that is already here.

One status group relates specifically to pelagic birds—the pelagic migrant (19 species). Some pelagic species breed on distant islands and disperse widely over pelagic waters after their breeding season, including the pelagic and coastal waters of Nicaragua. Their seasonality does not show a strong pattern.

The vagrant status (14 species) is reserved for species known only from very few records. Nicaragua lies outside of their normal distribution and they are not expected with any frequency.

Finally, in a few cases, when data is insufficient, we describe a bird's status as unknown (3 species).

Abundance. Abundance expresses the likelihood of encountering a species, by sight or sound, on any given day. For some birds, the expected abundance changes from location to location or from one time of the year to another. The abundance terms used are the following:

- abundant: Observed almost every day in the field, sometimes in large numbers.
- common: Observed on more than half of all days in the field; frequently encountered.
- uncommon: Observed on fewer than half of all days in the field; infrequently encountered.
- rare: Observed on fewer than 10% of all days in the field.
- very rare: Few records exist. Experienced observers may not encounter the species over the course of years in the field.
- accidental: Only a handful of records exist.
- local: Appears only at specific locations, often being absent from where one might expect to find it based on habitat and other preferences. Note the use of the term in such phrases as *locally common* or *uncommon and local*.

Distribution. We briefly describe the distributions within Nicaragua and also include range maps for each species. The country map on the inside front cover will help you locate most of the geographic features and place names that we mention in the text. Three main terrestrial ecoregions are used for reference—the Pacific, Northern Highlands, and Caribbean—along with two marine ecoregions—the Pacific Ocean and the Caribbean Sea. We also reference specific geographic features (e.g., lakes, rivers, lagoons, mountain ranges, volcano peaks, islands), protected areas, and private wildlife reserves within the ecoregions to make our descriptions of the distribution more precise.

The elevation range within which a bird occurs is an important aspect of its distribution. Elevation in Nicaragua ranges from sea level to 6,900 feet (2,100 meters). Generally, we refer to three categories of elevation:

- lowlands: sea level to 700 ft (200 m)
- foothills: 700 to 3,000 ft (200 to 900 m)
- highlands: above 3,000 ft (900 m)

But when the information is available—and relevant—we also offer more precise elevation ranges. We list the elevation information in feet first (rounded to the nearest 100 ft) and in meters in parenthesis (rounded to the nearest 50 m). When the elevation limits are different among the ecoregions, we list the limits separately. While these limits are very helpful, lack of data on some species means that they are not always perfect, and, on occasion, birds are found outside of their normal elevation range, either as accidentals or because of poorly understood seasonal movements.

Central America serves as a land bridge between North and South America. It is home to lineages of birds that either originated in North America before spreading south or that originated in South America before spreading north. As a result, several species reach their southernmost distribution (47 species) in Nicaragua, or their northernmost distribution (31 species). We note these cases in the text.

In many cases, we reference protected areas within the Nicaraguan government's protected area system managed by the *Ministerio del Ambiente y Recursos Naturales* (Environmental and Natural Resource Ministry). MARENA oversees 76 protected areas that make up 17% of the national terrestrial territory. The protected areas we reference are categorized as follows:

BR   biological reserve
GRR genetic resource reserve
NM   national monument
NP   national park
NR   natural reserve
WR   wildlife refuge

In a few cases, we reference privately owned preserves that are members of the *Reservas Silvestres Privadas* (Private Wildlife Reserve) system, which are also legally recognized by MARENA as protected areas. We abbreviate these as RSP. Seventy-one private reserves are members of this program, and some are well-known as birding hotspots, and as a source of valuable distribution records.

Protected Areas of Nicaragua

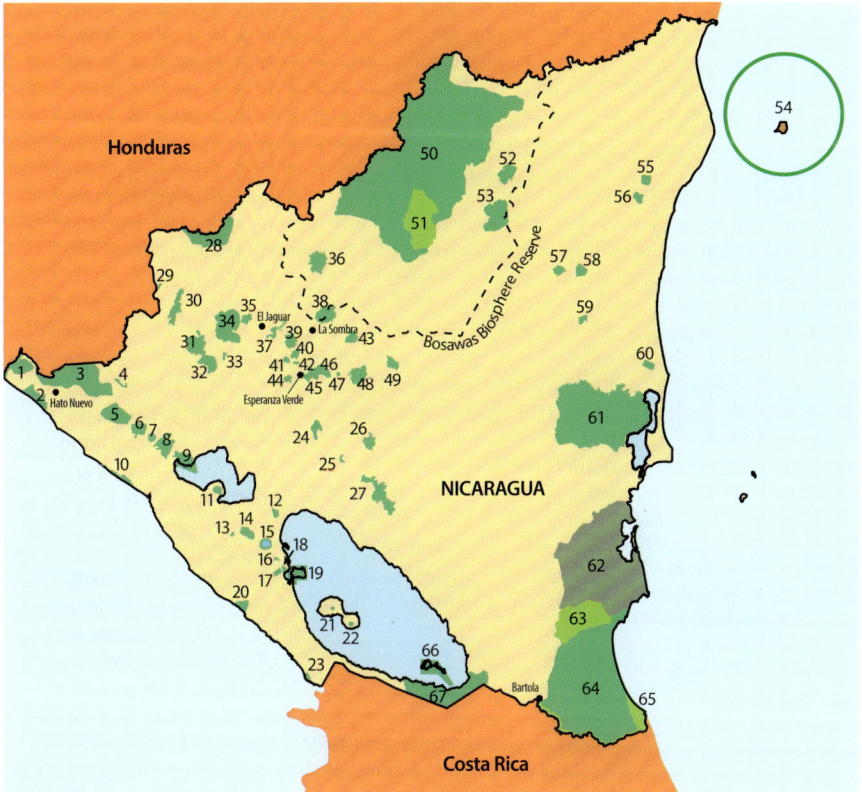

PACIFIC
1 – Cosigüina Volcano NR
2 – Padre Ramos NR
3 – Estero Real NR
4 – Apacunca GRR
5 – San Cristóbal-Casita NR
6 – Rota Volcano NR
7 – Telica Volcano NR
8 – Pilas-El Hoyo NR
9 – Momotombo NR
10 – Isla Juan Venado WR
11 – Chiltepe NR
12 – Tisma NR
13 – Chocoyero-El Brujo WR
14 – Masaya Volcano NP
15 – Laguna Apoyo NR
16 – Mombacho Volcano NR
17 – Mecatepe NR
18 – Río Manares NR
19 – Zapatera NP
20 – Escalante-Chacocente WR
21 – Concepción Volcano NR
22 – Maderas Volcano NR
23 – La Flor WR

24 – Cumaica-Alegre NR
25 – Mombachito-La Vieja NR
26 – Masigue NR
27 – Amerrisque NR

NORTHERN HIGHLANDS
28 – Dipilto-Jalapa NR
29 – Somoto Canon NM
30 – Tepesomoto-Pataste NR
31 – Tisey-Estanzuela NR
32 – Tisey-Estanzuela NR
33 – Tomabú NR
34 – Miraflor and Moropotente NR
35 – Yalí Volcano NR
36 – Kilambe NR
37 – Lake Apanás NR
38 – Peñas Blancas NR
39 – Dantanlí-El Diablo NR
40 – Frío-La Cumplida NR
41 – Arenal NR
42 – Salto Río Yasica NR
43 – Kuskawas NR
44 – Apante NR
45 – Yucul GRR

46 – Guabule NR
47 – Pancansán NR

CARIBBEAN
48 – Quirragua NR
49 – Musún NR
50 – Bosawas NR
51 – Saslaya NR
52 – Cola Blanca NR
53 – Bana Cruz NR
54 – Cayos Miskitos BR
55 – Yulu NR
56 – Kligna NR
57 – Alimikamba NR
58 – Limbaika NR
59 – Makantaka NR
60 – Karawala NR
61 – Wawashan NR
62 – Cerro Silva NR
63 – Punta Gorda NR
64 – Indio Maíz BR
65 – Río San Juan WR
66 – Solentiname NM
67 – Guatuzos WR

Range Maps. The range maps are intended to be used in conjunction with the distribution descriptions in the text. Given the scale of the maps, they are approximations only. And, it is important to keep in mind that birds will not occur within their depicted range in any regions that lack the appropriate habitat. Because of deforestation, birds reliant upon forest often are very local, in forest patches, or absent altogether unless large areas of forest remain. Needless to say, in the absence of data, in some cases we have been forced to make educated guesses about some distribution ranges.

Colors on the range maps represent the eight statuses previously described (p. 2). In a few cases, a map will have more than one color; in such cases, the total population of the species is composed of two different status categories. The following colors represent the eight statuses:

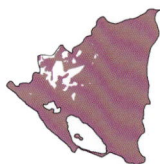

Purple = breeding resident. The species breeds in Nicaragua and remains present year-round. Some species perform short, seasonal movements within the country, and therefore parts of the shaded area may be occupied only on a seasonal basis. If the breeding resident population is joined by winter residents of the same species, it is not indicated on the range, but only in the text.

Blue = winter resident. A migrant between the Neotropics and North America, with a portion of the population residing in Nicaragua during the nonbreeding season (boreal winter), but returning to North America for the breeding season (boreal summer).

Yellow = passage migrant (NA). A migrant between the Neotropics and North America that passes through Nicaragua for a short span of time on its way to and from wintering grounds farther south.

Red = breeding migrant. A migrant that arrives from the south to breed in Nicaragua; returns south during the austral summer.

Orange = passage migrant (SA). A migrant that passes through Nicaragua, either north toward breeding grounds or south on its return trip to nonbreeding grounds in South America.

Teal = pelagic migrant. A pelagic species that breeds on distant islands and disperses widely over pelagic waters after its breeding season. Its seasonality does not show any strong pattern.

Green (dot) = vagrant. A species known only from very few records; Nicaragua lies outside of its normal distribution. Not expected with any frequency.

Gray = status unknown. Used when there is insufficient information to determine the species' status within Nicaragua.

Distributions described with an accidental status are not colored in on the range maps. However, dots are often used to show where accidental records have occurred. Dots are also used to indicate a bird's presence on the Corn Islands.

Habitat and behavior. Often, a bird's habitat and its behavior are critical clues to identification. Birds range from being habitat generalists to habitat specialists. And some behaviors are so distinctive that they can lead to a proper identification. For each species account, we describe the habitat that a species is expected to be found in, general or specific, along with behaviors that can be helpful in making an identification.

Behavior descriptions often include a description of where a bird is likely to be found within its habitat; for birds in forest habitats, this is defined by the level at which they forage. This includes the ground, understory (from ground to 10 ft [3 m] above ground), mid-canopy (from 10 to 30 ft [3 to 9 m] above ground), canopy (from 30 ft [9 m] above ground to the top of the tree crown), and above the canopy. Other behaviors that can be important to observe include whether the species joins same-species or mixed-species flocks; if it attends army ant swarms; and foraging techniques (e.g., sit-and-wait, stalk-and-strike, plunge-diving, gleaning). Some bodily movements are relatively unique, such as tail pumping, pendulous tail swinging, and wing flicking.

Vocalizations. Using vocalizations to identify birds is especially helpful in tropical forests, where some birds are often nearly impossible to see. In many species accounts, you will come upon the phrase "more often heard than seen." In addition, vocalizations are the best way to detect nocturnal birds and, in some cases, they are the only way to distinguish between two very similar species. We typically describe the full song and at least the most common call. Descriptions include the quality or characteristics of the sounds—e.g., length of song, pitch quality and changes, modulation of sound, speed of delivery, and intensity (loudness). When vocalizations are not overly complicated, we transcribe them, signified by italic font. It is important to note that variations on a given song or call should be expected. Vocalizations may differ to an extent between regions, between individuals, and depending on the time of day.

In describing the speed of delivery, dashes indicate a moderately paced delivery that allows for segments of the vocalization to sound distinct (e.g., *chik-der-vee*); absence of punctuation marks or spaces signifies a vocalization with a very fast delivery (e.g., *bzzrrrt*); and a comma indicates a pronounced pause between syllables (e.g., *pip, pip, pip*). In describing inflection and loudness, an accent mark denotes that a syllable is notably emphasized, but not louder than the rest of the vocalization (e.g., *whyáh*); an uppercase letter signals a stressed syllable that is noticeably louder than the rest of the vocalization (e.g., *feu-wEE*); and an exclamation point signals an abrupt and sharp end to a sound (e.g., *wyeea!*). When a vocalization is repeated at length, an ellipsis is used to denote the ongoing repetition (e.g., *chi-chi-chi…*).

Endemic status. An endemic species is one that occurs within a specific region. Depending on how the term is used, the scale of endemism varies, but birders are often interested in country or regional endemics. Nicaragua does not contain any country endemics, but it has several regional endemics, 18 in total. To organize the regional endemics, we subscribe to the Endemic Bird Areas (EBA) proposed by BirdLife International. Four EBAs, with a total of 16 regional endemics, are found within Nicaragua: North Central American highlands, North Central American Pacific slope, Lake Nicaragua marshes, and Central American Caribbean slope. We recognize 2 additional endemic species—a newly named species not yet added to the Central American Caribbean slope EBA and one species that does not fit into the BirdLife EBA scheme. Of course, these categories do not consider subspecies, which are genetically unique. The Mosquitia alone is home to several subspecies that are endemic.

North Central American highlands EBA	Central American Caribbean slope EBA
Ocellated Quail	Lattice-tailed Trogon
Green-breasted Mountain-Gem	Streak-crowned Antvireo
Bushy-crested Jay	Tawny-chested Flycatcher
Rufous-browed Wren	Snowy Cotinga
Blue-and-white Mockingbird	Gray-headed Piprites
North Central American Pacific slope EBA	Black-throated Wren
White-bellied Chachalaca	Canebrake Wren
Blue-tailed Hummingbird	Nicaraguan Seed-Finch
Pacific Parakeet	**Southern Nicaragua to western Panama**
Lake Nicaragua marshes EBA	Purple-throated Mountain-Gem
Nicaraguan Grackle	

Conservation status. When a species is categorized as threatened with extinction, as determined by the International Union for Conservation of Nature (IUCN), we indicate that in the species account. Threatened species are placed in one of four categories:

NT – near threatened (**22** species)
VU – vulnerable (**12** species)
EN – endangered (**3** species)
CR – critically endangered (1 hypothetical species)

Illustrations. We include more than one image for a species whenever there is a significant difference between male and female, adult and immature, or any other significant difference in plumages. Generally, all perched birds on a given page are to the same scale; when there is a change in scale, this is indicated by a horizontal line across the page. Birds in flight are often represented at a smaller scale than the corresponding perched bird or birds. The scale varies from page to page.

Biogeography and avian distribution. If there is one thing that is certain about biogeography and avian distribution in Nicaragua, it is that it is often a complicated affair! Patterns do emerge, however, often correlated to elevation and climate, which, in turn, play a critical role in determining habitat. In very simple terms, the country is separated into three terrestrial ecoregions—Pacific, Northern Highlands, and Caribbean— and two marine ecoregions—the Pacific Ocean and the Caribbean Sea.

The Pacific

The Pacific ecoregion is characterized by picturesque volcanic peaks, dry habitats, two large lakes, and a pronounced dry season. Although it is the driest part of the country, it experiences both a wet season and a dry season. The seasonal extremes of precipitation are such that the vegetation annually cycles between wet, verdant vegetation and dry, brown terrain. Here, drier habitats such as dry forest, thorn forest, and scrub dominate the landscape. But even here, small areas with wetter habitats can be found, namely in the form of gallery forest and cloud forest. Of course, the Pacific coastline is abundant with beach habitat and mangrove forest.

Despite the variety of habitats, dry forest is the dominant habitat in the Pacific. Although, the term *dry forest* can be strictly assigned to a very specific vegetation type based on environmental variables, for the sake of simplicity we use the term to describe a range of forest types, from deciduous to semi-deciduous forest, all

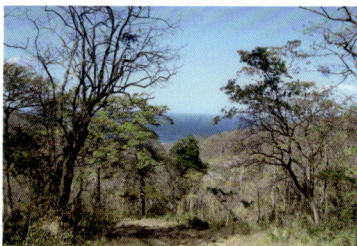

Dry forest as seen in the dry season.

Dry forest as seen in the wet season.

of which endure the intense dry season and respond at some level by dropping leaves. An even more arid forest, stunted in appearance, is the thorn forest. Vegetation is sparse in the thorn forest and the small trees often use thorns as a defense against predators. On the most extreme spectrum of the arid habitats is scrub, consisting of short, woody vegetation. Gallery forest is a unique forest type, consisting of taller trees that line river corridors that are surrounded by dry habitat. Especially in the dry season, this forest stands out, as its higher levels of moisture allow it to maintain its greenery. It can be an oasis for forest dwelling birds during the driest months of the year.

Although the Pacific has an elevation range from sea level to 5,700 ft (1,750 m), most of its terrain is low elevation. Volcanic peaks of the Sierra Maribios rise from the lowlands in the northern Pacific and several more prominent peaks lie in the southern Pacific. Their ascending elevations give way to foothills and small patches of highland terrain. At the northern section of the chain, San Cristóbal and Casita Volcanos reach 5,725 ft (1,745 m) and 4,610 ft (1,405 m) respectively; on these grow highland pine forest. At higher elevations on these peaks, the pine trees host a small number of species that are typically associated with the Northern Highlands. On the outskirts of Granada, Mombacho Volcano reaches 4,413 ft (1,345 m); it is one of two locations in the Pacific with cloud forest, where species that thrive in this habitat can be found in an isolated pocket atop the peak. The other Pacific location with cloud forest is Maderas Volcano, on Ometepe Island, which reaches 4,573 ft (1,394 m).

Gradually rising to the west of Managua are the Sierras Managua. They reach 3,000 ft (900 m) before sloping down to the Pacific coast. Receiving higher amounts of precipitation from the lake effect, these foothills house a slightly more humid forest preferred by some bird species. The peak of Sierras Managua, referred to as El Crucero, is a harsh, windblown environment.

The lakes of Nicaragua, referred to as the lakes region in the text, are a prominent biogeographic feature. Not only do they provide an abundance of aquatic habitats, they also lie in a depression that forms a biogeographic barrier between northern Nicaragua and Costa Rica. This depression, along with other factors, creates disjunct populations leading to speciation, and causes some species to reach distributional limits either in Nicaragua or Costa Rica. Also, these lakes and their surrounding freshwater marshes are home to hundreds of species of aquatic birds (e.g., grebes, ducks, rails, shorebirds, waders, gulls, and terns). Marshes are defined as herbaceous vegetation in soil that is flooded with water at least some times of the year. Of special note is Ometepe Island, in Lake Nicaragua, with prominent volcano peaks, marshes, dry forest, and cloud forest. It is a microcosm of climatic gradation across landscapes, with drier habitats on the north island and wetter habitats on the south island.

Habitats on the Pacific coastline also are a dominant feature. Beach types vary from sand to rocky outcroppings. Extensive mudflats are found where the waters of estuaries rise and fall with the tide. Pockets of mangroves dot the Pacific coastline, concentrated on the borders of estuaries.

In the northeast, the Pacific foothills transition into Northern Highland foothills, at approximately 1,300 ft (400 m). To the south, where the Pacific ecoregion transitions into the Caribbean ecoregion, the boundary between the two is less clear. Here the species typical of the Caribbean can also be found in Pacific forests. This is particularly true in the northern region of Sierra Chontaleña, Rivas Isthmus, and areas south of Lake Nicaragua.

Northern Highlands

This mountainous ecoregion is defined by the highest peaks in Nicaragua and is home to the southernmost stands of highland pine in Central America. The rapidly changing elevations, and the variation in climatic influences from the Pacific and the Caribbean, create a region with great variability in bird diversity and distributional patterns. The western slopes of the region that connect to Pacific foothills are drier, while the eastern slopes that connect to the Caribbean foothills are wetter. To the west, the intermontane valleys are dominated by arid scrub and thorn forest, while to the east forest more similar to humid lowland forest can be found. As would be expected, the western slopes have more Pacific birds than do the eastern slopes, where Caribbean birds are more prevalent. As the elevations climb higher, highland pine and pine-oak forest begin, and then, at the highest elevations, give way to cloud forest. Within these two fairly localized habitats, most of the highland specialty species are found.

The highland pine and pine-oak forest of Central America extend no further south than the Northern Highlands of Nicaragua. This pattern is mirrored in the avifauna; 34 bird species occur here but no farther south. This forest is dominated by pines and, in some cases, is intermixed with oaks; it generally occurs between 2,600 and 5,900 ft (800 and 1,800 m). Also primarily found in the Northern Highlands, cloud forest is a wet, broadleaf evergreen forest type known for the almost daily presence of clouds recycling moisture into the forest and maintaining high levels of humidity. Within the Northern Highlands, it either occurs at elevations above the highland pine and pine-oak forest, or, where this forest is not present, it may be the dominant forest type at as low as 3,280 ft (1,000 m).

The topography of the Northern Highlands is created by four primary mountain ranges, all of which have a west to east orientation. On the border with Honduras, and within the Segoviana Plateau, Sierra Dipilto-Jalapa boasts the two highest peaks in the country. Cerro Mogotón stretches 6,913 ft (2,107 m) above sea level; slightly to the west, Cerro El Volcán rises to 6,125 ft (1,867 m). This mountain range is home to the largest band of highland pine forest in the country and its highest peaks are covered with cloud forest. To the west of Sierra Dipilto-Jalapa are Sierra La Botija and Tepesomoto-La Pataste. The arid habitats in their foothills give way to patches of highland pine and pine-oak forest and cloud forest at the tallest peaks, which reach 5,675 ft (1,730 m). Combined, Sierra Isabelia and Sierra Dariense make up the large mountainous block of the southern half of the Northern Highlands. Both are notably influenced by Caribbean species on their eastern slopes, and bird diversity in the Northern

Highland pine forest.

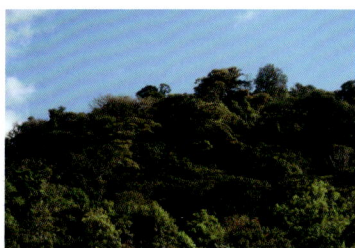

Cloud forest.

Highlands is at its highest at these locations. Between them is Lake Apanás, which adds to the diversity of avifauna in the region, offering unique aquatic habitat. Sierra Isabelia is demarcated by the Coco River to the north and the Tuma River to the south. The continuous expanse of peaks and valleys within the range is comprised of Cerro Tisey-Estanzuela and the highlands of Miraflor and Moropotente on the western fringe; moving east, Yalí Volcano, 5,059 ft (1,542 m), Cerro Kilambé, 5,741 ft (1,750 m), and the Peñas Blancas Massif, 5,725 ft (1,745 m), are the most dominating presences. Sierra Dariense is demarcated by Tuma River to the north and, to the south, Grande de Matagalpa River, which is the southern edge of the Northern Highlands. While most of the peaks in this range do not reach as high as peaks in other ranges, they are still high enough to support highland pine and pine-oak forest and cloud forest; these include Cerro Dantanlí, 5,085 ft (1,550 m), Cerro Chimborazo, 5,538 ft (1,688 m), and Cerro Apante, 4,731 ft (1,442 m).

Scrub in the foreground and thorn forest in the background.

Humid lowland forest.

Caribbean

The Caribbean ecoregion is the largest of the three terrestrial regions and receives the highest volume of rain. It is mostly represented by lowland elevations with humid lowland forest and lowland pine savanna. The terrain rises to foothill elevations in the north-central region (mainly in Bosawas Biosphere Reserve), the eastern slopes of Sierra Chontaleña, and the southeastern region (mainly Indio Maíz Biological Reserve); even rarer, highland elevations with cloud forest are present in the north-central region.

The Mosquitia is a distinct sub-region of the northeast Caribbean. It boasts some of the most intriguing bird species in the country. It is dominated by lowland pine savanna, which is natural grassland scattered with pine woodlands; it supports a number of species that only occur in this habitat. Amid the pine savanna are long rivers with lush gallery forest; and, where there are low-lying areas with poor drainage, rainforest islands persist, creating a patchwork of small, lush forests within the expansive savanna. Both the gallery forest and rainforest islands have vegetation similar to that of humid lowland forest. As a result, the Mosquitia has species unique to lowland pine savanna and also many species associated with humid lowland forest, although the latter are confined to the gallery forests and rainforest islands.

Where forest has not been altered by the human hand, the prominent habitat within the ecoregion is humid lowland forest. This is wet, evergreen forest, with a tall canopy and high species diversity. At higher elevations, humid lowland forest transitions into cloud forest. To the west, on Quirragua (4,478 ft [1,365 m]) and Cerro Musún (4,718 ft [1,438 m]), cloud forest begins at 3,900 ft (1,200). Further east, on Cerro El Toro (5,452 ft [1,662]) and Cerro Saslaya (5,449 ft [1,661 m]), cloud forest begins slightly lower, at 3,600 ft (1,100 m). The mountains tower above the lowlands, creating an exceptional elevational gradient from lowlands to highlands in a short distance. Interestingly, these highland peaks within the Caribbean support disjunct populations of many cloud forest species known primarily from the Northern Highlands.

A prominent feature of the Caribbean is the many river systems that collect water from far inland and drain it into the Caribbean Sea. They attest to the copious amounts of water that fall within the region throughout the year. In low-lying areas closer to the coastlines, where drainage becomes backed up, typical humid lowland forest is replaced by swamp dominated by palm trees. Like the coastline of the Pacific, the Caribbean coastline has sporadic mangroves throughout, particularly in the calmer waters of the larger bays, including Pearl Lagoon and Bluefields Bay.

Lowland pine savanna.

Swamp.

Pacific Ocean and Caribbean Sea

Nicaragua's two oceans create pelagic and coastal habitat for a variety of birds. Similar though they are in some ways, each coast has unique geographic features that create distinct foraging and breeding habitats. On the Pacific Ocean, the Gulf of Fonseca, shared by El Salvador, Honduras, and Nicaragua, is an area of calm waters that some species prefer for foraging. The Farallones Islets are rocky islands that jut out of the gulf, creating roosting and breeding surfaces for a large number of birds. On the Caribbean Sea, several lagoons and bays offer calm foraging waters, and the Miskito and Booby Cays, not far off the coast, create roosting and breeding habitat. Finally, the Big and Little Corn Islands, 43 miles (70 km) offshore of Bluefields, are an important refuge for migratory birds.

Anatomical Features

See the Glossary, p. 21, for definitions of terms not related to bird anatomy.

Upperparts: The upperparts are composed of the crown, nape, back, rump, and uppertail coverts. Distinguishing between the rump and uppertail coverts is sometimes a tricky affair, but note that the uppertail coverts cover the base of the uppertail.

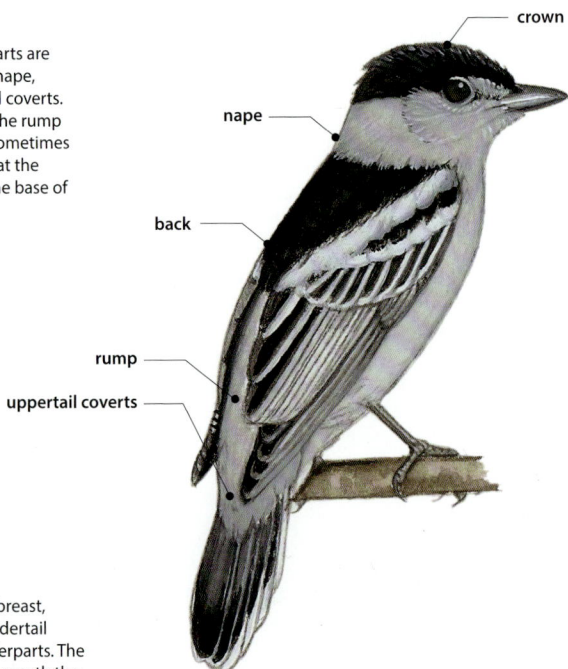

crown

nape

back

rump

uppertail coverts

Underparts: The throat, breast, flanks, belly, vent, and undertail coverts make up the underparts. The small area immediately beneath the lower mandible is referred to as the chin. The vent and undertail coverts lie close to each other and are sometimes difficult to distinguish, but the vent is the area of feathers covering the cloacal opening and the undertail coverts cover the base of the undertail.

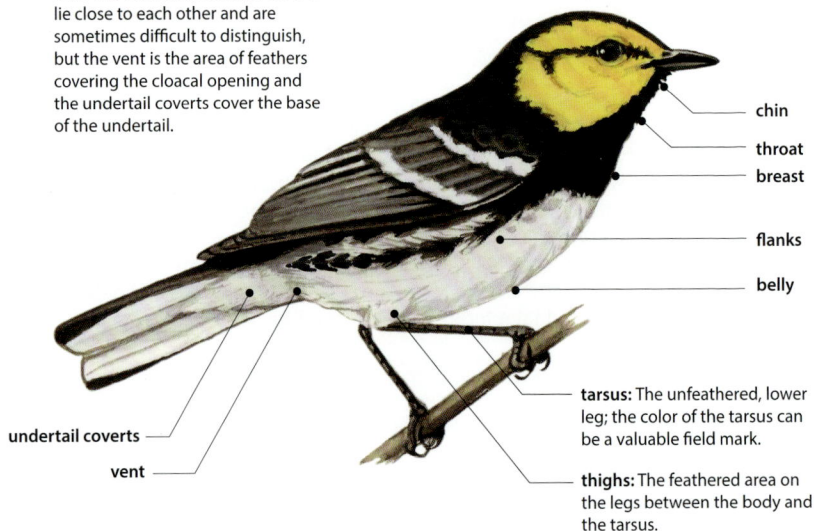

chin

throat

breast

flanks

belly

tarsus: The unfeathered, lower leg; the color of the tarsus can be a valuable field mark.

undertail coverts

vent

thighs: The feathered area on the legs between the body and the tarsus.

Wings: The wing feathers consist of two categories—flight feathers and covert feathers—both of which can aid in identification. Flight feathers are primarily designed for flight, while covert feathers provide protection for the flight feathers.

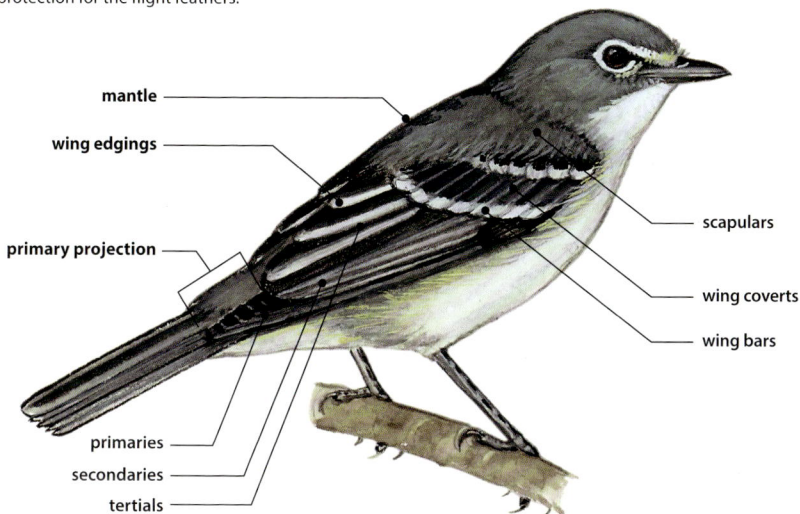

mantle

wing edgings

primary projection

scapulars

wing coverts

wing bars

primaries

secondaries

tertials

Flight feathers consist of three sets. The **primaries** are the 9–10 outermost, longest, and narrowest flight feathers. The **secondaries** are located between the primaries and tertials; they are shorter and broader than the primaries. The **tertials** are the 3–4 innermost feathers; when the wing is folded, they act as protective cover for the other flight feathers. On the folded wing, the **primary projection** is the length that the primary tips extend beyond the longest secondaries or tertials.

The covert feathers on the upperwing often aid in identification. The rows of feathers that cover the base of the flight feathers are called **wing coverts**. When wing coverts have pale tips, they form **wing bars**. The **scapulars**, corresponding to the shoulder, cover the area where the wings connect to the body. The upper back is referred to as the **mantle**. Paler colors on the edges of either flight feathers or wing coverts are called **wing edgings**, and create a finely streaked appearance.

leading edge of wing

axillar: An area corresponding to the armpit; covers the base of the underwing.

wing linings: Also known as underwing coverts, these are the feathers that cover the base of the flight feathers on the underwing.

speculum: A patch of distinctly colored, sometimes glossy, secondaries; can be useful in identifying ducks.

trailing edge of wing

Head: The head is sometimes the only thing you see on a bird, and typically provides important field marks for identification. They include the shape of the head; head plumage patterns; shape, size, and color of the bill; and the color of the eyes.

Bill Shapes

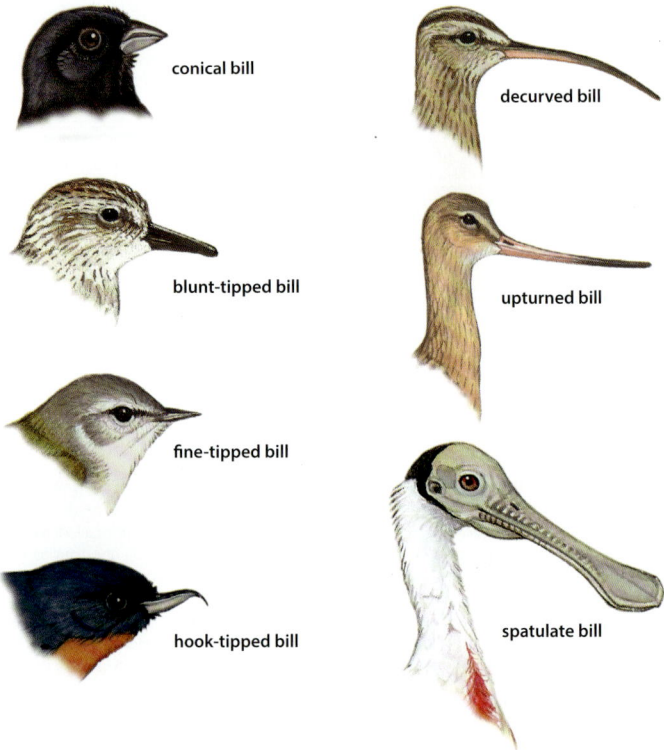

conical bill

decurved bill

blunt-tipped bill

upturned bill

fine-tipped bill

hook-tipped bill

spatulate bill

Bill Parts

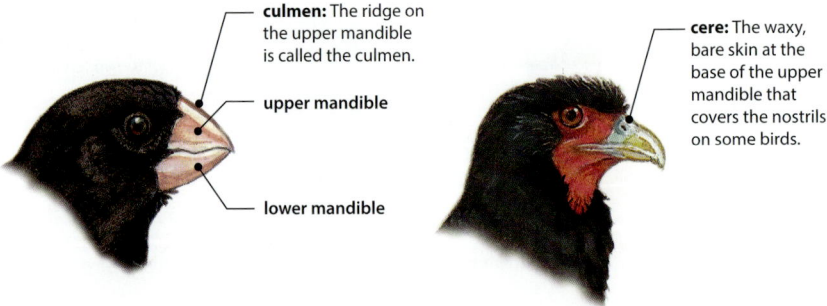

culmen: The ridge on the upper mandible is called the culmen.

upper mandible

lower mandible

cere: The waxy, bare skin at the base of the upper mandible that covers the nostrils on some birds.

midcrown

forecrown

hindcrown

nape

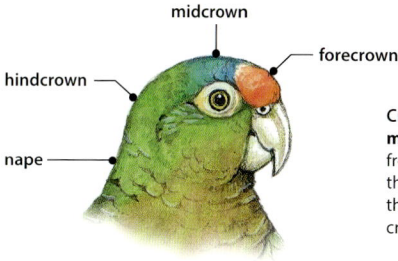

Crown: The crown is composed of the **forecrown**, **midcrown**, and **hindcrown**. The forecrown extends from the bill to the front of the eyes. The midcrown is the center of the crown. The **hindcrown** extends from the back of the eyes to the nape. Descending from the crown is the **nape**, an area corresponding to the neck.

crown patch: Sometimes there is a set of erectile feathers concealed below the crown; they are very rarely raised for display.

crest: Some crown feathers have the ability to be raised into a crest or remain fixed in a raised position.

hood: The hood is generally formed by feathers of the same color that go from the top of the head to the bottom (usually including the crown, nape, and throat), and often contrasting with the forehead, eye area, and chin.

half-hood: The half-hood includes the crown and cheeks, but not the chin and throat.

supraloral: The area above the lore.

lore: The area between the base of the bill and the eye.

superciliary: The superciliary is a differently colored line that extends from the base of the bill to behind the eye, which it passes over (includes the supraloral).

eye line: A differently colored line that crosses from the lore to behind the eye and appears to pass "through the eye."

postocular stripe: A differently colored line that begins behind the eye.

postocular spot: A differently colored small area behind the eye; can be round, square, or teardrop in shape.

gorget: An iridescent patch of feathers on the throat and upper breast of some hummingbirds.

eye ring: When the feathered circle around the eye is conspicuously colored it is called an eye ring. An eye ring that is not continuous is called a broken eye ring. If there is only half of an eye ring, above or below the eye, it is referred to as an eye crescent. (A broken eye ring is formed by two eye crescents.)

orbital skin: Featherless area of skin around the eye. It is called an orbital ring when it forms a circle immediately around the eye.

spectacles (formed with the lore)

When the color of the eye ring is continuous with the color of either the lore (sometimes supraloral) or the postocular stripe, it forms spectacles.

spectacles (formed with the postocular stripe)

malar stripe: A differently colored line descending from the base of the lower mandible along the lower edge of the ear coverts.

lateral throat stripe: A differently colored line bordering the throat; sits below the malar stripe.

moustachial stripe: A differently colored line that extends horizontally from the malar area to below the ear coverts.

mask: When a dark area surrounds the eye, usually extending from the base of the bill to ear the coverts, it is called a mask; a reduced mask may be little more than an area from the lore to the back of the eye.

ear patch (cheek): Ear coverts are small feathers covering the ear openings located behind the eye. When these feathers contrast in color with surrounding feathers, they form an ear patch (cheek).

Tail: Individual feathers on the tail are referred to as rectrices; the difference between the color pattern of the outer rectrices and the central rectrices can be important. Rectrices with paler outer tips are sometimes a distinguishing field mark.

outer rectrices

central rectrices

subterminal band: A band of contrasting color close to, but not reaching, the tip of the tail.

terminal band: A band of different color at the end of the tail.

Tail Shapes

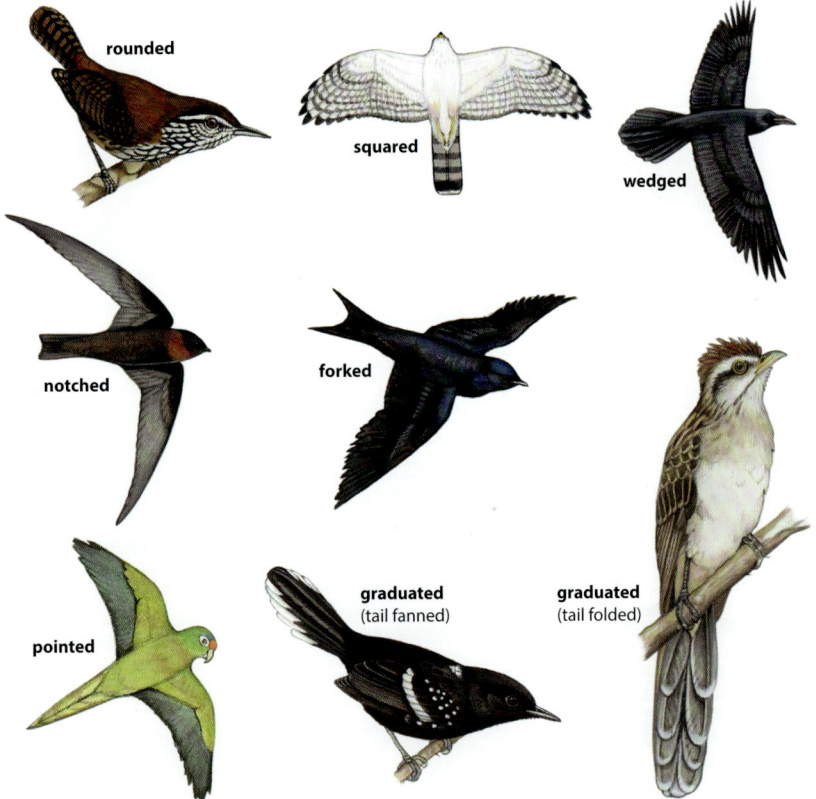

rounded

squared

wedged

notched

forked

pointed

graduated (tail fanned)

graduated (tail folded)

Glossary

See Anatomical Features (p. 14) for additional definitions.

army ant swarm. Massive groups formed by foraging army ants. Many bird species attend army ant swarms, feeding on arthropods as they attempt to flee the swarm.

austral. Pertaining to the southern hemisphere.

barred. The plumage pattern formed by horizontal bars.

disjunct. A biogeographic term used to describe a species with two or more populations that are not connected to one another.

boreal. Pertaining to the temperate region of the northern hemisphere; its climate is characterized by long winters and short summers.

brood parasite. A bird species that lays its eggs in the nest of another species, where they are incubated and raised, by the unwitting host.

breeding plumage. A plumage worn by birds during their breeding season; typically more colorful and distinctly patterned than the plumage of birds that are not breeding. Most often in reference to North American migrants, which may arrive to Nicaragua in this plumage or molt into this plumage prior to returning north.

buffy. A pale yellowish color (e.g., as seen on the rump of the Buff-rumped Warbler).

carpal patch. A patch of differently colored feathers on the underwing at the spot corresponding to the wrist.

chevroned. The zigzag plumage pattern formed by adjacent V-marks.

covey. A group, or flock, of quail.

cosmopolitan. Describes a species, genus, or family with a worldwide distribution.

cracid. A bird that belongs to the Cracidae family (guans, chachalacas, and curasows).

cryptic. Describes a plumage pattern or behavior that allows a bird to go unnoticed.

dabbling. In reference to ducks, the behavior of foraging for aquatic food by immersing the front half of the body while tipping the rear up.

dihedral. Describes the angle formed when the wings are held in a V-shape.

EBA. Abbreviation for Endemic Birds Areas, as described by BirdLife International.

endemic. Describes a species that is confined to a limited geographic region.

epiphyte. Plants that grow entirely on the substrates available within the canopies and trunks of trees, without rooting into the ground.

extirpated. Describes birds that no longer occur in a specific area; sometimes referred to as locally extinct.

facial disc. The concave, feathered area surrounding the eyes that acts as a parabolic device to direct sound waves, primarily in reference to owls.

frontal shield. A hard or fleshy plate extending from the base of the upper mandible to the forehead. Typically found on gallinules, moorhens, and jacanas.

frugivorous. Describes birds that feed on fruit.

gape (or **flange**). The point where the upper and lower mandible join.

genus (plural: **genera**). The taxonomic group below the level of family, usually containing multiple species.

glean. To catch invertebrates by plucking them from foliage or the ground.

gregarious. Describes the behavioral propensity to form groups, either to feed or during colonial roosting or nesting.

gular pouch (or **sac**). A bulge of fleshy skin on the throat.

hawking. Snatching food, usually insects, in flight, with only the bill, and subsequently consuming the prey without landing to perch.

immature. Term used to describe all ages prior to adult plumage.

insectivorous. Describes the feeding preference for insects.

irruptive. Refers to a species' irregular movement, temporarily and in large numbers, into an area that is outside its normal distribution.

kettle. Large groups of soaring birds, circling as a group. A kettle may be composed of several different species, and most often refers to migrating raptors.

kleptoparasitism. The feeding behavior of stealing food from another bird, not necessarily of the same species. Most often in reference to some pelagic species.

leaf litter. The accumulation of decomposing leaves and twigs on the forest floor.

lek. A congregation of males that perform breeding displays in order to attract females and compete for mating opportunities.

mixed-species flock. Flocks composed of multiple species; such flocks increase foraging efficiency and are better able to spot predators.

monotypic. Describes a genus or family that only consists of one species.

morph. Species sometimes occur in different colors, and these are referred to as morphs. A species with several morphs is polymorphic.

mottled. The plumage pattern formed by small, erratically spaced spots.

Neotropics. The biogeographic region extending from southern Mexico through South America.

nonbreeding plumage. A plumage worn by nonbreeding birds; typically less colorful and less distinctly patterned than birds in breeding plumage. Most often in reference to North American migrants, which retain this plumage for most of their time in Nicaragua.

nuchal collar. A contrasting band located on the nape, at the base of the hindcrown.

pantropical. Describes a distribution range that includes the tropics of both western and eastern hemispheres.

pelagic. Refers either to the open ocean, far from the coast, or to a species that primarily lives in this habitat.

polyandrous. The reproductive behavior in which one female pairs with several males.

polymorphic. When a species includes birds with two or more different plumage colors or forms.

primary forest. A relatively undisturbed forest.

raptor. Collectively describes birds equipped to hunt larger prey; includes osprey, vultures, hawks, kites, eagles, falcons, caracaras, and owls. A bird of prey.

rictal bristles. Stiff hairlike feathers at the base of the bill. Typically found on nightjars, flycatchers, and swallows.

roost. Where a bird sleeps.

rufous. A brownish-red color (e.g., the color on the head of Rufous-capped Warbler).

sally. To perform small flights to catch insects; after making a sally, the bird typically returns to the same branch or general area to perch.

scaled. The plumage pattern formed by what appear to be scales; caused by many small feathers overlapping one another.

second growth. Describes a forest with young successional growth after a heavy disturbance; typically with dense, shrubby vegetation that is under 10 ft (3 m) tall.

secondary forest. Describes a forest that is regenerating after significant disturbances. The resulting forest structure is less mature and less stable than that of primary forest.

sexual dimorphism. A characteristic of some species in which the male and female show different physical characteristics, often size, coloration, or adornment.

shorebird. Collectively describes birds that rely on aquatic shorelines to forage for food. Includes plovers, sandpipers, oystercatchers, jacanas, and stilts and avocets.

speckled. The plumage pattern formed by very small, scattered spots or specks.

spotted. The plumage pattern formed by large spots; the spots are less densely spaced than on mottled birds.

stitching. A shorebird foraging strategy of making rapid probes with the bill in mud or sand; such birds typically move in a line.

striped. The plumage pattern formed by vertical lines; less densely spaced than on a streaked pattern.

streaked. The plumage pattern formed by vertical lines; more densely spaced than on a striped pattern.

subspecies. A taxonomical level describing genetically distinct populations within a species; most often results from geographic isolation of breeding populations. May or may not show plumage, body size, or vocalization differences.

sympatric. Refers to two species that share, at least partially, the same distribution.

tawny. An orangish-brown to yellowish-brown color (e.g., the color seen on the crown of the Tawny-crowned Greenlet).

territorial. Describes birds that defend a resource from competing individuals or groups.

trapline. The hummingbird foraging strategy in which the birds move between nectar sources separated by considerable distances instead of defending a territory; primarily done by hermit hummingbirds.

IUCN. The International Union for Conservation of Nature is a global non-governmental organization responsible for assessing a species' conservation status.

vermiculated. The plumage pattern formed by thin, wavy lines.

wader. Collectively describes birds that primarily rely on the shallow waters of aquatic habitat to forage for food. Includes storks, bitterns, herons, egrets, and ibises.

wattle. Fleshy, ornamental skin hanging from different parts of the head; typically found on males.

Species Accounts and Illustrations

These terrestrial, secretive birds occur in forests throughout the Neotropics, where they roam the forest floor in search of fruit, seeds, and small invertebrates, and nest within exposed tree buttresses. Tinamous are recognized by the disproportionate size between their small dovelike head and plump chicken-like body; they have a very short tail and short, stocky legs. Leg color, body size, and vocalizations are important distinguishing characteristics for identification within the family. Tinamous are more often heard than seen because of their secretive behavior and cryptic plumage. The tremulous and pleasantly haunting vocalizations they make at dawn and dusk are unforgettable.

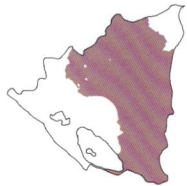

Great Tinamou (Tinamú Grande) *Tinamus major*
17 in (43 cm). **Largest** tinamou. **Large size** and **gray legs** distinguish it from the other, smaller tinamous. Generally common in the Caribbean but abundant in Bosawas Biosphere Reserve and Indio Maíz BR; rare and local on eastern slopes of Northern Highlands; to 4,600 ft (1,400 m). When foraging, walks quietly on ground of humid lowland forest and cloud forest. Shy and secretive; more often heard than seen as it vocalizes from within dense forest cover, primarily at dawn and dusk. Whistles a series of forlorn, quivering notes, often rising in pitch, strength, and undulation toward end of series; similar to that of Little Tinamou but lower pitched. **NT**

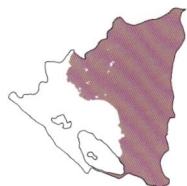

Little Tinamou (Tinamú Chico) *Crypturellus soui*
9 in (23 cm). **Small size** and **lack of barring** distinguish it from other tinamous. Only tinamou with **dull yellow legs**. Common throughout Caribbean lowlands and foothills; to 3,300 ft (1,000 m). Uncommon and local on eastern slopes of Northern Highlands; to 4,600 ft (1,400 m). Forages by gleaning seeds, berries, and insects from the ground in primary forest edge and second growth. Vocalizes during the day, unlike other tinamous. Gives a single clear whistle that rises in pitch before falling with undulating tones (sometimes emphatically repeats an accelerating whistle that rises in pitch); also whistles a series of 2 notes (with first note rising and second note falling). Call similar to that of Great Tinamou, but higher pitched.

Thicket Tinamou (Tinamú Canelo) *Crypturellus cinnamomeus*
11.5 in (29 cm). A medium sized tinamou with **extensively barred black and buffy-white** upperparts (black and cinnamon on female Slaty-breasted Tinamou); on male, **cinnamon breast** usually contrasts with slate-gray breast of similar Slaty-breasted Tinamou (difference not always detectable on respective females). Note **orange-red legs**. Common and local in Pacific lowlands and foothills (only tinamou species in that region); to 3,000 ft (900 m). Uncommon in Northern Highlands (no known records east of Lake Apanás); to 4,600 ft (1,400 m). Inhabits dry forest, thorn forest, cloud forest edge, and second growth. Feeds while walking on the ground; remains hidden within dense thickets and shrubbery of understory. Gives a single, 1-second, high-pitched whistle; there is a considerable amount of time between repetitions (seemingly nonconsecutive).

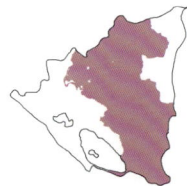

Slaty-breasted Tinamou (Tinamú Pizarroso) *Crypturellus boucardi*
11 in (28 cm). Named for **unpatterned slate-gray breast**, which differs from cinnamon breast of similar Thicket Tinamou (color of breast on respective females sometimes impossible to distinguish). Noticeably sexually dimorphic; back and wings are dark brown on male but extensively **barred with black and cinnamon** on female (black and buffy-white on female Thicket Tinamou). Note **red legs**. Common in Caribbean (more likely in lowlands and foothills); rare on eastern slopes of the Northern Highlands; to 4,600 ft (1,400 m). Prefers humid lowland forest and cloud forest, where it feeds on fruit and seeds on the ground. Intermittently repeats a hollow 2-note whistle; low-pitched in comparison to the other tinamou vocalizations.

Bright blue eggs
of the Great
Tinamou at its
nest site on the
forest floor.

Great Tinamou

Little Tinamou

female

**Thicket
Tinamou**

male

female

**Slaty-breasted
Tinamou**

male

These web-footed, aquatic birds are found worldwide. Their broad, flattened bills with rounded tips set them apart from other aquatic birds. The majority of Nicaraguan ducks are migratory (13 species), spending a portion of the year in boreal North America; only 4 breed in-country. Except for the whistling-ducks, all exhibit sexual dimorphism (and some of these molt from nonbreeding to breeding plumage while in Nicaragua, adding to identification challenges). Ducks are linked exclusively to aquatic habitats, where they feed on aquatic plants, insects, fish, mollusks, worms, and amphibians. Most ducks fall under one of two foraging categories: dabbling ducks (genera *Spatula*, *Mareca*, and *Anas*) feed on the water's surface, often raising their back end and foraging with their head submerged; diving ducks (genus *Aythya*) plunge their entire body underwater to forage at greater depths.

Black-bellied Whistling-Duck (Piche Piquirrojo) *Dendrocygna autumnalis*

20 in (50 cm). Only Nicaraguan duck with **red-orange bill** and **pink legs**; also note **black belly**. In all plumages, flying birds show a distinctive **white stripe across wing coverts**. Immature, with gray bill and legs, can be confused with immature Fulvous Whistling-Duck, but rump is dark and white stripe is still noticeable. Abundant countrywide; to 3,300 ft (1,000 m). Found in marshes, lakes, lagoons, and ponds; sometimes visits mangroves, estuaries, and salt ponds. Primarily feeds at night in shallow waters, foraging for plant material and invertebrates. Gives various high-pitched, screamlike whistles, sometimes leading to a *whit-WEE-whiwhiwhiwhiwhi* phrase; flocks often create a continuous cacophony.

Fulvous Whistling-Duck (Piche Canelo) *Dendrocygna bicolor*

19 in (48 cm). **White stripes on flanks** of perched or swimming birds are diagnostic; **white U-shaped band on rump** is diagnostic on birds in flight. Similar to immature Black-bellied Whistling-Duck, but note **black wings**. Common in the Pacific, Playitas-Moyúa-Tecomapa lagoons and Lake Apanás, and extreme southwestern Caribbean lowlands; to 3,300 ft (1,000 m). Prefers marshes, lakes, lagoons, and rice fields; less frequent in brackish waters. Forages in shallow waters for plant material, especially at night. Calls with a loud and raspy *pi-CHEW*.

Muscovy Duck (Pato Real) *Cairina moschata*

M 32 in (82 cm); F 25 in (64 cm). **Glossy black-green coloration** and obvious **white under- and upper-wing patches** make this duck unmistakable. Wing patches are significantly smaller on female and immature. (Beware of domesticated varieties, which vary in color and often have mostly white plumage with some black spots.) Common in Caribbean lowlands and uncommon and local in Pacific lowlands and foothills; locally very rare on Lake Apanás. Prefers wooded lakes, rivers, marshes, and mangroves, where it feeds on aquatic plants, grains, small fish, and insects. Very shy; mostly observed flying to and from feeding locations at dawn and dusk, likely a learned behavior resulting from extensive hunting pressure.

Masked Duck (Pato Careto) *Nomonyx dominicus*

13.5 in (34 cm). On breeding male, large **black mask contrasting with chestnut nape, neck, and breast** is distinctive; also note **broad blue bill with black tip**. Two prominent **horizontal dark stripes on face** of female separate it from similar female Ruddy Duck; nonbreeding male and immature resemble female. Rare to uncommon, in Pacific, Northern Highlands, and southern Caribbean; to 3,900 ft (1,200 m). Found on calm, shallow waters, within or near dense, emergent aquatic vegetation.

Ruddy Duck (Pato Cariblanco) *Oxyura jamaicensis*

15 in (38 cm). Distinctive **white cheek on male** contrasts with **black crown and nape**. Female and immature show a **single dark horizontal line on cheek**, which distinguishes it from similar female Masked Duck. Winter resident (Nov to early April); uncommon to common in Pacific lowlands and foothills; also occurs up to 3,900 ft (1,200 m) on western slopes of Sierra Isabelia and Sierra Dariense in Northern Highlands. Most likely at Salinas Grandes salt ponds, Moyúa Lagoon, and Lake Apanás. Recent records in May and June at Playitas-Moyúa-Tecomapa lagoons may indicate the colonization of breeding populations in Nicaragua. Forages in marshes by diving for seeds, roots of aquatic plants, insects, and small crustaceans.

Black-bellied Whistling-Duck

immature

A mixed group of Black-bellied and Fulvous whistling-ducks.

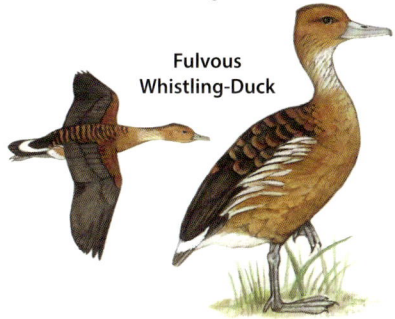

Fulvous Whistling-Duck

female

male

Muscovy Duck

immature

immature

female/immature/ nonbreeding male

breeding male

Masked Duck

female/ immature/ nonbreeding male

breeding male

Ruddy Duck

nonbreeding male

female

female

nonbreeding male

Blue-winged Teal (Cerceta Aliazul) *Spatula discors*
15 in (39 cm). On breeding male, **white facial crescent** contrasting with **dark head** is distinctive. Female has faint facial crescent with a distinct **dark eye line** and **broken white eye ring**; it is helpful to compare bill length with those on other similar females (shorter than Cinnamon Teal; longer than Green-winged Teal, p. 32); nonbreeding male and immature resemble female. In all plumages, note yellow legs (Green-winged Teal has dusky legs). In flight, on male and female green speculum contrasts with pale blue secondary wing coverts, as on Cinnamon Teal and Northern Shoveler. Winter resident (Sept to April); abundant countrywide with few individuals remaining year-round; to 4,300 ft (1,300 m). Found on lakes, ponds, rice fields, marshes, and mangroves, where it forages for insects, small invertebrates, and submerged aquatic vegetation.

Cinnamon Teal (Cerceta Castaña) *Spatula cyanoptera*
16 in (40 cm). Distinctive **chestnut-cinnamon plumage** of breeding male fades to a faint wash in nonbreeding plumage, but he retains **red iris**. Female has plain face with an indistinct **dark eye line**; richer coloration and **longer, slightly spatulated bill** is helpful in comparisons with female Blue-winged Teal and Green-winged Teal (p. 32), which have similar plumage. In flight, green speculum contrasts with pale blue secondary wing coverts, a pattern similar to that on Blue-winged Teal and Northern Shoveler. Also note yellow legs (Green-winged Teal has dusky legs). Rare winter resident (Nov to March) in Pacific lowlands; to 500 ft (150 m). Accidental in Caribbean lowlands, with 1 record (San Juan del Norte, Nov 1982). Frequents lakes, ponds, rice fields, marshes, and mangroves, where it forages for insects, small invertebrates, and submerged aquatic vegetation.

Northern Shoveler (Pato Cuchara) *Spatula clypeata*
19 in (48 cm). In all plumages, the diagnostic feature is its **large, spatulate bill that is longer than the head**. Breeding male is unmistakable; female and immature have buffy head and an orange bill with dusky smudging (nonbreeding male is similar, but note yellow eye and indistinct white patch at base of bill). In flight, green speculum with broad white border contrasts with pale blue secondary wing coverts, a pattern similar to that on the smaller bodied (and billed) Blue-winged Teal and Cinnamon Teal. Winter resident (Oct to April, occasionally to mid-May). Common in Pacific lowlands and Playitas-Moyúa-Tecomapa lagoons. Rare to uncommon on Lake Apanás; at 3,300 ft (1,000 m). Rare to uncommon in extreme southwestern Caribbean lowlands; otherwise accidental in Caribbean. Dabbles for plant material and aquatic invertebrates in shallow ponds, lakes, and coastal marshes.

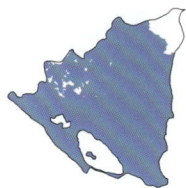

American Wigeon (Pato Calvo) *Mareca americana*
18.5 in (47 cm). Small **pale blue bill with black tip** separates this from other ducks. In flight, conspicuous white secondary coverts (grayer in female and immature) distinguish it from other dabbling ducks. Male in breeding plumage is unmistakable, with **white or buffy crown** and **green postocular stripe**; male in nonbreeding plumage resembles female, which has **speckled gray head with a faint dark mask** and cinnamon flanks. Uncommon to common winter resident (Nov to mid-April); countrywide, to 3,900 ft (1,200 m). Seen in estuaries, ponds, and along lake shores, where it forages by dabbling for plant material.

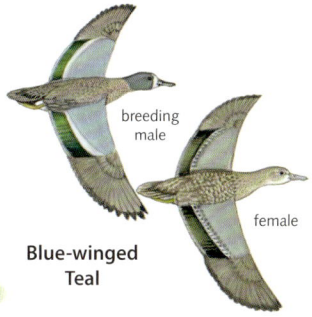

Blue-winged Teal

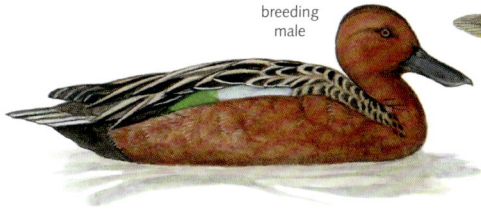

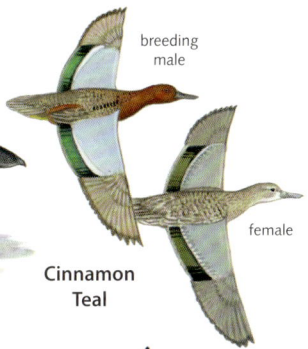

female

breeding male

female

breeding male

Cinnamon Teal

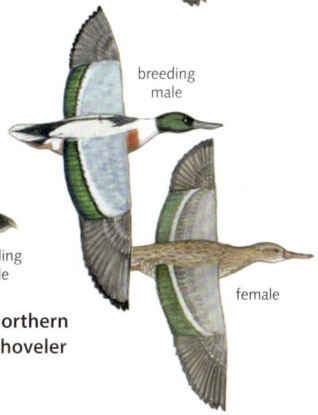

breeding male

breeding male

female

Northern Shoveler

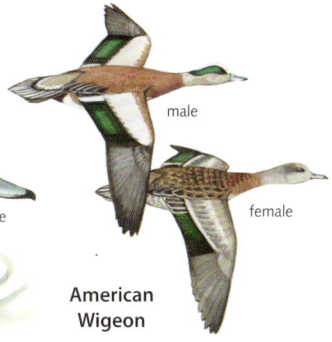

female

breeding male

breeding male

female

American Wigeon

female

female

male

male

female

Mallard (Pato Cabeciverde)   *Anas platyrhynchos*
23 in (58 cm). Largest dabbling duck. Male in breeding plumage has narrow **white collar**, **chestnut breast**, and **yellow bill**. Female is mottled brown overall, with a dark eye line and mostly **orange bill (dark central smudging)**; nonbreeding male and immature resemble female but with yellowish (male) and brown (immature) bills. In flight, note flashy **white underwings** and **blue speculum with white border**. Accidental winter resident, with records from Lake Apanás, Playitas-Moyúa-Tecomapa lagoons, Guayabo Wetlands, and Lake Nicaragua; to 3,300 ft (1,000 m). Forages for plant material and small invertebrates in shallow waters of freshwater marshes.

Northern Pintail (Pato Rabudo)   *Anas acuta*
22.5 in (57 cm). **Long neck, pointed tail,** and **slender body** make this attractive dabbling duck unmistakable. Male in breeding plumage has **dark brown head** with **white neck-stripe** and white breast. Female has an **unmarked buffy head**, but told apart from other dabbling ducks by unique body form; nonbreeding male and immature are similar to female. Uncommon to common winter resident (Oct to April) in Pacific and at Lake Apanás; to 3,300 ft (1,000 m). Uncommon in southwestern Caribbean lowlands. Found on lakes, freshwater and coastal ponds, lagoons, and brackish marshes.

Green-winged Teal (Cerceta Común)   *Anas crecca*
14 in (36 cm). Smallest dabbling duck. On breeding male, combination of **chestnut head, broad, green postocular stripe,** and **white vertical stripe** on side is distinctive. Female is distinguished from female Blue-winged Teal and Cinnamon Teal (p. 30) by **smaller body and bill, buffy undertail coverts**, and **white belly contrasting with mottled flanks**. Nonbreeding male and immature resemble female. In flight, both sexes show **green-and-black speculum** and **dusky legs**. Very rare winter resident (Nov to March). Occurs in Pacific lowlands and foothills and extreme southwestern Caribbean, otherwise accidental in Caribbean; to 1,600 ft (500 m) but potentially higher. Forages on lakes and ponds and in mangroves, for seeds and submerged vegetation and, to a lesser extent, for small invertebrates.

domestic
varieties

Mallard

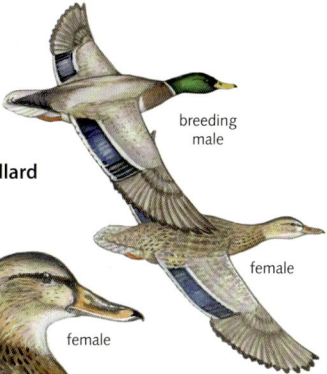

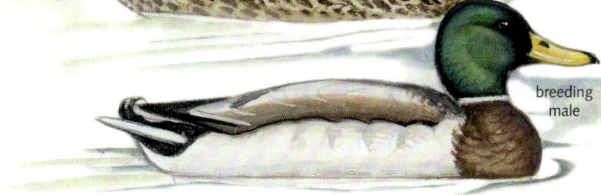

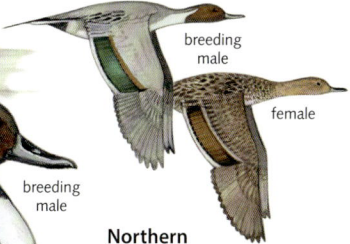

breeding
male

female

female

breeding
male

female

breeding
male

**Northern
Pintail**

breeding
male

female

breeding
male

**Green-winged
Teal**

female

breeding
male

female

Redhead (Porrón Cabecirrojo)   *Aythya Americana*
19 in (48 cm). **Round head** and **tricolored bill** are distinctive in all plumages. General plumage pattern of breeding male is similar to that of Canvasback but note red head and neck, **yellow iris**, **gray body**, and blue bill with black tip. Both nonbreeding male and female are tawny-brown overall. Very rare winter resident. Occurs in Pacific lowlands and foothills; often found in sizable groups (20+ individuals at Las Playitas Lagoon, Jan 2016; 4 individuals on Moyúa Lagoon, Feb 2017); to 1,500 ft (450 m). Might also occur in southwestern Caribbean lowlands.

Ring-necked Duck (Porrón Collarejo)   *Aythya collaris*
17 in (43 cm). Combination of **peaked head** (at rear of crown) and **white ring behind black-tipped bill** is diagnostic. Similar to Greater Scaup and Lesser Scaup, but told apart by bold bill pattern, **solid black back**, and **white vertical stripe** on side. Female has dark brown crown, grayish face, narrow white eye ring and pale postocular stripe; nonbreeding male and immature resemble female. Uncommon winter resident (mid-Oct to March), with patchy distribution records at Playitas-Moyúa-Tecomapa lagoons, Miraflor NR, Lake Apanás, and San Juan River headwaters; otherwise accidental in Caribbean lowlands; to 3,900 ft (1,200 m). Dives for mollusks, invertebrates, and aquatic vegetation in both deep and shallow waters of lakes, ponds, and rivers.

Canvasback (Porrón Picudo)   *Aythya valisineria*
21 in (53 cm). **Long sloping profile of head and bill** separates this species from other ducks. General plumage pattern of breeding male is similar to that of Redhead but note chestnut head and neck, **red iris**, white body, and **black bill**; nonbreeding plumage is duller. Female and immature have light brown head. Accidental winter resident in Pacific lowlands; only 1 record (NE Lake Nicaragua, Jan 1957).

Greater Scaup (Porrón Mayor)   *Aythya marila*
18 in (46 cm). Almost identical to Lesser Scaup but larger; note **more rounded head, which is held lower**. Also compare to Ring-necked Duck. In good light and at close range, note greenish-glossy head of breeding male. Female is dark brown overall, with bold white patch at the base of bill; nonbreeding male and immature resemble female, but male mostly lacks white at base of bill. In flight, shows **white stripe on secondaries and inner primaries** (outer primaries are pale gray). Accidental winter resident; countrywide, to 3,300 ft (1,000 m). Four country records: Papaturro River in Guatuzos WR, Feb 2001; Lake Apanás, Nov 2011; San Juan River, Jan 2013; Moyúa Lagoon, Feb 2017. Prefers saltwater habitats but can also be found in freshwater habitats.

Lesser Scaup (Porrón Menudo)   *Aythya affinis*
16 in (42 cm). Almost identical to Greater Scaup but note **smaller size**, **narrower bill**, and a **taller, peaked head** (best seen when bird is relaxed). Also compare to Ring-necked Duck. In good light and at close range, note purple-glossy head of breeding male. Female is dark brown overall, with white flecks; note white patch (of variable intensity) at base of bill; nonbreeding male and immature resemble female, but male mostly lacks white at base of bill. In flight, shows **bold white stripe only in secondaries**, while primaries are pale gray. Common winter resident countrywide (mid-Oct to May); to 3,900 ft (1,200 m). Frequents a variety of freshwater habitats; uncommon in salt water.

breeding
male

female

Redhead

female

breeding
male

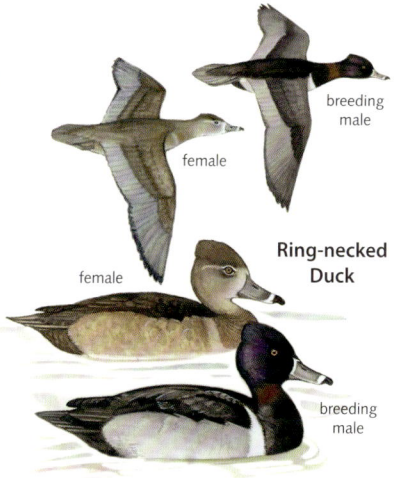

breeding
male

female

**Ring-necked
Duck**

female

breeding
male

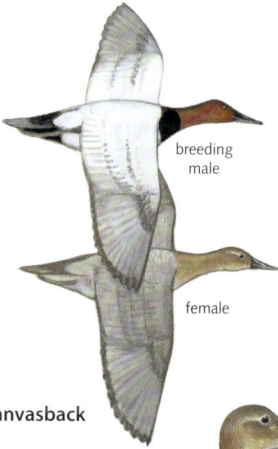

breeding
male

female

Canvasback

Greater Scaup

female

breeding
male

female

breeding
male

female

female

breeding
male

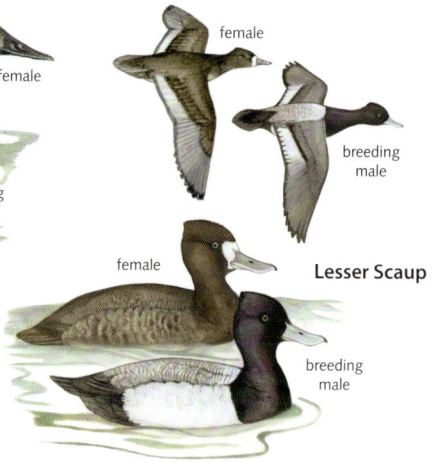

female

Lesser Scaup

breeding
male

These large birds are distant relatives of turkeys (family Phasianidae). They occur only in Neotropical forests. Although most members of the family depend on primary or mature forests, chachalacas prefer relatively open woodlands and forest edges. Chachalacas and guans spend much of their time in tree canopies, silently traversing large branches to find the fruit and leaves that they feed on, but, when forced into flight, they noisily crash through branches and foliage. Great Curassow, on the other hand, consistently roams the forest floor in search of food. Very shy and secretive, cracids are more often heard than seen; they often call at dawn and dusk. Due to deforestation and extensive hunting, many populations are in decline.

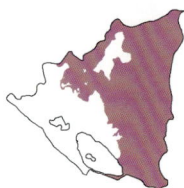

Gray-headed Chachalaca (Chachalaca Cabecigrís)   *Ortalis cinereiceps*
20.5 in (52 cm). **Rufous primaries** (visible on folded wing) and distinct vocalization separate it from very similar Plain Chachalaca. Common on eastern slopes of Northern Highlands, to 4,600 ft (1,400 m). Common in Caribbean lowlands and foothills, to 1,300 ft (400 m). Favors the edges of cloud forest and humid lowland forest, tall second growth, and gallery forest and rainforest islands in the Mosquitia. In flocks of up to 10 individuals, forages for fruit and leaves at mid-canopy and canopy; easiest to see at dawn. Repeats a high-pitched *weeUT!*, once or twice per second, sometimes crescendoing to a continuous and emphatic *WEEIT-WEEIT-WEEIT*... .

Plain Chachalaca (Chachalaca Lisa)   *Ortalis vetula*
20 in (51 cm). **Absence of rufous in primaries** and distinct vocalization distinguish it from very similar Gray-headed Chachalaca. Uncommon in Pacific lowlands but locally common in foothills, particularly in Sierras Managua, as well as in dry intermontane valleys of Northern Highlands; to 3,300 ft (1,000 m), but wanders to 4,400 ft (1,350 m) during the dry season. Prefers dry forest and edges, secondary forest, scrub, and shaded coffee plantations. During early morning hours, flocks of 4–7 individuals forage at mid-canopy, descending to ground to pick fruit. Incessantly repeats a raucous and raspy, 3-syllable, onomatopoeic *CHA-CHA-LAC*; group members return a similar call, only higher in pitch. Also calls with a variety of burry vocalizations, sometimes making a parrot-like chattering.

White-bellied Chachalaca (Chachalaca del Pacífico)   *Ortalis leucogastra*
20 in (51 cm). **Whitish underparts** from lower breast to undertail coverts distinguish it from the other chachalacas. Uncommon in Pacific lowlands and foothills, reaching as far south as Mombacho Volcano; to 1,800 ft (500 m). Occurs in dry forest and secondary forest, mangroves, swamps, and scrub. Mostly arboreal; in pairs or small flocks, forages for fruit, flowers, and leaves at mid-canopy. Repeats its loud scratchy call; a distinct *rhót* is followed by a jumbled *urdar*; group members return a similar call, only higher in pitch. Reaches southernmost distribution in Nicaragua. Endemic to North Central American Pacific slope EBA.

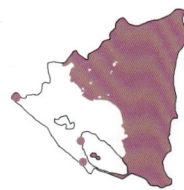

Crested Guan (Pava Crestada)   *Penelope purpurascens*
34 in (87 cm). **Large size**, prominent **crest**, and **extensive bluish gray orbital skin contrasting with red iris** combine to make it distinctive. Also note **red wattle and coral-red legs**; white flecking on dark brown neck and breast distinguishes it from smaller female Highland Guan (p. 38). Hunting and deforestation have largely extirpated it from historical distribution. Uncommon in Caribbean, but locally common in Saslaya NP, Bosawas NR, and Indio Maíz BR; to 3,300 ft (1,000 m). Rare throughout Pacific (recent reports from Rivas Isthmus, Mombacho Volcano NR, and Cosigüina Peninsula), but locally common on Maderas Volcano NR; to 3,900 ft (1,200 m). Rare on eastern slopes of Northern Highlands; to 4,600 ft (1,400 m). Mainly arboreal, preferring mid-canopy; in pairs or groups, walks along branches in search of fruit. Very vocal; pipes a rapid, loud, high-pitched *whut-whut-whut*..., usually making 2 notes per second; the sequence often lasts for minutes, with waves of increased intensity and pitch. Also utters a soft, nasal, low-pitched *whúaan*.

Gray-headed
Chachalaca

Plain
Chachalaca

White-bellied
Chachalaca

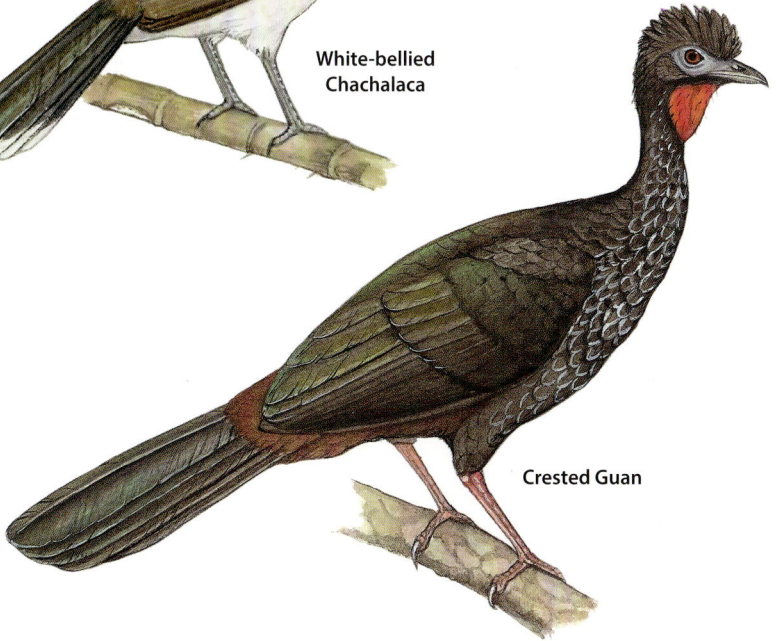

Crested Guan

Highland Guan (Pava de Altura)   *Penelopina nigra*
24.5 in (62 cm). On male, combination of **glossy black plumage, orangish-red bill and legs, and red wattle** is diagnostic. Female has extensive **black and rufous barring** and paler legs than male; immature resembles female. Generally rare, but locally common on eastern slopes of Northern Highlands, and locally common in Musún NR (probably in nearby Quirragua NR) and in Saslaya NP; above 3,600 ft (1,100 m). Reaches southernmost distribution in Nicaragua. Inhabits cloud forest with little disturbance and highland pine-oak forest. Forages in pairs or groups, plucking fruit at all levels. Male whistles an unmistakable, high-pitched, sliding song that rises slowly before rapidly accelerating to a piercing quality at the end, lasting 2–3 seconds; alarm call is an incessant, squeaky *piá-piá-piá…*, followed by scrambled chatter. Male also produces a mechanical flight noise of 2 raps immediately followed by an electronic, descending trill (only in breeding season, March to May). **VU**

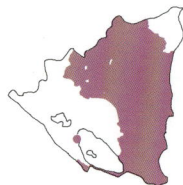

Great Curassow (Pavón Grande)   *Crax rubra*
35 in (89 cm). Largest of all cracids. **Curly-feathered crest** is prominent on both sexes. Male is unmistakable, with gaudy **yellow knob** on base of upper mandible; **glossy black plumage** is only interrupted by **white vent and undertail coverts**. Female varies in color, but most individuals have **rufous upperparts and chest, black and white barring on crest, and speckling on head**. Hunting and deforestation have extirpated it from most of its historical distribution (it once occurred countrywide). Rare and local in Pacific (recent reports from Rivas Isthmus and Mombacho Volcano NR), on eastern slopes of Northern Highlands, and in Caribbean (locally common in Musún NR, Saslaya NP, Bosawas NR, and Indio Maíz BR; uncommon in the Mosquitia); to 4,400 ft (1,300 m). Mostly dependent on mature forest, where pairs or small flocks feed primarily on the ground, seeking fruit and small invertebrates; usually shy and difficult to observe. Recognized by both a thin, high-pitched, descending whistle (2 seconds in length) and a punctuated, yippy *wheet!* call. Male produces a slow and deep *whuuUU* boom call; the resonant sound resembles a mammal-like growl or grunt. **VU**

female

male

Highland Guan

female
barred morph

female

Great Curassow

male

Occurring in many parts of the New World, these quail are characterized by short, curved bills; strong legs; a compact body; and short, rounded wings. All dwell exclusively on the ground, in forests, except for the Crested Bobwhite, which nests in trees and prefers open habitat. They travel in small groups, called coveys, in search of seeds, fruit, small invertebrates, and the occasional reptile or amphibian. When alarmed, new world quail either freeze or flush, flying away a short distance, low to the ground. Cryptically colored and elusive, these birds are more often heard than seen; their presence inside of dense forests is often announced with explosive dawn choruses.

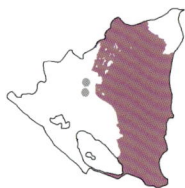

Tawny-faced Quail (Codorniz Carirrufa)   *Rhynchortyx cinctus*
7.5 in (19 cm). Small and compact. Male has **tawny-orange face, dark brown eye line,** and gray breast. On female, **white postocular stripe and throat** contrast with **brown head and breast**. Uncommon in Caribbean lowlands and foothills, locally common in Saslaya NP; to 2,300 ft (to 700 m). Status unknown on eastern slopes of Northern Highlands, but there are three historical records: Peñas Blancas NR, ca. 1926; Tuma River, April 1909 and ca. 1926. In small coveys, roams on the floor of humid lowland forest, in search of seeds, fruit, and small invertebrates. Delivers a sequence of 5–7 languid notes: *whuu whú whú whú whú-whú-whú*. The first note is lower-pitched than the rest of the series and is repeated at 2-second intervals for an extended period before the rest of the song is delivered; the final 2 or 3 notes are given more quickly.

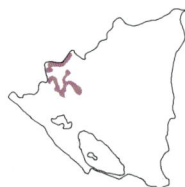

Buffy-crowned Wood-Partridge (Perdiz Cariblanca)   *Dendrortyx leucophrys*
13 in (33 cm). Large. Taken individually, each of these traits is unique among Nicaraguan quail: **long tail, red orbital skin, and red legs**. Further distinguished by **buffy-white forecrown**, superciliary, and throat. Uncommon in Northern Highlands, as far south as Arenal NR (no known records from Kilambé NR and Peñas Blancas NR); above 4,000 ft (1,200 m). Occurs in highland pine and pine-oak forests, cloud forest edges and adjacent coffee plantations, gallery forest, and second growth. Usually in pairs, foraging on the ground for seeds, fruit, leaves, and small invertebrates. Very shy and wary; when scared, walks toward the cover of dense vegetation. Delivers a continuous and scratchy *quuAH!* phrase, repeated once a second for lengthy periods of time; begins with a slight downslur before quick ascension. Also communicates within covey using soft, jittery chirps.

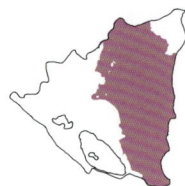

Black-eared Wood-Quail (Codorniz Pechicastaña)   *Odontophorus melanotis*
9.5 in (24 cm). No other Nicaraguan quail has **solid rufous breast and belly** and **black face and throat**. Short crest is only sometimes raised. Uncommon in Caribbean lowlands and foothills and locally common in Saslaya NP; to 3,300 ft (1,000 m). Coveys of up to 10 birds roam the ground of humid lowland forest and adjacent second growth, where they scratch the leaf litter for seeds, fruit, and small invertebrates. Proclaims a fast, oscillating song: *whút-u-whút-u-whút-u…*, seemingly without stopping; maintains communication within covey with short, slightly electronic sounding, chirps.

Spotted Wood-Quail (Codorniz Moteada)   *Odontophorus guttatus*
10 in (25 cm). **White streaks on black cheeks and throat** and **bold white spots on rufous underparts** are distinctive. Displays orange crest when excited. Locally common in Saslaya NP and Bosawas Biosphere Reserve; from 1,000 to 5,900 ft (350 to 1800 m). Rare in Northern Highlands; above 3,300 ft (1,000 m). Found in humid lowland forest and cloud forest. Moves about in coveys on the ground, scratching for seeds, fruit, and small invertebrates. Repetitively sings a fast, 4-syllable phrase: *wét-u-u-u…*, sometimes interrupted by 2 descending notes.

female   male

**Tawny-faced
Quail**

**Buffy-crowned
Wood-Partridge**

**Black-eared
Wood-Quail**

**Spotted
Wood-Quail**

Black-throated Bobwhite (Codorniz Gorginegra)   *Colinus nigrogularis*
8 in (20 cm). On male, combination of extensive **black throat and eye line** and **white superciliary and lateral throat stripe** is diagnostic; also note boldly scaled underparts. Female is very similar to female Crested Bobwhite, but has **scaled underparts** (not spotted); no range overlap. Locally common in the Mosquitia, where it is the only quail; to 700 ft (200 m). Reaches southernmost distribution in Nicaragua. In coveys, moves over the ground of lowland pine savanna; scratches in leaf litter for seeds and small invertebrates, and searches for fallen fruit. Intermittently whistles the onomatopoeic *bob whITE!* and maintains communication within the covey with short, sweet whistles delivered in quick succession.

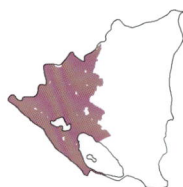

Crested Bobwhite (Codorniz Crestada)   *Colinus cristatus*
8 in (20 cm). Small crest. Male has black throat and **brown eye line** bordered by **white superciliary and cheek** (on female, superciliary, cheek, and throat are buffy); note spotted underparts. Female is very similar to female Black-throated Bobwhite, but has **spotted underparts** (not scaled); no range overlap. Common to abundant in Pacific lowlands and foothills; to 3,000 ft (900 m). Common in dry intermontane valleys of Northern Highlands (range expanding southeast into Sierra Chontaleña); to 5,200 ft (1,600 m). Occurs in thorn forest, scrub, dry forest, and savannas; scratches ground in search of seeds, fruit, and small invertebrates. Intermittently whistles a 3-note, onomatopoeic *bob bob-WHITE!*, with the final note abruptly rising in pitch and intensity; maintains communication within the covey with soft, sweet whistles delivered in quick succession.

Ocellated Quail (Codorniz Ocelada)   *Cyrtonyx ocellatus*
8 in (20 cm). No other Nicaraguan quail has the intricate **black-and-white facial pattern**, which is bold on male and faint on female and immature. Limited distribution: locally uncommon on Sierra Dipilto-Jalapa; status unknown throughout rest of Northern Highlands, with only 1 historical record (San Rafael del Norte, April 1917); from 3,300 to 5,300 ft (1,000 to 1,600 m). Occurs in highland pine and pine-oak forests, where it forages on the ground in small coveys; scratches for seeds and small invertebrates, and searches for fruit. Sings a single high-pitched, downslurred, buzzy whistle that abruptly ends with an up or down note; also gives a short series of 5–7 dulcet whistles (lasting 3 seconds) and repeats a soft but gravelly *feeur* call. Reaches southernmost distribution in Nicaragua. Endemic to North Central American highlands EBA. **VU**

female

male

Black-throated Bobwhite

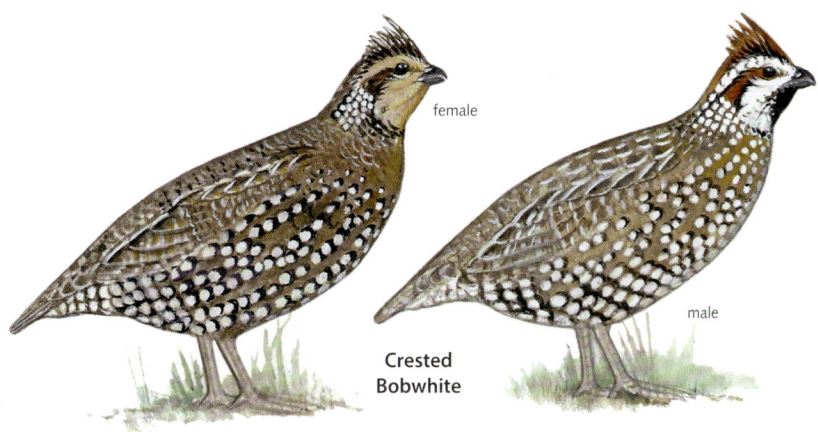

female

Crested Bobwhite

male

female

male

Ocellated Quail

A relatively small family of aquatic birds with worldwide distribution. Grebes superficially resemble ducks (family Anatidae), but have lobed toes, pointed bills, and extremely short tails; they are also generally more cautious than ducks. Associated with tranquil freshwater habitats, grebes are frequently seen sitting on the water's surface, but proficiently dive underwater to hunt prey (small fish, invertebrates, and crustaceans); their diving skills are due in part to feet that are set far back on their bodies. When startled, grebes prefer to dive rather than fly, disappearing underwater only to emerge far away from the point of entry. Males and females are similar; breeding birds look slightly different from nonbreeding birds.

Least Grebe (Zampullín Enano)   *Tachybaptus dominicus*
9 in (23 cm). Petite body, vivid **yellow eye**, and **slender dark bill** distinguish it from larger Pied-billed Grebe. Common countrywide; to 4,600 ft (1,450 m). Occurs in marshes and swamps and on ponds, lakes, and slow-moving rivers bordered by dense vegetation; dives for small fish, invertebrates, and crustaceans. Found in pairs or small groups, often near Pied-billed Grebes. Utters an 8–9 second, low-pitched trill that trails off at the end; very similar to that of Ruddy Crake (p. 96) and White-throated Crake (p. 96), but drier and more nasal. Also barks a loud, emphatic, nasal *ERRH-ERRH-ERRH…*, which often continues for a long time.

Pied-billed Grebe (Zampullín Piquipinto)   *Podilymbus podiceps*
13 in (33 cm). Larger size, thin white eye ring, **dark iris**, and **thick whitish bill** distinguish it from Least Grebe. Note **black ring on bill** of breeding birds. Common in Pacific lowlands and foothills; to 1,500 ft (500 m). Locally common at Lake Apanás and ponds in Miraflor NR; otherwise uncommon in Northern Highlands; to 4,800 ft (1,450 m). Uncommon in Caribbean. Inhabits marshes, lakes, ponds, and slow-moving rivers; can also be found on brackish waters. Dives for larger prey than does Least Grebe, including frogs, large crustaceans, fish, and aquatic insects. Found in pairs, but also in small groups, often near Least Grebe. Whistles a series of pure toot notes (*huh!-huh!*), followed by a soft, trilling *bidibidi*.

The common name comes from their lobately webbed feet, which are adapted for locomotion in aquatic habitats. This family has only 3 species: 1 in the Neotropics and 1 each in Africa and southeast Asia. Although they resemble grebes at first glance, they are more closely related to rails. Males exhibit a peculiar parental-care behavior, in which they transport helpless hatchlings to safety in underwing skin pockets. They feed on insects, snails, spiders, frogs, and lizards gleaned from vegetation hanging over the water's edge.

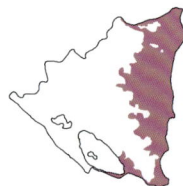

Sungrebe (Pato Cantil)   *Heliornis fulica*
11 in (28 cm). Distinguished by a combination of plumage (**black-and-white striped head and neck**), behavior (pumps its head while swimming), and habitat (found close to overhanging vegetation on calm waters). Breeding female has reddish bill and tawny cheek. Locally common in Caribbean lowlands; to 700 ft (200 m). Northernmost populations extend south on tributaries of the Coco River. Accidental on eastern slopes of Northern Highlands. Prefers calm waters (lakes and ponds, rivers and streams, lagoons, mangroves, and places within flooded forests) where there is prevalent overhanging vegetation at the water's edge. Can be found roosting in low hanging branches during midday and at night; very shy and wary and easily overlooked. Yelps a resonating *WUH!* Also gives a variety of weak, nasal yips and ticking noises.

nonbreeding

Least Grebe

breeding

nonbreeding

Pied-billed Grebe

breeding

male/
nonbreeding
female

Sungrebe

breeding
female

Members of this family occur on every continent except Antarctica, in a variety of habitats, including deserts and rainforests. They are characterized by robust bodies and relatively small heads; short necks, legs, and bills enhance their compact appearance. The members of the genera that occur in Nicaragua typically feed either in trees (mainly fruit) or on the ground (mainly seeds). Although it can be a challenge to identify species, determining genera is often straightforward: *Patagioenas* pigeons perch high in treetops of forest and forest edge and are often only observed in flight; *Columbina* ground-doves are small-bodied residents of open country, and all flash rufous primaries in flight; *Claravis* ground-doves are medium-bodied residents of forests, and relatively arboreal; *Geotrygon* quail-dove are large-bodied birds that inhabit the forest floor; *Leptotila* doves have large bodies and pale coloration and reside in forests and at forest edges, and in flight, all flash white-tipped outer rectrices; and *Zenaida* doves are large-bodied, pale-colored residents of open areas.

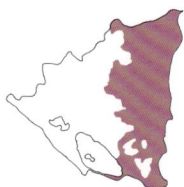

Pale-vented Pigeon (Paloma colorada) — *Patagioenas cayennensis*

12 in (31 cm). Multi-colored overall, but underparts are **whitish, from lower belly to undertail coverts**. **Grayish tail** has a pale broad terminal band, a characteristic used to distinguish it in flight from Scaled Pigeon (black tail), p. 48; also note **black bill**. Common in Caribbean lowlands and foothills; to 900 ft (300 m). Perches atop tall trees. Found at the edges of humid lowland forests, second growth, lowland pine savanna, and mangroves. Alone or in flocks, feeds on berries of fruiting trees. Coos a rhythmic 4-phrase song: *wuuUUu HU-hu-huu HU-hu-huu HU-hu-huu*.

Red-billed Pigeon (Paloma Piquirroja) — *Patagioenas flavirostris*

13.5 in (34 cm). Plumage is vinaceous-purple and blue-gray; **mostly white bill** with red base (red best seen up close) distinguishes it from similar, but smaller, Short-billed Pigeon (all-dark bill). In flight, blue-gray underparts (belly to undertail coverts) distinguish it from the other *Patagioenas* pigeons and contrast with **dark tail**. Common in Pacific and in Northern Highlands; to 6,500 ft (2,000 m). Rare in Caribbean lowlands and foothills; to 2,600 ft (800 m). Inhabits open areas with scattered trees; scrub; and thorn forest. Alone or in pairs, feeds on seeds, nuts, and figs. Coos a *wuuú wut!-wu-wu-wuú*; the first note ascends slowly, and the following notes are delivered quickly—and often repeated several times.

Band-tailed Pigeon (Paloma Collareja) — *Patagioenas fasciata*

14 in (36 cm). No other pigeon has a **yellow bill with a black tip**, yellow legs, or a **white nuchal color**. In flight, note **whitish underparts (lower belly to undertail coverts)** and **broad, pale terminal band** (Scaled Pigeon has a black tail). Uncommon in Northern Highlands (formerly common); above 3,300 ft (1,000 m). Perches atop tall trees, often out of sight. Occurs throughout highland pine and pine-oak forests; occurs locally in cloud forests. Feeds on pine nuts, acorns, and berries. Seasonal migrations follow acorn masts. Coos a throaty 1-note *whuU* and a 2-note *whú-uu*; both are sometimes repeated at length and interspersed with a *whuU whú-uu*. Also emits a nasally growl.

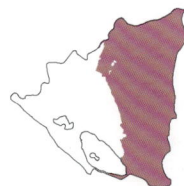

Short-billed Pigeon (Paloma Piquicorta) — *Patagioenas nigrirostris*

11 in (28 cm). Has **unpatterned, vinaceous-purple plumage**; also note **short, black bill** (similar, but larger, Red-billed Pigeon has white bill with red base). Common in Caribbean, to 3,300 ft (1,000 m); accidental on eastern slopes of Northern Highlands. Perches inconspicuously in humid lowland forest and edges, but will venture out to semi-open areas and second growth to forage. Feeds on berries (including mistletoe) and *Cecropia* fruit at mid-canopy and canopy. More often heard than seen, calling loudly and repetitively from canopy; makes a somewhat rigid, 4-syllable *whUT-hu-huhuu*, the second note jumping to the highest pitch of the phrase.

Pale-vented
Pigeon

Red-billed
Pigeon

Band-tailed
Pigeon

Short-billed
Pigeon

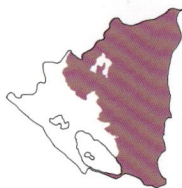

Scaled Pigeon (Paloma Escamosa) — *Patagioenas speciosa*

12.5 in (32 cm). **Heavy scaling on neck and underparts** is unique; also note **red bill** with white tip. In flight, **whitish underparts (lower belly to undertail coverts)** contrast with **black tail**; Pale-vented Pigeon (p. 46) has gray tail, Band-tailed Pigeon (p. 46) has pale tail. Uncommon throughout Caribbean lowlands and foothills, including on eastern slope of Sierra Amerrisque; to 1,600 ft (500 m). Rare on eastern slopes of Northern Highlands; to 4,900 ft (1,500 m). Prefers primary and secondary forest (often overlooked in dense canopy), but is also seen perched atop tall trees at forest edge, river edge, and in semi-open areas; feeds alone or in flocks on berries of fruiting trees. Coos 2 mournful renditions with the same notes but different phrasing, *huú-hu-huú* and *hu-hu-huú*.

White-crowned Pigeon (Paloma Gorriblanca) — *Patagioenas leucocephala*

13.5 in (34 cm). Conspicuous **white crown** (mostly absent on immature) is diagnostic; it contrasts sharply with **dark gray plumage**. Locally common on Pearl Keys and Corn Islands; most likely rare along coastal Caribbean lowlands, but not yet recorded there. Prefers islands but also visits mainland forests and flies between the two in search of fruiting trees (observed flying from Little Corn Island directly to mainland Nicaragua). Feeds on fruit and berries, sometimes hanging upside down to reach them. Coos a 3-note *hÚT-hu-huu*, sometimes followed by a soft, sweet growl. **NT**

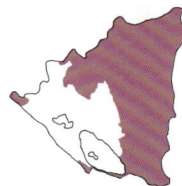

Blue Ground-Dove (Tortolita Azulada) — *Claravis pretiosa*

8 in (21 cm). **Largest** of the ground-doves; **yellowish bill** is diagnostic on both sexes. **Blue-gray** male is unmistakable. Female (often with male) is similar to female Maroon-chested Ground-Dove (not yet recorded, but expected in Northern Highlands), p. 413, but note all-rufous tail and rump, yellowish bill, and wing covert pattern; has **rufous spots on wing coverts** (smaller female Ruddy Ground-Dove has black spots). Common throughout Caribbean lowlands and foothills. Rare on eastern slopes of Northern Highlands, on Cosigüina Peninsula, and in Rivas Isthmus; to 6,200 ft (1,900 m). Prefers forest, forest edge, second growth, and adjacent forest patches. Feeds on seeds and small insects on the ground. Perches at mid-canopy and canopy (more arboreal than other ground-doves); found singly and in pairs. Repeats a loud, double-noted *HUUP HUUP*; the double note lasts a second, with a two-second interval between the double notes. Sometimes repeats a single *HUUP* note in a continuous series.

**Scaled
Pigeon**

immature

adult

**White-crowned
Pigeon**

female

male

**Blue
Ground-Dove**

Inca Dove (Tortolita Colilarga)   *Columbina inca*

8 in (21 cm). Combination of **extensive scaling** and **long, slender tail** distinguishes it from other ground-doves. In flight, note **white outer rectrices** and rufous primaries; **wing coverts lack spots**. Abundant in Pacific, common in Northern Highlands, and uncommon on Sierra Chontaleña (following deforestation) and extreme southwestern Caribbean; to 4,600 ft (1,400 m). Prefers arid and semi-arid habitat. Seems to tolerate the presence of humans. In pairs or flocks, forages for seeds on the ground, and perches in understory of nearby vegetation; sometimes joins Common Ground-Dove and Ruddy Ground-Dove. Continuously whistles a 2-note *hú hú* (repeated every 1–2 seconds), the second note slightly lower in pitch than the first. In flight, produces a distinct mechanical rattling noise.

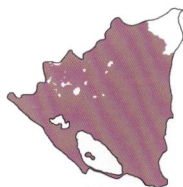

Common Ground-Dove (Tortolita Común)   *Columbina passerina*

6.5 in (17 cm). Has **less scaling** (limited to head and breast) than Inca Dove; all other small doves lack scaling. Only dove with **black-tipped, pink-orange bill** (flesh-colored on female). In flight, flashes black outer rectrices (inconspicuously white-tipped) and rufous primaries. Common in Pacific and Northern Highlands; to 4,600 ft (1,400 m). Rare in Caribbean lowlands and foothills. Occurs in arid and semi-arid habitat. Found in pairs or small flocks; sometimes joins Inca Dove and Ruddy Ground-Dove. Forages for fruit and seeds it picks from the ground, often on unpaved roads. Delivers an almost constant, clear, melancholic *huuá huuá huuá…*, at a rate of 1 note per second, with each note smoothly rising in pitch at end. Gives a slower delivery than does Ruddy Ground-Dove.

Plain-breasted Ground-Dove (Tortolita Menuda)   *Columbina minuta*

6 in (15 cm). **Smallest** of the ground-doves; notably petite. Male (grayish) and female (brown) both similar to female Ruddy Ground-Dove, but **dark purple spots on wing coverts** and **gray bill** on Plain-breasted help distinguish them. In flight, flashes black outer rectrices (inconspicuously white-tipped) and rufous primaries (less extensive and paler than in other ground-doves). Uncommon in Northern Highlands; to 4,600 ft (1,400 m). Rare in Caribbean (local in the Mosquitia). Very rare in Pacific lowlands and foothills; to 1,900 ft (600 m). Prefers grasslands and semi-arid areas, where it forages for seeds and fruit on the ground of open understory; found singly or in pairs. Delivers an incessant *huúp!-huúp!-huúp!…*, at a rate of 2 notes per second, with each note abruptly and upwardly inflected at end.

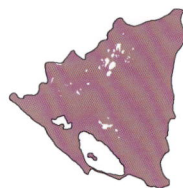

Ruddy Ground-Dove (Tortolita Rojiza)   *Columbina talpacoti*

7 in (18 cm). Only ground-dove with black axillaries (seen in flight). **Rufous plumage** of male is unmistakable. Female (often with male) is similar to Common Ground-Dove, but has **plain breast** (not scaled); she is distinguished from smaller Plain-breasted Ground-Dove and larger female Blue Ground-Dove (p. 48) by **black spots on wing coverts**; also note **dark bill**. In flight, flashes black outer rectrices (inconspicuously white-tipped) and rufous primaries. Abundant in Pacific and common in Northern Highlands; to 4,900 ft (1,500 m). Common in Caribbean; to 1,600 ft (500 m). Forages for seeds on exposed ground of open areas, second growth, woodland edge, shrubby areas, and gardens; found in pairs and sometimes joins Inca Dove and Common Ground-Dove. Delivers a constant, rapid, hoarse *huá-huá-huá…*, at a rate of 2 notes per second, each note quickly rising in pitch at end; gives a faster delivery than that of Common Ground-Dove.

Inca Dove

male

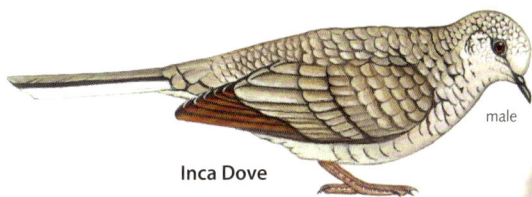

Common Ground-Doves and an Inca Dove (middle) feeding by a gravel road.

Common Ground-Dove

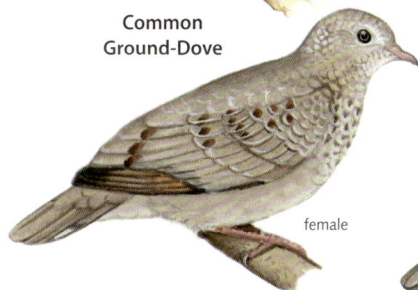

male

male

female

Plain-breasted Ground-Dove

male

female

Ruddy Ground-Dove

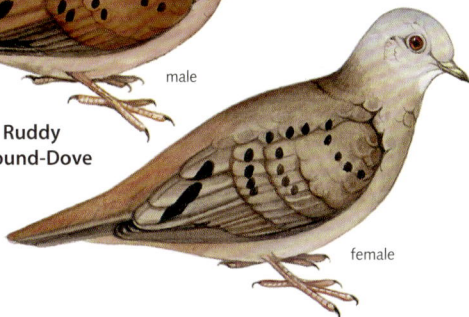

male

female

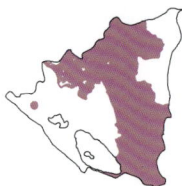

Ruddy Quail-Dove (Paloma Perdiz Rojiza)   *Geotrygon montana*
9 in (23 cm). Male is only quail-dove with warm **rufous-brown upperparts and a moustachial stripe**; female distinguished by **olive-brown** plumage. Uncommon in Caribbean lowlands and foothills (locally common in foothills of Bosawas Biosphere Reserve). Uncommon in Northern Highlands and locally very rare on Casita Volcano (two historical records: April 1907 and Sept 1996); to 4,600 ft (1,400 m). Inhabits primary and secondary forest. Singly or in pairs, walks on the ground in search of seeds, insects, and fallen fruit. When startled, freezes in place or flies to a low perch. Coos a quiet, mournful, and mostly monotone *wuuu* (1 second in length) that is repeated every 2–3 seconds. Call is similar to that of Violaceous Quail-Dove but noticeably lower pitched; also compare with call of Gray-chested Dove (p. 54).

Violaceous Quail-Dove (Paloma Perdiz Violácea)   *Geotrygon violacea*
8.5 in (22 cm). Note **glossy-violet nape and back** and **chestnut tail**. Has a **deep red bill with a black tip** (White-faced Quail-Dove has a black bill). Very rare on eastern slopes of Northern Highlands, but a total of 6 historical records from Peñas Blancas NR (May 1909) and Arenal NR (April 1996) suggest it was common. Very rare in Caribbean, but uncommon locally in foothills of Saslaya NP; to 4,600 ft (1,400 m). Reaches northernmost distribution in Nicaragua. Inhabits dense understory of cloud forest and humid lowland forest; forages on the ground for seeds, insects, and fallen fruit. Perches high in understory, where it is easily overlooked; shy and more arboreal than other quail-doves. Coos a mostly monotone *wuuu* (1 second in length) that is repeated every 2–3 seconds. Song similar to those of Ruddy Quail-Dove and White-faced Quail-Dove but noticeably higher pitched; also compare to song of Gray-chested Dove (p. 54).

Olive-backed Quail-Dove   *Leptotrygon veraguensis*
(Paloma Perdiz Bigotiblanca)
9 in (23 cm). **White forecrown and cheek stripe** contrast strongly with otherwise **dark plumage**. Probably rare to uncommon in Indio Maíz BR, with only 2 records (1999, 2004). Reaches northernmost distribution in Nicaragua. Forages singly or in pairs on the ground of humid lowland forest with dense understory. Feeds on seeds, insects, and fallen fruit. Coos a low-pitched, quiet, and lethargic *wuuuuuu* (2 seconds in length) that is repeated every 2–3 seconds; note a slight mid-note spike in pitch. Also utters a short, amphibian-like *boyee*, with a muffled quality.

White-faced Quail-Dove (Paloma Perdiz Cariblanca)   *Zentrygon albifacies*
11.5 in (29 cm). Has a distinctive head and neck pattern; note **white face** and **dark furrows on sides of neck**. Violet coloration on back is similar to that on Violaceous Quail-Dove, but note black bill. Locally common on eastern slopes of Northern Highlands; above 4,000 ft (1,200 m). Locally uncommon in Caribbean highlands of Musún NR (probably in nearby Quirragua NR) and Saslaya NP; above 4,300 (1,300 m). Reaches southernmost distribution in Nicaragua. Prefers cloud forest with dense understory, but also found in highland pine forest, where it is uncommon. Shy and inconspicuous. Walks quietly on ground in search of seeds, insects, and fallen fruit; roosts in vegetation above ground. Usually alone, but can be in small groups. Coos a very quiet, mournful, and monotone *huuaw* (1 second in length) that is repeated every 3 seconds, similar to that of Violaceous Quail-Dove but noticeably lower pitched; also compare to song of Gray-chested Dove (p. 54).

female

Ruddy
Quail-Dove

male

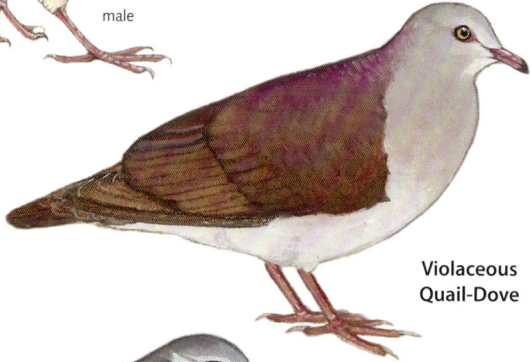

Violaceous
Quail-Dove

Olive-backed
Quail-Dove

White-faced
Quail-Dove

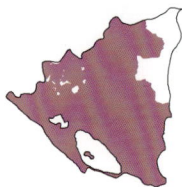

White-tipped Dove (Paloma Coliblanca)   *Leptotila verreauxi*
11 in (28 cm). Has a **pink wash on hindcrown and nape** (wash sometimes blends into breast); of the *Leptotila*, has the most extensive white-tipped outer rectrices. Orbital skin color varies from red (in the north) to blue (in the south). Common in Pacific and uncommon in Northern Highlands; to 4,600 ft (1,400 m). Rare in Caribbean (except for extreme southwestern corner, where it is common). Prefers dry forest, second growth, and scrub, but also found in cloud forest edge and highland pine forest. Walks alone or in pairs on ground, gathering seeds, grit, and small insects. Calls from low branches; repeats a mournful and hollow 2-note w*hu-huuú* (1 second in length) with the second note slightly ascending at end; repeated every 7–8 seconds. Sometimes it only gives the second note.

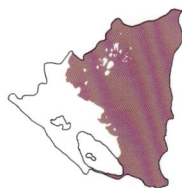

Gray-chested Dove (Paloma Pechigís)   *Leptotila cassini*
10 in (25 cm). Has **warm brown hindcrown and nape** and **gray breast**; shows least extensive white-tipped outer rectrices of the *Leptotila*. Uncommon in Caribbean lowlands and foothills, but locally common in foothills of Saslaya NP; to 2,100 ft (650 m). Rare on eastern slopes of Northern Highlands; to 4,600 ft (1,400 m). Found in humid lowland forest edge and adjacent second growth and in semi-open areas such as cacao plantations. Wanders alone or in pairs on ground, collecting seeds, grit, and occasionally insects; also picks fruit from shrubs and trees. Repeats a mournful and gravelly w*huuu* (1 second in length), repeated every 3 seconds with consistent pitch and tone; compare to song of Quail-Dove (p. 52).

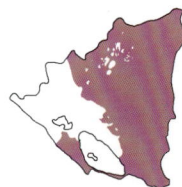

Gray-headed Dove (Paloma Cabecigrís)   *Leptotila plumbeiceps*
10 in (26 cm). Has **blue-gray hindcrown and nape** and **buffy cheeks and breast**; shows less extensive white-tipped outer rectrices than on White-tipped Dove. Uncommon in Caribbean lowlands and foothills; to 2,000 ft (600 m). Locally rare on eastern slopes of Northern Highlands; to 4,600 ft (1,400 m). Locally rare in Pacific lowlands and foothills, from Sierras Managua and south; to 1,600 ft (500 m). Inhabits forest edge, second growth, and semi-open areas such as cacao plantations. Singly or in pairs, picks seeds, grit, fallen fruit, and sometimes small insects off the ground. Makes a mournful and slightly nasal, single-noted w*uóh* (1 second in length), repeated every 2 seconds.

White-tipped
Dove

Gray-chested
Dove

Gray-headed
Dove

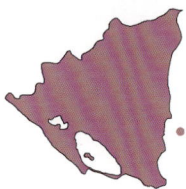

White-winged Dove (Tórtola Aliblanca)   *Zenaida asiatica*
11 in (28 cm). Only dove with **white wing stripe**, which looks like a crescent when the wing is folded. Also note broad, white-tipped outer rectrices. Breeding resident population is joined by NA migrants (Sept to April). Abundant in Pacific, common on western slopes of Northern Highlands, rare on eastern slopes of Northern Highlands and in Caribbean. Prefers thorn forest, scrub, mangroves and edge, gardens, and urban areas. Forages on ground for seeds and grit, and also picks fruit from shrubs; not wary of humans. Sings a fast and rythmic 7-note *huh-hú-hu-huú-huu-huó-huuh* in a variety of tones; also makes a 4-note *hu-húh-húh-huuh*. Both are raspy but calm, with the final note slightly drawn out.

Mourning Dove (Tórtola Rabuda)   *Zenaida macroura*
11.5 in (29 cm). Has a **long, pointed tail, white-tipped outer rectrices**, and a slim body. Larger than Inca Dove (p. 50); also lacks extensive scaling and primaries are black (not rufous). Locally common breeding resident in Pacific; winter resident (Oct to March), when it is common throughout Pacific and rare on western slopes of Northern Highlands, but locally common in Mesa del Moropotente in Miraflor NR. Found in thorn forest and scrub, mangroves, and salt ponds. Forages on ground, alone or in small flocks, for seeds, grit, and insects. As name implies, coos a mournful 4-note *huuUU-hu-hu-huu*, with the first note ascending in pitch and volume.

Eurasian Collared-Dove (Tórtola Turca)   *Streptopelia decaocto*
13 in (33 cm). **Black collar** on nape is diagnostic. In flight, square tail distinguishes it from smaller Mourning Dove (which has graduated tail). A non-native species recently documented for the first time in Nicaragua (Cosigüina Peninsula, June 2017); this suggests that it has started colonizing the country, moving south from El Salvador or Honduras. Native to subtropical Asia, it was introduced to the US in the mid-1970s. Found in open areas, agricultural land with scattered trees, gardens, and parks in urban areas. Perches on exposed telephone lines and other human structures. Coos a hollow, hoarse, frequently repeated *hu-hooo huh!* Call is a nasal, gravelly *wyáuh*.

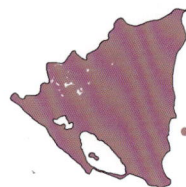

Rock Pigeon (Paloma de Castilla)   *Columba livia*
12.5 in (32 cm). No other pigeon has **two dark wing bars**. Four color morphs exist, including an all-white morph (rare) with no wing bars. Common to abundant countrywide; to 4,900 ft (1,500 m). Introduced pigeon (native to N. Africa and Eurasia), most often found in parks and streets in urban areas. Feeds on seeds and discarded food.

White-winged
Dove

Mourning
Dove

Eurasian
Collared-Dove

Rock Pigeon
(various color morphs)

The members of this cosmopolitan family are characterized by slender bodies and long, graduated tails. While some species are solely terrestrial (Lesser Roadrunner and Rufous-vented Ground-Cuckoo), most are partly or completely arboreal. Although this family is notorious for brood parasitism, only 2 species exhibit this behavior in Nicaragua, the Striped Cuckoo and Pheasant Cuckoo, who lay their eggs in the nests of other species, which unknowingly incubate and raise the cuckoo hatchlings. Cuckoos are normally solitary (with the exception of anis). They are inconspicuous and often only sing during the breeding season, which means that it is often difficult to spot them the rest of the year.

Dark-billed Cuckoo (Cuclillo Piquioscuro)   *Coccyzus melacoryphus*
11 in (28 cm). Very similar to larger Mangrove Cuckoo, but bill is dark (no yellow on lower mandible), mask is dusky (not black), and the whitish-gray on sides of the face extends down to upper flanks. Vagrant, with 1 country record from mangroves on Indio River in Indio Maíz BR (April 2017). Occurs in gallery forest and mangroves, where it feeds on grasshoppers, caterpillars, cicadas, spiders, and lizards. Calls include a low, growl-like wooden rattle.

Yellow-billed Cuckoo (Cuclillo Piquigualdo)   *Coccyzus americanus*
12 in (30 cm). Similar to Black-billed Cuckoo and Mangrove Cuckoo but note **mostly yellow bill** (Mangrove has some yellow on lower mandible) and **rufous primaries**; further distinguished from Black-billed Cuckoo by broad white-tipped outer rectrices that appear as **large white spots on undertail**. Note yellow eye ring. Rare to uncommon passage migrant (Sept to Nov and mid-April to May). Occurs countrywide; to 4,400 ft (1,350 m). Favors forest and forest edge, mangroves, coastal scrub, and shrubby vegetation along lakeshores. Searches for caterpillars, larvae, cicadas, and other insects in mid-canopy and canopy. Furtive, inconspicuous, and generally silent in migration.

Mangrove Cuckoo (Cuclillo Enmascarado)   *Coccyzus minor*
12 in (30 cm). **Buffy underparts** and **black mask** distinguish it from both Yellow-billed Cuckoo and Black-billed Cuckoo. Also note yellow eye ring. Yellow on lower mandible distinguishes it from smaller Dark-billed Cuckoo. Common in Pacific lowlands and foothills and southeastern Caribbean lowlands (mid-Oct to May), but not clear whether population is of winter residents or breeding residents. Very rare in Northern Highlands (most likely only a passage migrant). Breeding resident on Corn Islands and potentially along Caribbean coast. Found in scrub, marsh thickets, and mangroves. Perches motionless for long periods and walks or hops very slowly from branch to branch in search of grasshoppers, caterpillars, cicadas, spiders, and lizards. Sings in breeding season with a 2- to 8-second series of dry, staccato notes; trails off in speed and intensity at end; somewhat resembles croaking sound of frogs.

Black-billed Cuckoo (Cuclillo Piquinegro)   *Coccyzus erythropthalmus*
12 in (30 cm). Similar to Yellow-billed Cuckoo and Mangrove Cuckoo but distinguished by **black bill** and narrow white-tipped outer rectrices that appear as **small white spots on undertail**; further distinguished from Yellow-billed Cuckoo by lack of rufous on primaries. Note **red eye ring**. Rare to uncommon passage migrant (mid-Sept to Nov and April to May), in Pacific, Northern Highlands, Sierra Chontaleña, southeastern shores of Lake Nicaragua, and San Juan River headwaters; to 4,400 ft (1,350 m). Prefers forest, forest edge, second growth, and scrub. Forages for caterpillars, cicadas, and other insects. Very secretive, elusive, and generally silent in migration; possibly more common than records suggest.

**Dark-billed
Cuckoo**

undertail

**Yellow-billed
Cuckoo**

**Mangrove
Cuckoo**

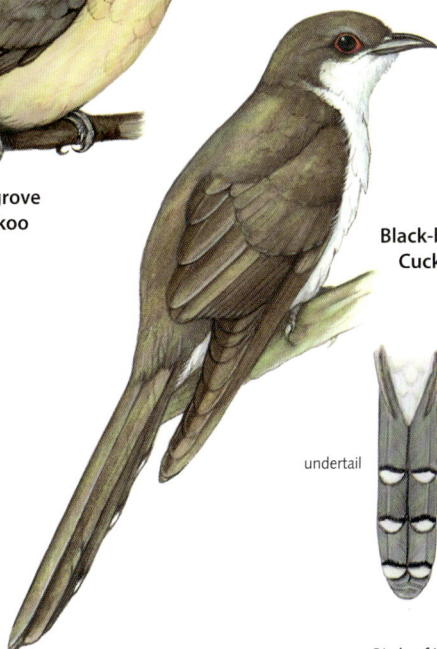

**Black-billed
Cuckoo**

undertail

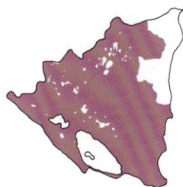

Striped Cuckoo (Cuclillo Listado)   *Tapera naevia*
11.5 in (29 cm). Has **buffy upperparts** with **bold black streaking**; smaller than Pheasant Cuckoo and **lacks spots on throat and breast**. Common in Pacific lowlands and foothills; to 2,800 ft (850 m). Uncommon in Northern Highlands; to 4,600 ft (1,400 m). Rare in Caribbean lowlands and foothills (expanding range eastward with deforestation); to 2,000 ft (600 m). Occurs in second growth, scrub, and a variety of open areas with scattered trees and thickets. Forages within thickets or on ground for grasshoppers, caterpillars, spiders, and other insects. Known nest parasite of Rufous-and-White Wren (p. 318) and Cabanis's Wren (p. 318). Very vocal; sings a variety of combinations of clean, short, high-pitched notes, sometimes piercing in quality, while simultaneously raising and lowering crest. Most common call is a 2-note *fe-fi* (second note with noticeable rise in pitch); this is sometimes extended into a crescendoing 3–7 note phrase: *fe-fi-fi-feeAW*.

Pheasant Cuckoo (Cuclillo Faisán)   *Dromococcyx phasianellus*
15 in (38 cm). **Very long, broad, graduated tail** is distinctive; larger than Striped Cuckoo and has **dark spots on throat and breast**. Upperparts are dark brown. Common in Northern Highlands, locally common in Caribbean foothills and highlands, and accidental in Pacific foothills (San Cristóbal Volcano, May 1907); generally above 2,000 ft (700 m) but sometimes lower. Forages on ground of primary forest for insects, lizards, and nesting birds. Very secretive and furtive, it is more likely to be heard than seen. Vocalizes only during breeding season (Feb to early Aug); sings from mid- to upper canopy (terrestrial when not singing). Whistles 2 songs. The first (*fuh-fee-fweeáw*) ends with a tremulous and drawn out note; the second (*fuh-fee-fi-fi-fi*) ends with 3 rapidly delivered notes.

Smooth-billed Ani (Garrapatero Piquiliso)   *Crotophaga ani*
14 in (35 cm). **Higher-arched culmen** (without grooves on upper mandible) than on smaller Groove-billed Ani. Occurs only on Corn Islands, where it is common; to 200 ft (50 m). Found in open areas, grassland, second growth, and roadsides. Most often in small flocks; forages on ground for insects (associates with cattle). Repeats a strained *weeeUP!*

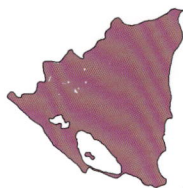

Groove-billed Ani (Garrrapatero Común)   *Crotophaga sulcirostris*
12.5 in (32 cm). Has **lower-arched culmen** than on Smooth-billed Ani. Grooves on upper mandible are difficult to observe. Abundant countrywide; to 5,000 ft (1,500 m). Ubiquitous in open areas, grassland, second growth, and roadsides. Forages on ground for insects; often feeds on prey disturbed by grazing cattle, agricultural machinery, or army ant swarms. Produces a sharp, high-pitched *TEEah, TEEah*; also delivers a constant *petah-petah-petah… .*

Striped
Cuckoo

Pheasant
Cuckoo

Smooth-billed
Ani

Groove-billed
Ani

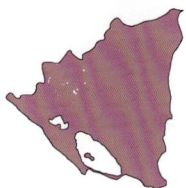

Squirrel Cuckoo (Cuco Ardilla)   *Piaya cayana*

18 in (46 cm). **Very long, graduated tail** with **broad, white-tipped rectrices** makes it unmistakable; rufous upperparts. Common countrywide; to 4,900 ft (1,500 m). Found in all forest types and forest edge, scrub, and second growth. Forages for insects, lizards, eggs, and small fruits at mid-canopy and canopy. Glides from one tree to the next; hops from branch to branch in a squirrel-like manner. Displays an impressive repertoire of vocalizations. Gives an explosive *PIP!-PIP!* or *PIP!-yeur*; a raspy, nasal 3-note *whért-ter-der*; and a lengthy, agitated, squirrel-like rattle. Sings a high-pitched, monotone *whip-whip-whip…*, at a rate of 2 notes per second, carrying on for a prolonged period of time.

Lesser Ground-Cuckoo (Cuclillo Sabanero)   *Morococcyx erythropygus*

10.5 in (27 cm). Unmistakable. **Colorful orbital skin** (yellow in front and blue behind) is surrounded by black that extends to sides of neck; also note **cinnamon underparts**. Common in Pacific lowlands and foothills; to 2,600 ft (800 m), and higher to peak of Casita Volcano. Uncommon in dry intermountain valleys of Northern Highlands; to 4,300 ft (1,300 m). Prefers arid habitat such as thorn forest, scrub, and dry forest. Mainly terrestrial, foraging on ground within thickets for insects; freezes when alarmed. Often sings while exposed on perch. Carols with multiple, short rolling whistles (lasting 7 seconds) that begin quickly but slow down at the end. Also repeats a plaintive, double-noted *wee-der* (every 4 seconds) and makes a similar, single *wher*.

Lesser Roadrunner (Correcaminos Menor)   *Geococcyx velox*

18 in (46 cm). Unmistakable. **Extensive white streaking on brown upperparts** and buffy underparts; long tail. Common in Northern Highlands; very rare and local on Sierra Maribios; from 1,600 to 5,200 ft (500 to 1,600 m). Rare on western slopes of Sierra Chontaleña, as far south as Amerrisque NR; above 700 ft (200 m). Reaches southernmost distribution in Nicaragua. Found in arid habitat such as highland pine forest and edge, thorn forest, and scrub. Mostly terrestrial but often seen perched atop rocks with its tail characteristically cocked upward. Furtive; forages for insects and small vertebrates. Sings a slow, mournful *UUh uuh uuh eeuu eeuu eeuu*, with the final notes noticeably sliding down in pitch. Length of song varies from 2 to 6 notes, with 1 second between notes. Unique sound resembles a whimpering dog.

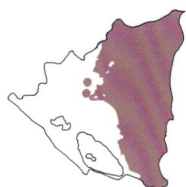

Rufous-vented Ground-Cuckoo (Cuco Hormiguero)   *Neomorphus geoffroyi*

19.5 in (50 cm). Difficult to find but unmistakable. Note **large body, long-tail**, prominent **erect crest**, and **dark breast band**. Rare in Caribbean lowlands and foothills, locally common in Saslaya NP; to 2,900 ft (900 m). Likely extirpated from eastern slopes of Northern Highlands (Peñas Blancas NR, May 1909; Tuma River, March 1909), Inhabits primary forest, where it inconspicuously follows army ant swarms to catch fleeing large insects and spiders. Gives a low-pitched, dove-like *whuuuú* (2 seconds in length) repeated every 4 seconds; regularly exhibits bill clacking. **VU**

Squirrel
Cuckoo

Lesser
Ground-Cuckoo

Lesser
Roadrunner

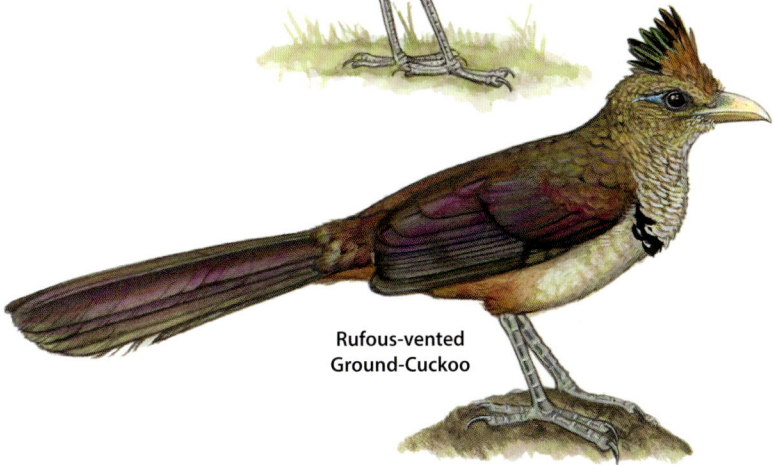

Rufous-vented
Ground-Cuckoo

Although cosmopolitan, this family is most diverse in the Neotropics. A large gape, large eyes, and rictal bristles aid in detecting prey, and allow these birds to hunt insects in low-light situations. The members of this family are grouped into nighthawks and nightjars. Nighthawks forage on the wing at twilight; nightjars sally from the ground or a low perch, strictly at night. Both are cryptic in coloration, with complex plumage patterns. This, combined with their silent disposition and motionless horizontal posture, makes it difficult to find them on their exposed day roosts. Identification without vocalizations can be quite challenging; it is helpful to note throat coloration, presence or absence of nuchal collar, and tail pattern (which is often sexually dimorphic).

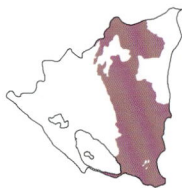

Short-tailed Nighthawk (Añapero Colicorto)   *Lurocalis semitorquatus*
8 in (20 cm). **Dark overall**, with mostly chestnut body and a **white throat patch**. Flight silhouette is distinctive because of **long, blunt-tipped wings** and a **short, squared tail**. Rare in Caribbean lowlands and foothills; to 1,600 ft (500 m). Forages on the wing and with erratic flight, within and above the canopy of humid lowland forest and edges; also flies above forest rivers. Roosts on horizontal branches in mid-canopy and canopy. Most often heard vocalization is an upslurred, liquid *whit!*, which it makes in flight; this is sometimes repeated in fast succession. Also repeats a *ewIT-ewIT-ewIT…* (each note upslurred).

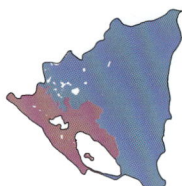

Lesser Nighthawk (Añapero Menor)   *Chordeiles acutipennis*
8.5 in (22 cm). Very similar to Common Nighthawk, which also has long wings with a white band on the primaries and a white (male) or buffy (female) throat patch, but distinguished by the following subtle characteristics. At rest, the **wingtips are short of, or barely reach, the tip of the tail**. In flight, **primary band is closer to the wingtip**, and wingtips look slightly rounded because the outermost primary is shorter than the next primary feather. There are 3 subspecies. The *littoralis* subspecies is a breeding resident throughout Pacific lowlands and foothills. The *texensis* and *micromeris* subspecies are winter residents (mid-Aug to mid-May); they join the breeding resident population in the Pacific, and are rare in the Northern Highlands and the Caribbean; to 4,300 ft (1,300 m). Flies erratically at low heights while foraging on the wing; roosts on horizontal branches or sometimes on the ground. While foraging, delivers a nasal and choppy *weer-e-e-e-e-e*, which almost sounds synthetic. Song in breeding season is a soft, hollow trilling that extends for long periods of time. Sounds toadlike.

Common Nighthawk (Añapero Zumbón)   *Chordeiles minor*
9.5 in (24 cm). Very similar to Lesser Nighthawk, which also has long wings with a white band on the primaries and a white (male) or buffy (female) throat patch, but distinguished by the following subtle characteristics. At rest, the **wingtips project beyond tip of the tail**. In flight, **primary band is farther from the wingtip** than on Lesser, and wingtips look pointed because the outermost primary is longer than other primary feathers. Two separate populations spend time in Nicaragua. The first is a common breeding migrant (*panamensis* subspecies); it arrives from SA to breed in the Mosquitia (April to early July), inhabiting lowland pine savanna. The second is a winter resident from NA (Sept to early May); uncommon in Caribbean lowlands and rare in Pacific lowlands and foothills; to 1,600 ft (500 m). Flies erratically high above the ground while foraging on the wing; roosts on ground or perches close to the ground. While foraging, aggressively screams a scratchy *beent!* in flight; during breeding season, male produces a deep and vibrating buzz created by air passing through flight feathers during display dives.

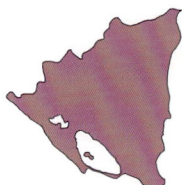

Common Pauraque (Pocoyo Tapacaminos)   *Nyctidromus albicollis*
11.5 in (29 cm). At rest, **tips of wings only reach halfway to tip of the tail**. Cinnamon mask is noticeable on brown body. In flight, quick flashes of the tail pattern show white stripes bordered with black outer rectrices (on male) or white-tipped outer rectrices (on female); **bold white band on primaries** (seen in flight) sets it apart from other ground foraging nightjars. Abundant countrywide; to 5,900 ft (1,800 m). Practically found in all terrestrial habitats but prefers to forage in open areas; often flushed from roads. Commonly seen sallying from the ground for flying insects; roosts on ground under dense vegetation or forest cover. Routinely sings well into the night with a far-carrying *weeOWW!* and a burrier *wee-úl*, both of which have a sharp, whiny quality. Also delivers a *wu-wu-wu-wu-WHHIR!*; the stuttering opening terminates with an explosive sound.

**Short-tailed
Nighthawk**

male

**Lesser
Nighthawk**

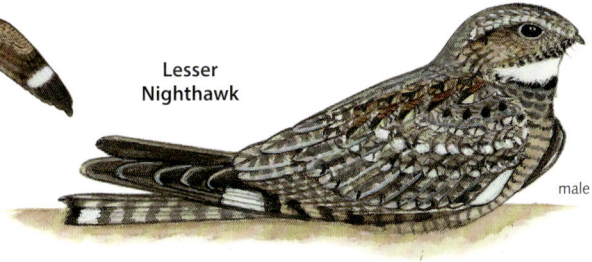

male

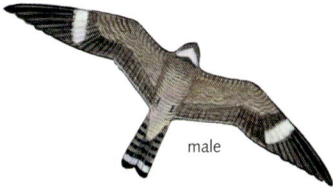

male

**Common
Nighthawk**

male

male

**Common
Pauraque**

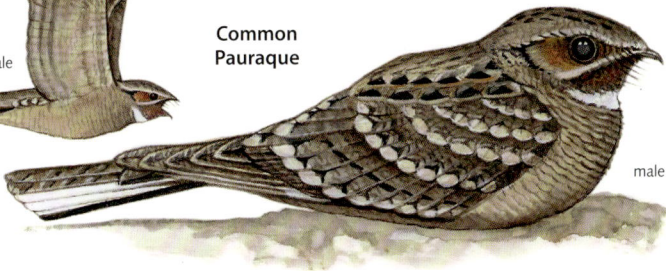

male

Chuck-will's-widow (Pocoyo de Carolina)   *Antrostomus carolinensis*
12 in (30 cm). **Largest of the nightjars**. Individuals range from gray to brown overall, with a **buffy-brown throat** bordered by thin buffy-white throat collar. Overall typically browner than the whip-poor-wills and head is larger and flatter. In flight, white inner webs of outer 3 rectrices are seen on male (tawny-tipped on female). Winter resident (Sept to April); countrywide; to 4,300 ft (1,300 m). Forages in a variety of forested and open habitats and roosts on horizontal branches. Sometimes sings its onomatopoeic *chuck-wills-widow*, with a cheery and rolling quality, but usually silent; may emit an abbreviated *chuk* when flushed.

Rufous Nightjar (Pocoyo Rojizo)   *Antrostomus rufus*
11 in (28 cm). Generally darker than the very similar Chuck-will's-widow. **Rufous throat** is bordered by thin, buffy-white throat collar. In flight, white inner webs of outer 3 rectrices are seen on male (female has no white). Very rare in extreme southeastern Caribbean lowlands, in San Juan River region. Reaches northernmost distribution in Nicaragua. Forages by sallying for flying insects from an understory perch; roosts in dense understory. Energetically sings a *chuk-wit-weeó…* (phrase delivered rapidly, once per second or even faster).

Eastern Whip-poor-will (Pocoyo Gritón Norteño)   *Antrostomus vociferus*
10 in (25 cm). Individuals range from gray to brown overall; **dark throat** bordered by a buffy-white throat collar. Typically grayer overall than Chuck-will's-widow, and head is smaller and more rounded. On male in flight, **outer rectrices have broad white tips that extend to over half the length of the tail** (showing more white than Mexican Whip-poor-will); on female, note only narrow buffy tips. Lack of obvious nuchal collar, as well as the distinct tail pattern, separates it from all nightjars except Mexican Whip-poor-will, which is only safely eliminated by distribution or vocalization. Very rare winter resident (mid-Nov to March) in Pacific lowlands and foothills and extreme southwestern Caribbean lowlands; to 1,300 ft (400 m), but could show up at higher elevations. Hunts nocturnally for flying insects by sallying from ground or low perch in open or semi-open areas. Roosts by day on ground or horizontal branches in dense vegetation of understory; can be found in virtually all terrestrial habitats. Generally silent while in Nicaragua.

Mexican Whip-poor-will (Pocoyo Gritón Sureño)   *Antrostomus arizonae*
9 in (23 cm). Individuals range from gray to brown overall; **dark throat** bordered by a buffy-white throat collar. Typically grayer overall than Chuck-will's-widow, and head is smaller and more rounded. On male in flight, **outer rectrices have broad white tips that extend to only half the length of the tail** (showing less white than Eastern Whip-poor-will); on female, note only narrow buffy tips. Lack of obvious nuchal collar, as well as the distinct tail pattern, separates it from all nightjars except for Eastern Whip-poor-will, which is only safely eliminated by distribution or vocalization. Rare and local on Sierra Dipilto-Jalapa; one breeding population is known from the base of Cerro Mogoton, at 4,600 ft (1,400 m). Prefers highland pine and pine-oak forests, shaded coffee plantations, and cloud forest, where it hunts nocturnally by sallying from the ground or understory perch for insects. Incessantly sings the burry onomatopoeic *whip-poor-will* (sometimes only the *poor-will* carries far enough to be heard).

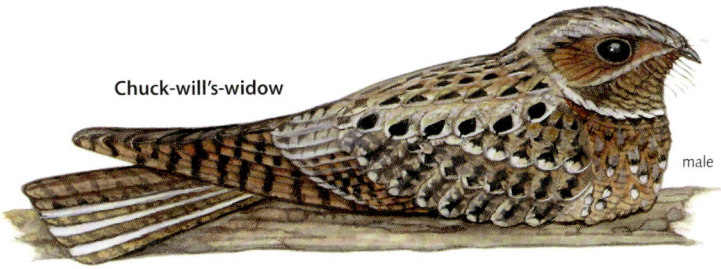

Chuck-will's-widow

male

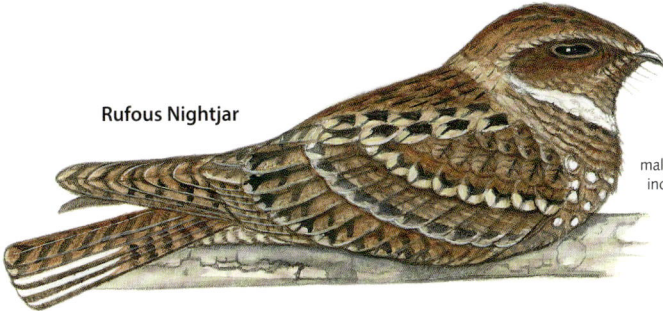

Rufous Nightjar

male rufous
individual

male

**Eastern
Whip-poor-will**

male

male

**Mexican
Whip-poor-will**

male

Ocellated Poorwill (Pocoyo Ocelado)   *Nyctiphrynus ocellatus*
8.5 in (22 cm). A **small, dark** nightjar. Has a clearly defined silvery-gray breast, **two rows of black spots on scapulars**, and a **white-spotted belly**; female generally shows more rufous in plumage than male. In flight, note white-tipped outer rectrices. Very rare, with only 2 populations recorded (Peñas Blancas NR, 1909; Indio Maíz BR, 2011). It is not known whether populations in Honduras and Nicaragua are completely disjunct or if the species simply occurs very locally. It is possible that other local populations occur throughout Caribbean lowlands and up into the eastern slopes of Northern Highlands. Reportedly forages by sallying for insects in second growth understory close to older forest; roosts in understory vegetation. Makes a soft, downslurred *prill*, with a slightly trilling quality (repeated every 3 seconds).

Tawny-collared Nightjar (Pocoyo Mexicano)   *Antrostomus salvini*
10 in (25 cm). **Cinnamon nuchal collar** noticeable on dark body. **Heavily mottled black throat and underparts** distinguish it from Buff-collared Nightjar and Spot-tailed Nightjar. In flight, outer 3 rectrices have broad white tips (on male) or narrow buffy tips (on female). One Nicaraguan record (Matagalpa Dept., April 13, 1917), at 2,100 ft (650 m); most likely a vagrant from northeastern Mexico. Not illustrated.

Buff-collared Nightjar (Pocoyo Collarejo)   *Antrostomus ridgwayi*
8.5 in (22 cm). **Cinnamon nuchal collar** noticeable on grayish body. **Dusky throat and gray underparts** distinguish it from Tawny-collared Nightjar and Spot-tailed Nightjar. In flight, inner webs of outer 3 rectrices have broad white tips (on male) or narrow buffy tips (on female). Very rare in the far eastern foothills of the Pacific and dry intermontane valleys of Northern Highlands; from 300 to 2,600 ft (100 to 800 m). Reaches southernmost distribution in Nicaragua. Prefers arid habitat such as thorn forest, scrub, and grassland, where it sallies from the ground for insects; roosts on ground. Unmistakable song is a rapid series of 10–12 staccato-like clucks that crescendo to a final screech (entire song is 1 second in length); song is repeated at length.

Spot-tailed Nightjar (Pocoyo Colimaculado)   *Hydropsalis maculicaudus*
8.5 in (22 cm). Has the most unique field marks among the nightjars. Cinnamon nuchal collar is noticeable on overall dark body. **Buffy superciliary** and **black triangular malar patch** distinguish it from Tawny-collared Nightjar and Buff-collared Nightjar. **Several rows of buffy spots on wing coverts** are conspicuous. In flight, male has white-tipped outer rectrices, while female shows no white. Common breeding migrant in the Mosquitia (March to July). Sallies from the ground or a low perch for flying insects in lowland pine savanna; also hunts on the wing. Whistles a sharp but sweet *pit-peeu*, with an upslurred second note; repeated with variable intervals of spacing, but often several seconds.

Ocellated
Poorwill

male

male

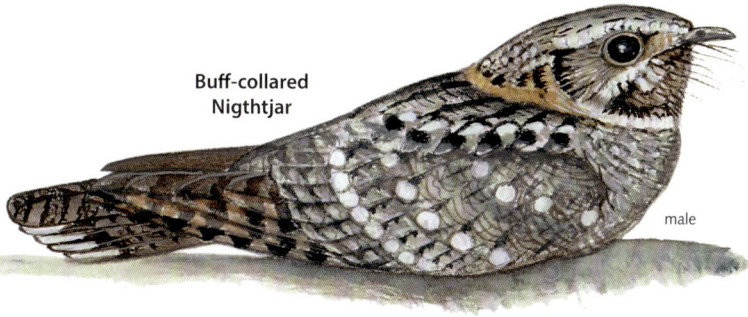

Buff-collared
Nigthtjar

male

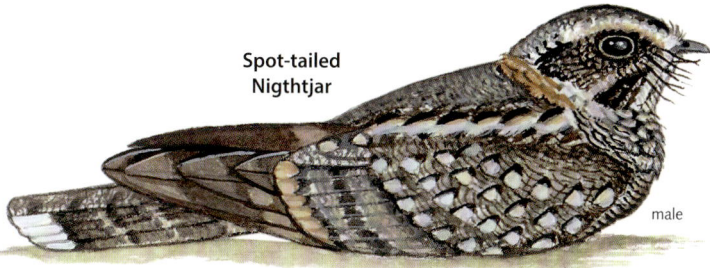

Spot-tailed
Nigthtjar

male

Potoos are found only in the Neotropics. This small family is closely related to nightjars, and both share a large gape, large eyes, and rictal bristles that aid in hunting insects at night. From an exposed snag or branch, where they sit in a vertical posture, they implement a sit-and-wait hunting strategy. Cryptic coloration, and the habit of extending their head and bill upward to make themselves resemble an extension of a snag or broken branch, makes them difficult to detect, and they are more often heard than seen. Long wings and tail can help to prevent confusion with any owl.

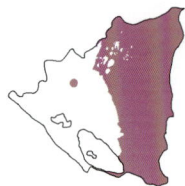

Great Potoo (Estaquero Grande)   *Nyctibius grandis*
20.5 in (52 cm). Has pale coloration and a **large head**. Large size, lack of moustachial stripe, and **chestnut iris** further distinguish it from Common Potoo and Northern Potoo. Note fine vermiculation on upperparts and underparts. Rare throughout Caribbean lowlands and foothills; to 2,000 ft (600 m). Very rare and local in Northern Highlands (records from El Jaguar RSP). Resides in humid lowland forest and forest edge, preferring mid-canopy. At night, sallies for large flying insects and bats; roosts on exposed perches high within the canopy. Makes a deep, raspy, ominous *WAOOOW*; similar to call of Northern Potoo but lower pitched and more robust. Sometimes in flight, makes a *wuOH*, similar to its other call but more nasal.

Common Potoo (Estaquero Común)   *Nyctibius griseus*
15 in (38 cm). Very little range overlap with virtually identical Northern Potoo, and Common generally has darker plumage. Smaller size, **black moustachial stripe**, and **yellow iris** distinguish it from Great Potoo. Underparts are finely streaked. Rare throughout southern Caribbean lowlands (reaches Bluefields at the northern limit of its range); to 1,000 ft (300 m). Reaches northernmost distribution in Nicaragua. Inhabits humid lowland forest and edge. Roosts on exposed mid-canopy or canopy perches. Gives a low-pitched, mournful series of whistles that descends in pitch and volume: *WHEE-U uuh uuh*.

Northern Potoo (Estaquero Norteño)   *Nyctibius jamaicensis*
16 in (40 cm). Very little range overlap with virtually identical Common Potoo, and generally has paler plumage. Smaller size, **black moustachial stripe**, and **yellow iris** distinguish it from Great Potoo. Underparts are finely streaked. Uncommon in Pacific, rare in Northern Highlands, and rare in the Mosquitia; to 4,900 ft (1,500 m). **Found** in a variety of forested habitats and semi-open areas, including in urban areas. Sallies for flying insects from both high and low perches; roosts on exposed mid-canopy or canopy perches. Delivers a wail-like, raspy *WHHAAAA*, sometimes followed by a *wa-wa-wa-wa*; calls similar to those of Great Potoo but higher-pitched and thinner.

**Great
Potoo**

alert
pose

**Common
Potoo**

cryptic
pose

**Northern
Potoo**

The members of this cosmopolitan family of fast flying, insectivorous aerial acrobats live up to their name. Erratic, batlike flight and swept-back wings help distinguish swifts from the superficially similar swallows (Family Hirundinidae). Rapid flight and generally dark plumage can make identification quite challenging. Sifting through a flock of high flying and fast moving swifts requires patience, but persistence and practice are rewarded with the gradual increase in confidence to determine subtle, yet important, differences in size, morphology, and flight patterns. Noting flight calls can also aid in a positive identification. Nonetheless, some observations will inevitably end with identification only to the genus level. Making observations from higher viewing points increases the ability to see field marks on both upper- and underparts.

[Genus *Cypseloides* is characterized by medium-sized bodies, longer tails, and broad bases on pointed wings. Flight is typically stiff, with straighter and down-tilted wings.]

Black Swift (Vencejo Negro) *Cypseloides niger*

7 in (18 cm); WS 15.5 in (39 cm). Relatively longer, **distinctly notched-tail** helps to separate it from White-chinned Swift, but depth of notch varies in female and immature and sometimes appears squared. Body size is important to note when comparing to Chestnut-collared Swift (smaller) and White-collared Swift (larger); broader tail further distinguishes it from Chestnut-collared Swift. White wash on forecrown and faint white scaling on underparts are only seen under the best viewing conditions. *Borealis* subspecies: rare passage migrant (Sept to Oct and March to April), in Pacific, in Northern Highlands, and in extreme southwestern Caribbean lowlands. *Costaricensis* subspecies: records from dates outside of migratory period suggest that it arrives to Nicaragua as a breeding migrant; current records support its occurrence in the Pacific, south of Lake Managua, but distribution is undetermined. Forages high within mixed swift and swallow flocks; seen soaring between wingbeat bursts and erratic twists, and exhibits banking movements more often than other swifts. Roost and nest sites are associated with waterfalls. In flight, sputters a continuous, erratic series of fast, high-pitched twittering notes, mainly characterized by a *twit* sound.

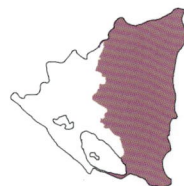

White-chinned Swift (Vencejo Barbiblanco) *Cypseloides cryptus*

6 in (15 cm); WS 13.5 in (34 cm). Relatively shorter and more **squared tail** helps to distinguish it from larger Black Swift and smaller Chestnut-collared Swift, but beware of squarish tales in some individuals of these 2 species. White wash on forecrown and spot on chin are only seen under the best viewing conditions. Rare but locally common in Caribbean. Forages with mixed swift and swallow flocks, within which it often soars less frequently than does Black Swift. Roost sites are associated with waterfalls or moisture-laden cliffs. Flight calls consist of sweet but squeaky chirps, often delivered in a series of 5–6 ascending notes that become sharper at the end, and sometimes devolving into a twittering trill. Also uses a clicking contact call within flocks.

[Genus *Streptoprocne* is characterized by medium- to large-sized bodies and long tails. Flight is typically powerful with deep wingbeats from swept-back wings.]

Chestnut-collared Swift (Vencejo Cuellicastaño) *Streptoprocne rutila*

5.5 in (14 cm); WS 12.5 in (32 cm). Although often difficult to see, the **chestnut collar** is diagnostic (reduced or completely lacking on female and immature). Longer, **slightly notched tail** helps to distinguish it from White-chinned Swift. Smaller and with narrower tail than Black Swift. Rare in Northern Highlands; above 2,500 ft (750 m), but sporadically moves to lower elevations in eastern foothills. Possible breeding population in Sierra Amerrisque; accidental in Caribbean. Forages in small- to medium-sized flocks. Typically flies higher than *Chaetura* swifts, and makes more shorter glides than those species. Roosts are associated with steep terrain along mountain streams. Flight calls consist of erratic, dry, buzzy chatter.

White-collared Swift (Vencejo Collarejo) *Streptoprocne zonaris*

8.5 in (22 cm); WS 20 in (51 cm). Large size and obvious **white collar** (absent on immature) is diagnostic. Only swift with a **forked tail**. Possible to see countrywide but more abundant at high elevations; common in Northern Highlands and Caribbean, and uncommon in Pacific. Prefers forested landscape, where it pursues insects high above canopy with powerful, precise flight, utilizing very little flapping and often diving or turning in acrobatic fashion. Roost and nest are associated with caves, wet crevices, and waterfalls. In flight, produces a robust and shrill call note that becomes a cacophony of screeches when large flocks are vocalizing in unison.

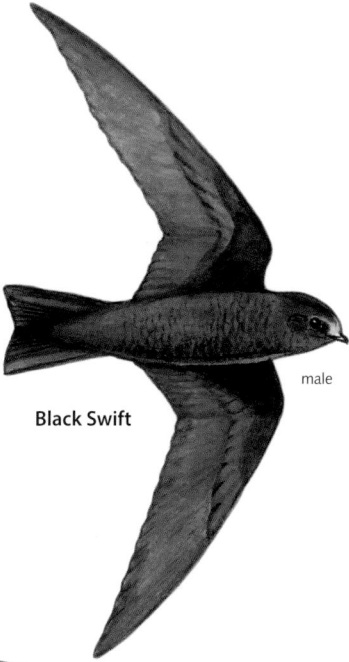

Black Swift

male

White-chinned
Swift

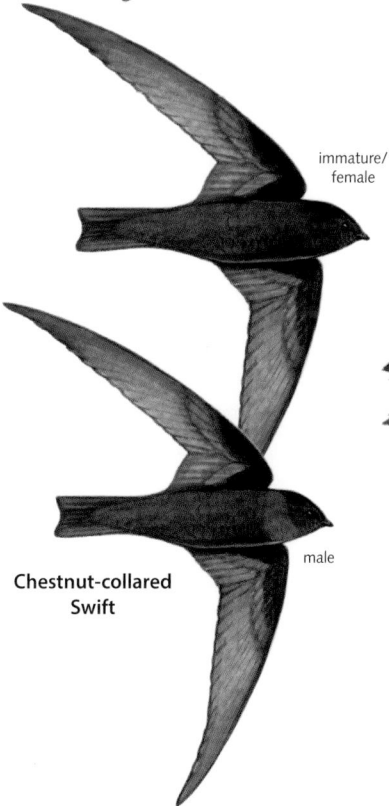

immature/
female

Chestnut-collared
Swift

male

White-collared
Swift

[Genus *Chaetura* is characterized by small- to medium-sized bodies with spine-tipped, short tails. Wings have a distinct shape. Note rounded leading edge of wing; also note that variation in the length of feathers on the trailing edge of wing results in varied feather shapes—hooked outer primaries, bulbous inner primaries, and attenuated secondaries. Flight is marked by erratic, short, fast, batlike movements, on down-tilted wings.]

Chimney Swift (Vencejo Pasajero)   *Chaetura pelagica*

5 in (13 cm); WS 12.5 in (32 cm). Dark overall, with **slightly lighter throat and upper breast**. Very difficult to distinguish from Vaux's Swift (which is shorter-tailed and shorter-winged), but note that **rump and uppertail coverts are only slightly paler than dark body**, with little contrast. Short, squared-tail can look rounded because of projecting spines. Passage migrant (late Sept to early Nov and March to mid-May); uncommon throughout Caribbean lowlands; rare throughout rest of country. Erratically twists and banks while maintaining almost constant wingbeats; will forage at lower heights but in migration is often seen flying in a single direction. Generally silent in migration, but in flight may emit a rapid twittering sound that oscillates in pitch. **NT**

Vaux's Swift (Vencejo Grisáceo)   *Chaetura vauxi*

4.5 in (11 cm); WS 10.5 in (27 cm). Dark overall, but **lighter plumage extends from the throat to lower breast**. Very difficult to distinguish from Chimney Swift (which is longer-tailed and longer-winged), but note that **pale rump and uppertail coverts noticeably contrast with dark body**. Short, squared-tail often shows projecting spines. Abundant in Northern Highlands and Caribbean; common in southern Pacific; rare in northern Pacific. Forages at low heights, sometimes in large flocks. Erratically twists and banks with short glides while maintaining almost constant wingbeats. In flight, emits a liquid chatter that rises and falls, sometimes interspersed with buzzy notes.

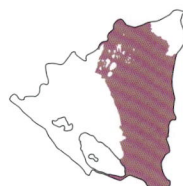

Gray-rumped Swift (Vencejo Lomigrís)   *Chaetura cinereiventris*

4 in (10 cm); WS 10.5 in (27 cm). In flight, note **gray rump and uppertail coverts**. Lighter throat and upper breast contrast more sharply with dark body than in Vaux's Swift. Short, squared-tail can look rounded because of projecting spines. Common in Caribbean lowlands and foothills; to 2,000 ft (600 m). Forages in small- to medium-sized flocks over humid lowland forest canopy and adjacent open areas, sometimes mixing with other swifts and swallows. In flight, emits a rapid, reedy twittering.

[Genus *Panyptila* is characterized by pied coloration; small- to medium-sized, slender bodies; forked tails; and long, slender wings. Flight is more graceful than that of other swifts.]

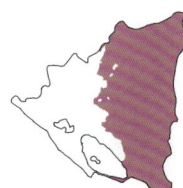

Lesser Swallow-tailed Swift (Macuá Menor)   *Panyptila cayennensis*

5 in (13 cm); WS 12 in (30 cm). Sleek **black-and-white plumage** and deeply forked tail (often held tightly together to form a point) can only be confused with Great Swallow-tailed Swift. **Smaller size** is best determined by comparison with other species in the air; portion of collar on the nape is narrower than on Great Swallow-tailed Swift, and often with a slight dusky smudge (only visible in the best viewing conditions). Rare throughout Caribbean lowlands and foothills. Forages high, either alone or in pairs, and typically not in flocks. Flies with rapid wingbeats; flies more erratically and is less likely to soar than Great Swallow-tailed Swift. In flight, gives a constant, rapid, thin chatter, with brief upticks in pitch.

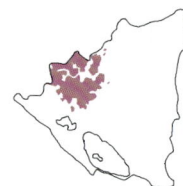

Great Swallow-tailed Swift (Macuá Mayor)   *Panyptila sanctihieronymi*

8 in (20 cm); WS 18 in (46 cm). Sleek **black-and-white plumage** and deeply forked-tail (often held tightly together to form a point) can only be confused with Lesser Swallow-tailed Swift. **Larger size** is best determined by comparison with other species in the air; portion of collar on the nape is broader than on Lesser Swallow-tailed Swift. Rare throughout Northern Highlands, but annual movement and status is not well understood; above 2,600 ft (800 m). Forages high, either alone or in pairs, and sometimes in small flocks; is known to join mixed swift and swallow flocks. Flies with powerful wingbeats; is less erratic and more likely to soar than Lesser Swallow-tailed Swift. Flight calls consists of scratchy screams, rapid chatter, and descending shrill whistles.

Chimney
Swift

Vaux's Swift

Gray-rumped
Swift

Great
Swallow-tailed
Swift

Lesser
Swallow-tailed
Swift

Hummingbirds occur only in the Americas. They are small birds with long, slender bills specialized to access nectar. Light refraction, rather than pigmentation, creates the striking iridescent colors on their gorgets and crowns. Body size, bill length and shape, and tail pattern and shape are keys to identification. Hummingbirds are powerful flyers and can rotate their wings 180°, allowing them to fly backwards or hover at a fixed point in front of flowers to feed. All are pollinators, and some have bills evolved to perfectly access specific flower structures. Some species perform seasonal movements in search of reliable nectar sources; protein-rich insects and small spiders supplement their diets. Hummingbirds display diverse nectar-feeding behaviors, including territorialism, marauding, filching, trap lining, and nectar robbing.

White-tipped Sicklebill (Pico de Hoz) *Eutoxeres aquila*
5 in (13 cm). Unmistakable. Note **long, dramatically decurved bill** and **heavily streaked underparts**. Only 2 records in Nicaragua (Indio Maíz BR, 2006 and 2007); probably uncommon in southeastern Caribbean lowlands and foothills; to 1,300 ft (400 m). Reaches northernmost distribution in Nicaragua. Found in understory of humid lowland forest, where it perches to feed on the curved flowers of heliconias: *Heliconia pagonantha*, *H. longa*, *H. trichocarpa*, *H. reticulata*, and *Centropogon granulosus*. Try observing these flowers at dawn to see this hummingbird. Song consists of a medley of sharp, very high-pitched, erratic notes, sometimes followed by individual *tseet* calls.

Bronzy Hermit (Ermitaño Bronceado) *Glaucis aeneus*
4.5 in (11 cm). Broad **rufous base on outer rectrices** and **cinnamon throat** distinguish it from similar Band-tailed Barbthroat. Uncommon in Caribbean lowlands and rare in foothills. Rare on eastern slopes of Northern Highlands (population has been diminished by deforestation); to 4,300 ft (1,300 m). Prefers understory of lowland humid forest, forest edge, and second growth (frequently along streams and sometimes at swamps). Traplines *Heliconia* and banana and plantain flowers. Sings a repeating, high-pitched *tse-tse-se-se-su-su-sui…*, comprised of 2 stacatto notes followed by a descending chatter-like trill.

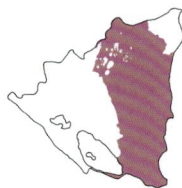

Band-tailed Barbthroat (Ermitaño Barbudo) *Threnetes ruckeri*
5 in (12 cm). **White base on outer rectrices** and **black chin contrasting with rufous breast** separate it from similar Bronzy Hermit. Uncommon in Caribbean lowlands and foothills; to 2,000 ft (600 m). Accidental on eastern foothills of Northern Highlands, with only 3 historical records on Sierra Isabelia and Sierra Dariense (1 in June 1909; 2 in Oct 1907); to 3,300 ft (1,000 m). Traplines *Heliconia*, bananas, and plantains in understory of humid lowland forest, forest edge, and secondary growth (occasionally nectar robbing). More likely to be found inside forest than Bronzy Hermit. Song starts with a shrill descending trill followed by sweet, high-pitched warbles.

Long-billed Hermit (Ermitaño Colilargo) *Phaethornis longirostris*
6.5 in (16 cm). Larger than other hermits. **Extremely elongated central rectrices** (protruding from **long, graduated tail**) set it apart from much smaller Stripe-throated Hermit. Common on eastern slopes of Northern Highlands and in Caribbean; locally rare on Sierra Chontaleña (northern extent and eastern slopes); locally very rare in Rivas Isthmus; to 4,900 ft (1,500 m). Found in the understory of cloud forest and lowland humid forest, forest edge, and second growth, especially near water; traplines and also eats small insects and spiders. Pumps tail while singing incessantly from an inconspicuous forest understory perch. Delivers a monotonous, fast, high-pitched *sueet-sueet-sueet…*; harsher than song of Stripe-throated Hermit. Males gather at leks and noisily sing simultaneously, but birds are surprisingly hard to locate.

Stripe-throated Hermit (Ermitaño Enano) *Phaethornis striigularis*
3.5 in (9 cm). **Very small** (bill and tail account for most of its size). **Short, elongated central rectrices** on **buff-tipped, graduated tail** distinguish it from much larger Long-billed Hermit. Common in Caribbean; locally common in Northern Highlands (mainly eastern slopes); locally uncommon from Lake Managua south; rare on Sierra Maribios and Sierra Chontaleña (northern extent and eastern slopes); to 4,600 ft (1,400 m). Found in the understory of cloud forest, humid lowland forest, gallery forest, dry forest, and forest edge. Traplines very small flowers and robs nectar from large flowers. Sings a fast, continuous, high-pitched, oscillating *sueet-sú-sueet-sú-suett-sú…*; sweeter in quality than song of Long-billed Hermit. Males will gather, in loose association, at leks sites.

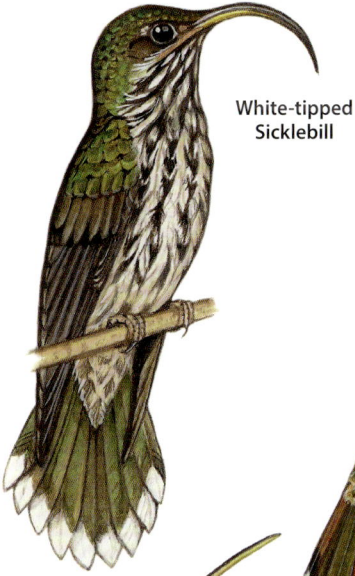

White-tipped
Sicklebill

Bronzy
Hermit

Band-tailed
Barbthroat

Long-billed
Hermit

Stripe-throated
Hermit

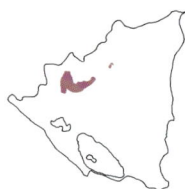

Brown Violetear (Colibrí Orejivioláceo Pardo)   *Colibri delphinae*
4.5 in (11 cm). Combination of **violet ear patch** and **brown plumage** is distinctive. Patchy distribution; common in highlands of Saslaya NP, uncommon on Sierra Isabelia and Sierra Dariense in the Northern Highlands; above 3,300 ft (1,000 m). Possibly occurs in Tepesomoto-Pataste NR (not yet documented). Found in cloud forest and—less commonly—in highland pine and pine-oak forests. Generally visits epiphyte flowers in canopy and flowering trees (usually insect-pollinated species such as those in the genus *Inga*). Sometimes visits roadside verbena flowers (genus *Stachytarpheta*) and *Salvia purpurea*, when it is easiest to see. Delivers a rapid, raspy *chí-chu-chu-chí-chu-chu-chí-chu-chí* (the final note has a noticeable upward inflection).

Mexican Violetear (Colibrí Orejivioláceo Mexicano)   *Colibri thalassinus*
4.5 in (11 cm). Combination of **violet ear patch** and **iridescent green plumage** is distinctive; also note **broad, dark subterminal band** on bluish tail. Rare to locally uncommon in the Northern Highlands: on Sierra La Botija, in Tepesomoto-Pataste NR, and on Sierra Isabelia and Sierra Dariense; 1 record on Casita Volcano (Dec 1983); above 4,100 ft (1,250 m). Reaches southernmost distribution in Nicaragua. Generally a trapliner in mid-canopy and canopy, but also descends to understory depending on availability of preferred flowers. Sits motionless while singing from canopy, which sometimes makes it easy to hear but difficult to see. Incessantly proclaims a rhythmic cadence with a consistent up-and-down pitch change: *chét-chechep-chét-chechep… .*

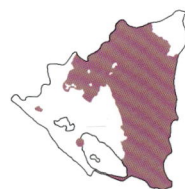

White-necked Jacobin (Jacobino Nuquiblanco)   *Florisuga mellivora*
4.5 in (11 cm). Male, with **bright blue head**, **snow white underparts**, and **mostly white tail**, is unmistakable. Female can be confused with slightly larger Scaly-breasted Hummingbird (p. 82), but note bolder throat and breast scaling (bluish-green), **all-black bill**, and evenly white-tipped rectrices. Common in Caribbean, rare on eastern slopes of Northern Highlands, and very rare in Pacific foothills (Casita Volcano, Dec 1961; between Chonco Volcano and San Cristóbal Volcano, Dec 2016; Mombacho Volcano, March 1962); to 3,000 ft (900 m), but occasionally wanders to 4,200 ft (1,300 m). Forages at mid-canopy and canopy of humid lowland forest, tall second growth, and sometimes visits flowering trees of coffee and cacao plantations. In flight, male fans tail completely in territorial display. When perched, calls with a repeated, harsh, loud *CHIT! CHIT! CHIT!…* (1 note per second) or a fast, soft *tsup-tsup-tsup… .* In flight, produces a single *tsup* note.

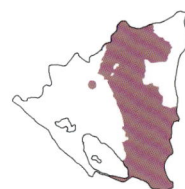

Purple-crowned Fairy (Espadachín Enmascarado)   *Heliothryx barroti*
5 in (13 cm). Note **immaculate white underparts** and **white outer rectrices on long, graduated tail**; **short, needle-like black bill** enhances its dainty appearance. Locally uncommon in Caribbean lowlands and foothills, rare south of Lake Nicaragua; one historical record from the eastern slopes of Northern Highlands (Peñas Blancas, June 1909); to 2,300 ft (700 m). Found in humid lowland forest, tall second growth, and forest edge; usually near streams. Gracefully flits in mid-canopy and canopy, sometimes descending to visit *Heliconia* flowers. Pierces flowers at base, robbing nectar; this species is more insectivorous than most hummingbirds. Often curious; frequently hovers at eye-level, making it easy to observe. Emits a soft *twip* call.

Brown Violetear

Mexican Violetear

female

female

male

White-necked Jacobin

male

Purple-crowned Fairy

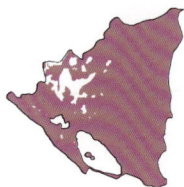

Green-breasted Mango (Manguito Colipúrpura)   *Anthracothorax prevostii*
4.5 in (11 cm). On male, **frontal stripe is black on chin and throat, becoming bluish-green on breast and upper belly** (consider width, length, and color of frontal stripe to rule out male Black-throated Mango). Female has **dark green frontal stripe**; immature is similar to female but with rufous stripes bordering the underparts. Common in Caribbean lowlands and foothills. Locally common to uncommon in Pacific lowlands and foothills. Uncommon in Northern Highlands; to 3,300 ft (1,000 m), rarely to 4,300 ft (1,300 m). Prefers open areas with remnant trees, dry forest, humid lowland forest and edge, and tall second growth. Visits canopy of flowering trees (e.g., *Erythrina* sp. and *Inga* sp.). Sings a fast, high-pitched warble; calls with a variety of short, high-pitched *chep* and *chet* notes.

Black-throated Mango (Manguito Gorginegro)   *Anthracothorax nigricollis*
4.5 (11 cm). On male, **extensive black frontal stripe (from throat to lower belly) is bordered by glossy blue** (consider width, length, and color of frontal stripe to rule out common male Green-breasted Mango). Female has **black frontal stripe**. Vagrant in Caribbean lowlands; one record from San Juan River headwaters (Dec, 2005) and 2 undocumented records on Pearl Lagoon shoreline (June, 2012).

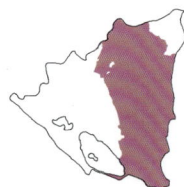

Bronze-tailed Plumeleteer (Colibrí Patirrojo)   *Chalybura urochrysia*
4.5 in (11 cm). **Red-pink feet** on both sexes (darker on male) are diagnostic. Plumage appears blackish under poor light within forest interior, but bronze rump leading into dark purple-hued tail offers a hint of color; female with inconspicuous white-tipped outer rectrices. Local and common in Saslaya NP, Bosawas Biosphere Reserve, and Indio Maíz BR, otherwise rare to uncommon in Caribbean lowlands and foothills; to 2,600 ft (800 m). Prefers interior of humid lowland forest, where it forages from understory to mid-canopy; very shy and seldom visits forest edge. Male is territorial. Song consists of a bizarre-sounding electronic trill: *tér-tididididididi*; introductory note is distinct. Calls with a rapid, repeated, forcefully delivered *tset!-tset!-tset!…* .

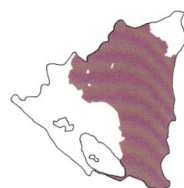

Crowned Woodnymph (Ninfa Violeta y Verde)   *Thalurania colombica*
M 4 in (10 cm); F 3.5 in (9 cm). On male, **iridescent green throat and violet-blue breast and belly** are distinctive; also note **deeply forked tail**. Drab female is distinguished from other females by combination of **grayish throat, vent, and undertail coverts** and **greenish belly and flanks**. Common in Caribbean, but rare south of Lake Nicaragua; uncommon on eastern slopes of Northern Highlands; to 4,300 ft (1,300 m). Favors humid lowland forest, cloud forest, tall second growth, and coffee plantations. Male forages from mid-canopy to canopy; female stays in understory. Sings a continuous and monotone *suit-suit-suit…*, given at variable speeds.

adult
female

**Green-breasted
Mango**

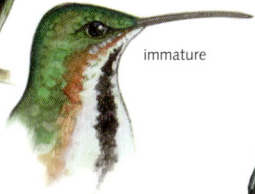

immature

female

**Black-throated
Mango**

adult
male

male

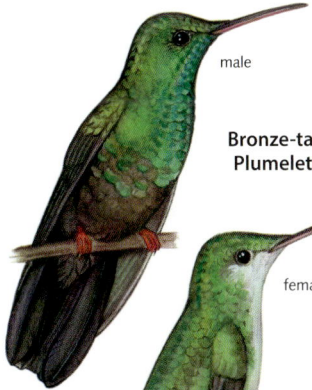

male

**Bronze-tailed
Plumeleteer**

female

male

**Crowned
Woodnymph**

female

Rivoli's Hummingbird (Colibrí de Rivoli)   *Eugenes fulgens*
5 in (13 cm). **Large**, with **long, straight bill (slightly decurved towards tip on female)**. Female and immature can be confused with smaller male Green-breasted Mountain-gem (p. 88), but distinguished by white postocular spot and outer rectrices with white tips only. Rare in Northern Highlands and Saslaya NP; above 4,300 ft (1,300 m). Reaches southernmost distribution in Nicaragua. Traplines in highland pine and pine-oak forests, cloud forest edge, and tall second growth. Continuously sings a series of high-pitched notes: *ch-ít ch-uh ch-ít-uh…*, each with quick pitch changes. Call is a dry, rapid *chit* note.

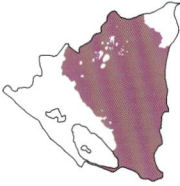

Scaly-breasted Hummingbird (Colibrí Pechiescamado)   *Phaeochroa cuvierii*
5 in (12 cm). Resembles female White-necked Jacobin (p. 78), but note white postocular spot; further distinguished by **yellowish-green, scaled breast**, pinkish basal half of lower mandible, and **white-tipped outer rectrices**. Straight bill and lack of purple on throat distinguishes it from larger female Violet Sabrewing. Common in Caribbean lowlands and foothills; to 2,000 ft (600 m). Uncommon on eastern slopes of Northern Highlands and in northern region of Sierra Chontaleña; to 4,300 ft (1,300 m). Prefers forest edge, second growth, and semi-open areas with scattered trees; also visits flowering trees in coffee and cacao plantations. Forages at mid-canopy and canopy. Sings a diverse series of short, high-pitched, rhythmic notes; may include *tseet-tseet-tseet-tseet-su* or a buzzy *zwee-zwee*, and sometimes with short trills intermixed.

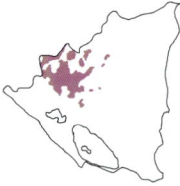

Violet Sabrewing (Sable Violáceo)   *Campylopterus hemileucurus*
5.5 in (14 cm). **Large**; both sexes flash **broad white-tipped outer rectrices**. **Striking purple plumage** on male is unmistakable. On female, decurved bill and **purple on throat** distinguish it from smaller Scaly-breasted Hummingbird. Locally common in Northern Highlands and locally very rare in northern region of Sierra Chontaleña; above 3,000 ft (950 m). Common in Saslaya NP and uncommon in Musún NR (probably in nearby Quirragua NR); generally above 3,300 ft (1,000 m), but in Saslaya NP it descends to 1,300 ft (400 m) outside of breeding season. Favors understory and mid-canopy of cloud forest, humid lowland forest, and forest edge; visits feeders. From perch, sings a loud, high-pitched, continuous series of mixed notes: *tsee tsee tsee tsát…*, 1 note delivered every second.

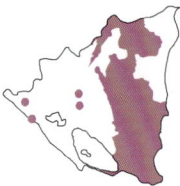

Long-billed Starthroat (Colibrí Piquilargo)   *Heliomaster longirostris*
5 in (12 cm). **White postocular spot** (not stripe) and **blue forecrown** distinguish it from Plain-capped Starthroat; gorget is magenta (appears black in poor light). Rare to uncommon in Caribbean lowlands and foothills; to 1,300 ft (400 m), but occasionally wanders to 3,000 ft (900 m). Accidental in Pacific lowlands (Chinandega Dept., 1908; Isla Juan Venado WR, 1983; Cumaica-Cerro Allegre NR, Nov 2015) and Northern Highlands (Finca Esperanza RSP, Jan 2005). Occurs in humid lowland forest and edge, gallery forest, second growth, semi-open areas with scattered trees, and coffee, banana, and plantain plantations (generally occurs in more humid habitats than does Plain-capped Starthroat). Forages in canopy, favoring *Erythrina* sp. and other tubular flowers. Repeats at length an assertive but monotone *peek! peek! peek!…*, with the time between calls evenly spaced.

Plain-capped Starthroat (Colibrí Pochotero)   *Heliomaster constantii*
5 in (12 cm). Combination of **dark auricular patch, white postocular stripe** (not spot), and **absence of blue on forecrown** distinguishes it from Long-billed Starthroat; gorget is red (appears black in poor light). Uncommon to locally common in Pacific and uncommon in dry intermontane valleys of Northern Highlands; to 4,300 ft (1,300 m). Favors dry forest and edge, gallery forest, scrub, second growth, semi-open areas with scattered trees, coffee plantations, and cloud forest edge (generally found in habitats that are more arid than those preferred by Long-billed Starthroat). Forages from mid-canopy to canopy, favoring *Erythrina* sp. and other tubular flowers; perches on exposed twigs. Repeats at length a monotone *pik! pik! pik!…*, interspersed with an occasional higher-pitched *peek!*

male

female

Rivoli's
Hummingbird

Scaly-breasted
Hummingbird

female

male

Violet
Sabrewing

Plain-capped
Starthroat

Long-billed
Starthroat

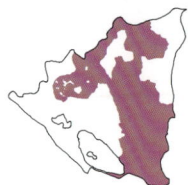

Black-crested Coquette (Coqueta Crestinegra)   *Lophornis helenae*
3 in (7 cm). No other Nicaraguan hummingbird has **complete white rump band**; **greenish-buffy spots on whitish underparts** are unique, if subtle. On male, note combination of red bill, **elongated black feathers** on green crown, and green gorget. Female can be confused with female Sparkling-tailed Hummingbird (partially sympatric) but is separated by rufous on tail and small black mask (and white rump band). Uncommon in Northern Highlands (excluding Sierra La Botija, Tepesomoto-Pataste NR, and Sierra Dipilto-Jalapa); to 4,300 ft (1,300 m). Uncommon and local in Caribbean (in Saslaya NP and Indio Maíz BR); to 1,300 ft (400 m). Traplines in canopy of cloud forest, humid lowland forest, and forest edge; sometimes descends to visit roadside verbena flowers (genus *Stachytarpheta*). Fluttery flight and plumage pattern make it possible to confuse this hummingbird with sphinx moths (family Sphingidae), which visit shared nectar sources. Mostly silent; emits a somewhat languid *chuup* call note.

Sparkling-tailed Hummingbird (Florín Gorgiazul)   *Tilmatura dupontii*
M 3.5 in (9 cm); F 2.5 (6 cm). **Square white patches on sides of rump** are distinctive. On male, **extremely forked tail (black with white spots)** makes it unmistakable. Female could be confused with female Black-crested Coquette, but note **cinnamon face and underparts**. Locally uncommon on Sierra Isabelia and Sierra Dariense of Northern Highlands; above 3,300 ft (1,000 m). Reaches southernmost distribution in Nicaragua. Occurs in cloud forest edge, highland pine and pine-oak forests, and second growth. Traplines at all levels of forest, and visits roadside verbena flowers (genus *Stachytarpheta*). Characteristically flies in a slow, bumblebee-like manner; male with tail cocked upward. Sings a series of thin, twittering, high-pitched notes (entire series lasts up to 10 seconds); the first 3 or 4 notes are strongly metallic.

Snowcap (Copete Nevado)   *Microchera albocoronata*
3 in (7 cm). Tiny. On male, combination of **white crown** and **purple-bronze plumage** is unique. Female is distinguished from similar small-bodied females by **white tail with dark subterminal band**. Common in foothills of Musun NR, Saslaya NP, and throughout Bosawas Biosphere Reserve; otherwise rare in Caribbean lowlands and foothills; to 2,600 ft (800 m). Local and rare in Peñas Blancas NR (eastern slope of Northern Highlands); to 4,000 ft (1,200 m). Forages at all levels within humid lowland forest, cloud forest, adjacent forest patches, and tall second growth. Male defends nectar-rich flower clusters, but is easily driven off by larger hummingbirds. Song consists of a high-pitched, metallic *tee-tú-tee-tú-tee-tú…*, repeated at length.

Black-crested Coquette

male

female

male

male

female

Sparkling-tailed Hummingbird

female

male

Snowcap

Ruby-throated Hummingbird (Colibrí Gorgirrubí)   *Archilochus colubris*
3.5 in (9 cm). On both sexes, note **white postocular spot**. On male, **ruby-red gorget** is distinctive (appears black in poor light); tail is forked. On female, **white-tipped outer rectrices** and **white breast extending up sides of neck** distinguish it from similar females. Winter resident (Oct to April); common in Pacific, uncommon in Northern Highlands; to 4,600 ft (1,400 m). Rare in Caribbean lowlands and foothills. Ubiquitous throughout range, but favors more arid habitat such as dry forest and edge, scrub, and highland pine and pine-oak forests. Rapidly and sporadically calls with a soft *tyup* or *tew*.

Canivet's Emerald (Esmeralda Rabihorcada)   *Chlorostilbon canivetii*
3.5 in (9 cm). On male, combination of **black-tipped, red bill** and **forked tail** is distinctive. On female, **short, white postocular stripe** distinguishes it from similar females. Common in Pacific, uncommon in dry intermontane valleys of Northern Highlands, and rare eastward into Amerrisque NR; to 4,600 ft (1,400 m). Recently discovered to be breeding on the Mosquitia coast; abundance is unknown. Prefers dry forest and edge, open areas with scattered trees, second growth, scrub, gardens, and urban parks. Traplines in understory and mid-canopy, where it visits small, insect-pollinated flowers. Sings a short, insect-like trill ending with a punctuated rise in pitch.

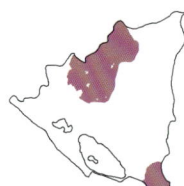

Violet-headed Hummingbird (Colibrí Cabeciazul)   *Klais guimeti*
3 in (8 cm). **Squarish, white postocular spot** is diagnostic on both sexes. On male, note **iridescent, purple crown and gorget** (iridescent feathers can appear blue) in good light; on female, note purplish-blue crown. Patchy distribution; rare on eastern slopes of Northern Highlands, where it can wander to 4,600 ft (1,400 m); rare to uncommon in Caribbean (locally common in foothills of Saslaya NP); to 3,600 ft (1,100 m). Inhabits humid lowland forest and edge; frequently found along streams. Forages in understory; male is typically territorial while female traplines, often nectar robbing. Sings a fast and continuous *tsu-tsít-tsu-tsít-tsu-tsít...*; notes are sweet but piercing and follow a distinct pattern of up-and-down pitch changes.

Emerald-chinned Hummingbird (Colibrí Gorgiverde)   *Abeillia abeillei*
3 in (7 cm). Tiny. On both sexes, note **very short, needle-like black bill** and **white postocular spot**. Male has **iridescent green gorget** (appears black in poor light). Locally common on Northern Highlands and in highlands of Saslaya NP; above 3,600 ft (1,100 m). Reaches southernmost distribution in Nicaragua. Forages in understory of cloud forest and edge, tall second growth, and gallery forest. Almost never ventures out from forest cover, but occasionally visits roadside verbena flowers (genus *Stachytarpheta*). Song is a sweet, but piercing, 7-syllable phrase (*see-út-si-siút-siút*) rapidly repeated for 4–8 seconds.

male

Ruby-throated Hummingbird

female

male

Canivet's Emerald

female

male

Violet-headed Hummingbird

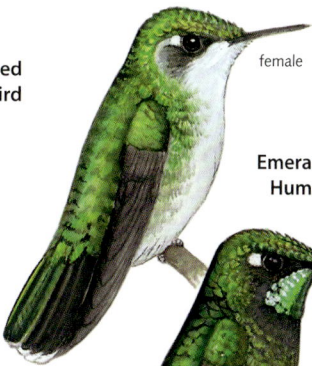

female

female

female

Emerald-chinned Hummingbird

male

Green-breasted Mountain-gem (Montañés Pechiverde)   *Lampornis sybillae*
4.5 in (11 cm). Note **white outer rectrices** and **white postocular stripe over dark cheek**, which help to distinguish it from larger female Rivoli's Hummingbird (p. 82). On male, throat and breast show bluish-green scales; note pale buffy throat on female. Common in Northern Highlands, Musún NR, and Saslaya NP; above 3,300 ft (1,000 m). Prefers cloud forest and edge but also occurs, less commonly, in highland pine and pine-oak forests. Forages from understory to mid-canopy. Weakly sings a fast series of scratchy squeaks followed by a squeaky trill. Calls with a dry and buzzy *bzzt*, sometimes repeated very rapidly. Reaches southernmost distribution in Nicaragua. Endemic to North Central American highlands EBA.

Purple-throated Mountain-gem   *Lampornis calolaemus*
(Montañés Gorgipúrpura)
4 in (10 cm). Both sexes have **white postocular stripe**; male has **purple gorget**, female has **pale buffy throat and underparts**. No similar hummingbird in its restricted range. Common on Mombacho and Maderas volcanos; above 3,600 ft (1,100 m). Found in steeply-sloped cloud forest. Feeds from mid-canopy to canopy. Male defends nectar-rich flower clusters; female may defend territory (less aggressively) but often also traplines. Sings a bizarre, mechanical-sounding compilation of buzzy and nasal notes, interspersed with short trills (entire song is 4–10 seconds in length). Call is a harsh *twit!* Reaches northernmost distribution in Nicaragua. Endemic from southern Nicaragua to western Panama.

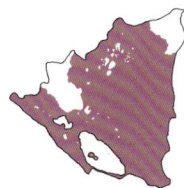

Blue-throated Goldentail (Zafiro Colidorado)   *Hylocharis eliciae*
3.5 in (9 cm). Combination of **black-tipped, bright red bill (thick at base)** and **greenish-golden tail** is diagnostic; upper mandible is dark on female. Common in Caribbean lowlands and foothills; to 2,000 ft (600 m). Locally common in Pacific lowlands and foothills; to 2,600 ft (800 m). Rare to uncommon on eastern slopes of Northern Highlands; to 4,300 ft (1,300 m). Feeds from mid-canopy to canopy in all types of forest and edge, shade coffee plantations, gardens, and open areas with scattered trees. Great songsters; in the dry season males gather in leks and sing from exposed twigs at mid-canopy. Sings a fast, strained *which-í-dituh* or *which-í-dit*, often repeated; song variations include the use of different trills. Call is a sharp *tsee*.

White-eared Hummingbird (Zafiro Bicejudo)   *Hylocharis leucotis*
3 in (8 cm). Unmistakable. **Bold white postocular stripe** contrasts with **black cheek**; bicolored bill is notably black-tipped on male but mostly dark on female (base of lower mandible is red). On male, note purple forehead and gorget (gorget appears black in poor light); on female, forehead is brown. Common in Northern Highlands; above 3,300 ft (1,000 m). One historical record from San Cristóbal Volcano (June 1917) suggests it may be present on Casita Volcano. Reaches southernmost distribution in Nicaragua. Occurs in highland pine and pine-oak forests, cloud forest, and adjacent open areas with flowering plants. Forages in understory and perches at mid-canopy. Sings a thin *tsee-tsaw tsee-tsaw tsee-tsaw…*, with incessant fervor.

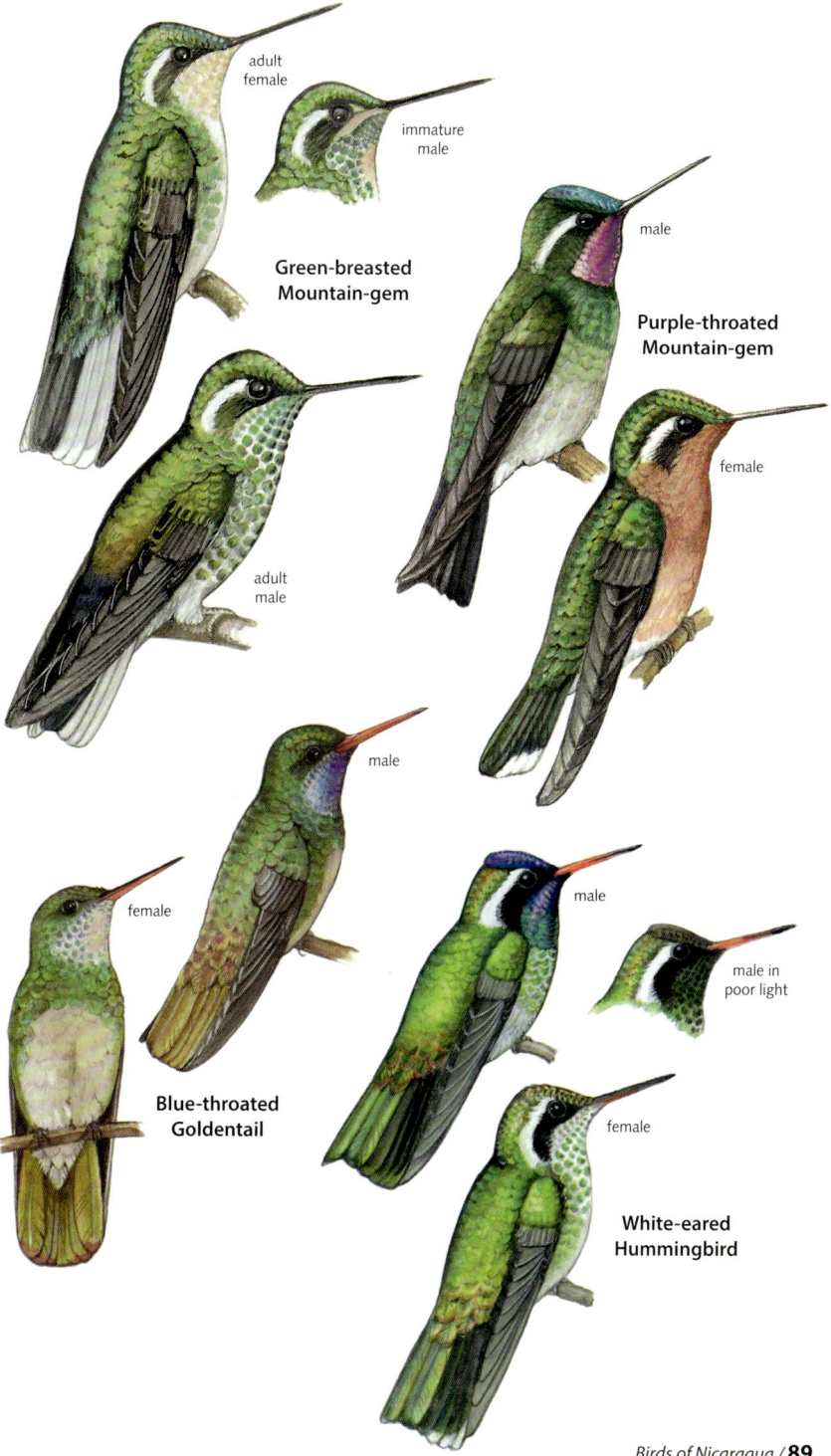

adult female

immature male

Green-breasted Mountain-gem

male

Purple-throated Mountain-gem

adult male

female

male

female

Blue-throated Goldentail

male

male in poor light

female

White-eared Hummingbird

Stripe-tailed Hummingbird (Colibrí Colirayado) *Eupherusa eximia*
3.5 in (9 cm). Conspicuous **rufous wing patch** is similar to that on Blue-tailed Hummingbird, but **white outer rectrices** distinguish it. Common on eastern slopes of Northern Highlands and in highlands of Saslaya NP; above 2,600 ft (800 m). Favors cloud forest, forest edge, and second growth. In canopy, male visits flowers of epiphytes (*Ericaceae* sp.) and trees and aggressively defends nectar-rich territories; female feeds in understory. Both sexes fan tail while feeding on nectar. Sings a series of scratchy and metallic notes (each 1 second in length). Calls with a sharp *chit!*

Blue-tailed Hummingbird (Amazilia Coliazul) *Amazilia cyanura*
3.5 in (9 cm). **Rufous wing patch** distinguishes it from very similar and partially sympatric Steely-vented Hummingbird; **bronzy rump** and **all-dark tail** distinguishes it from Stripe-tailed Hummingbird. Local and common on Sierras Managua, Casita Volcano, and Cosigüina Peninsula, but uncommon in other regions of the Pacific; locally uncommon in Northern Highlands; to 4,600 ft (1,400 m). Accidental in Caribbean. Prefers dry forest and edge, shaded coffee plantations, second growth, and gallery forest; forages from mid-canopy to canopy. Patchy seasonal distribution is caused by altitudinal migrations that are not well understood but may be timed with flowering of trees; appears to be absent in lowlands in June and July. Squeaky, twittering song has 5 notes, with the second note abruptly rising in pitch and the final 2 notes delivered more quickly than rest of the song: *zwee-é-pi-titi*. Calls with a buzzy and abrupt *twezt!* Endemic to North Central American Pacific slope EBA.

Steely-vented Hummingbird (Amazilia Rabiazul) *Amazilia saucerrottei*
3.5 in (9 cm). Almost identical to Blue-tailed Hummingbird. Both have **bronzy rump** and **all-dark tail**, but Steely-vented **lacks rufous wing patch**. Abundant on Mombacho Volcano, common elsewhere in Pacific lowlands and foothills (particularly on Sierras Managua), rare on western slopes of Sierra Isabelia and Sierra Dariense in Northern Highlands; to 4,300 ft (1,300 m). Reaches northernmost distribution in Nicaragua. Favors dry forest and edge, scrub, second growth, gallery forest, open areas with scattered trees, and shade coffee plantations. Aggressive and territorial; forages at all levels of vegetation.

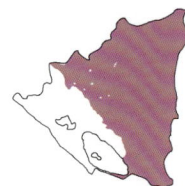

Rufous-tailed Hummingbird (Amazilia Rabirrufa) *Amazilia tzacatl*
4 in (10 cm). Note striking **rufous uppertail coverts and tail**; coloration on upperparts is similar to that of Cinnamon Hummingbird, but note iridescent green throat and breast (greener on male), blending into grayish-whitish belly. Abundant on eastern slopes of Northern Highlands; to 5,000 ft (1,500 m). Abundant in Caribbean lowlands and foothills; to 3,900 ft (1,200 m). Occurs in second growth, coffee plantations, gardens, and humid lowland forest edge. Most aggressive and territorial hummingbird in its range; defends feeders and flowers clusters at all vegetation levels. Most often sings a high-pitched, repeated *swee su-su*, often with a variety of additional notes added at end.

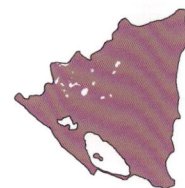

Cinnamon Hummingbird (Amazilia Canela) *Amazilia rutila*
4 in (11 cm). Coloration of upperparts similar to that of Rufous-tailed Hummingbird, but distinguished by **uniform cinnamon underparts**. Abundant in Pacific, common on dry intermontane valleys of Northern Highlands; 4,600 ft (1,400 m). Rare to uncommon in Caribbean lowlands and foothills; to 3,000 ft (900 m). Prefers open areas with scattered trees, dry forest and edge, gallery forest, and shrubby areas. Most aggressive and territorial hummingbird in Pacific; defends flower clusters. Sings a buzzy, 3-note *zeeut zeeut zeeut* and a 4–7 note series that descends in pitch and accelerates slightly: *see-su-sit-sit-sit-sit*.

Stripe-tailed
Hummingbird

male

female

Steely-vented
Hummingbird

Blue-tailed
Hummingbird

Rufous-tailed
Hummingbird

Cinnamon
Hummingbird

White-bellied Emerald (Amazilia Pechiblanca)   *Amazilia candida*
3.5 in (9 cm). Has immaculate white underparts and black-tipped bill (with red lower mandible); **dark subterminal band on outer rectrices** distinguishes it from female Blue-chested Hummingbird and larger Azure-crowned Hummingbird. Performs seasonal movements; common on eastern slopes of Northern Highlands, from April to May and Aug to Nov, but rare at other times; rare to uncommon in Caribbean, to Amerrisque NR (southwest) and Bluefields (southeast); to 4,600 ft (1,400 m). Reaches southernmost distribution in Nicaragua. Forages at all levels in primary forest, forest edge, second growth, forested river edge, and coffee plantations. Sings 2 songs. Although each has a distinct cadence, both display clear spacing between notes and have a similar tonal quality. One typically has 4 notes (the second falls in inflection and pitch) that are repeated 3 times in succession: *swí-u-swí-swí*. The other song has 5 notes (the second rises in inflection and pitch): *tew-chí-tu-tu-tu* (sometimes only the first 2 notes are sung).

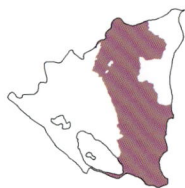

Blue-chested Hummingbird (Amazilia Pechiazul)   *Amazilia amabilis*
3.5 in (9 cm). On male, note distinctive iridescent **violet-blue lower throat and upper breast** contrasting with green plumage. Female has **greenish iridescent specks** (also flashing some blue and purple) on upper underparts; all-black tail additionally distinguishes it from partially sympatric White-bellied Emerald. Uncommon in Caribbean lowlands and foothills; to 1,000 ft (300 m). Accidental in Northern Highlands. Visits scattered flowers in humid lowland forest edge, second growth, forested river edge, old plantations, and open areas with scattered vegetation. Repeats a sharp, high-pitched *peet!-peet!-peet!…* (up to 4 notes per second).

Azure-crowned Hummingbird (Amazilia Frentiazul)   *Amazilia cyanocephala*
A. c. guatemalensis 4.5 in (11 cm); A. c. chlorostephana 4 in (10 cm). Two subspecies occur in Nicaragua; both are distinguished from smaller White-bellied Emerald by all-black tail (no subterminal tail band) and **small postocular spot**. Cyanocephala subspecies: **iridescent blue crown** (only visible in good light). Common in Northern Highlands; above 3,300 ft (1,000 m). Reaches southernmost distribution in Nicaragua. Prefers highland pine and pine-oak forests, but also found in cloud forest edge and scrub. Forages at all canopy levels. Delivers a harsh *síu-síu-síu-síu*, of 4–7 notes, each note with a sharp rise and fall in pitch. Chlorostephana subspecies: smaller, with **iridescent green crown**. Abundant in the Mosquitia; to 300 ft (100 m). Prefers lowland pine savanna, associated gallery forest, and humid lowland forest edge. Endemic to the Mosquitia of Honduras and Nicaragua.

White-bellied Emerald

female

Blue-chested Hummingbird

male

Blue-chested Hummingbird feeding on *Stachytarpheta* (porterweed) flowers.

cyanocephala ssp.

Azure-crowned Hummingbird

chlorostephana ssp.

LIMPKIN   Aramidae

The Limpkin is the only extant species in this Neotropical family. It occurs throughout the Neotropics, where it is associated with a variety of freshwater habitats. It hunts large aquatic snails. Although heron-like in appearance, the limpkin is more closely related to the rails. The combination of a forward and rising head movement and a jerky stride gives a limping appearance, thus its English common name.

Limpkin (Carao)   *Aramus guarauna*

26 in (66 cm). A large wader with long legs and neck and a large bill. Brown overall, with **extensive white streaking** on head and neck (streaking extends onto upper- and underparts). Similar in body shape and plumage to both immature and nonbreeding Glossy Ibis (p. 164), which is smaller and has a pronounced decurved bill. Plumage resembles that of immature night-herons, but those have shorter, stout bills. Locally common in lowlands, with known concentrations on shoreline of Lake Nicaragua, Playitas-Moyua-Tecomapa lagoons, and San Juan River banks; accidental throughout rest of Caribbean; to 1,500 ft (450 m). Usually seen foraging alone in freshwater marshes, riverbanks, lakeshores and lagoons, where it slowly walks in shallow water or on floating aquatic plants looking for prey. Gives a resonant scream (*krAOow*) and a subtler, nasal *koaw*. Also yelps a high-pitched *AOWH!* or *KUH KUH*.

RAILS, CRAKES, and GALLINULES   Rallidae

This cosmopolitan family includes endemic species on islands around the world. All members of the family live near or on bodies of water, in both open and wooded areas. Rails and crakes secretively weave through dense reeds and grasses, a behavior aided by their narrow body design; they sometimes forage in open edges of vegetation, at dawn and dusk. Their preference for foraging in dense vegetation and their retiring, skulking behavior often make it a challenge to see these birds. Gallinules and coots are easier to see, often on open water or at the edge of vegetation. Body size, leg color, and bill size and color are keys to identification, but vocalizations are often the only means of detecting and identifying these birds.

Rufous-necked Wood-Rail (Rascón Cuellirrufo)   *Aramides axillaris*

11.5 in (29 cm). Smaller than Russet-naped Wood-Rail and easily distinguished from other rails and crakes of similar size by **rufous head and neck** and **gray, upper back patch**. Uncommon and local in the mangroves of Isla Juan Venado NR. However, historical records of breeding individuals from San Cristóbal and Mombacho volcanos suggest they perform short-distance, altitudinal migrations to the foothills of Pacific volcanos. Gives a sharp and grating *CHO CHO CHO*…, sometimes devolving into a rattling *cheeoo*; when duetting, loud sound carries far. Calls with a low *tek*; also makes a deep, drumlike sound on breeding territory.

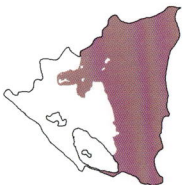

Russet-naped Wood-Rail (Rascón Poponé)   *Aramides albiventris*

15 in (39 cm). Larger than Rufous-necked Wood-Rail. Combination of size, **gray head and neck**, and **rufous underparts** is diagnostic. Common throughout Caribbean lowlands and foothills. Uncommon on eastern slopes of Sierra Isabelia and Sierra Dariense in Northern Highlands; accidental in the lakes region of the Pacific; to 4,600 ft (1,400 m). Frequents vegetation bordering freshwater aquatic habitats. Also forages on ground by picking through leaf litter, in the dense understory of cloud forests and at their edges and in humid lowland forests. Often communicates with a low, guttural, clucking *ka-ka-ka*… . Far-carrying song is a combination of the monotone *whut-whut-whut*…, and more explosive *waHA-waHA*, which becomes raucous when sung in duet.

Limpkin

Rufous-necked
Wood-Rail

Russet-naped
Wood-Rail

Ruddy Crake (Polluela Colorada)   *Laterallus ruber*
6 in (16 cm). Combination of **unbarred rufous underparts** and a dark gray half-hood distinguishes it from other small crakes and rails. Locally uncommon on eastern slopes of Northern Highlands; to 4,600 ft (1,400 m). Very rare in Pacific lowlands and Playitas-Moyúa-Tecomapa lagoons. Abundance is relatively undocumented in the Caribbean, with records from the Mosquitia and as far south as Pearl Lagoon. While foraging in marshes, flooded fields, and adjacent forest edge, moves with jerky body movements and tail pumping. Delivers a loud, gradually descending, low-pitched trill, practically identical to that of White-throated Crake, but more likely to trail off in intensity at end. Calls with an electric and snappy *chup!* as well as a *típ típ típ…*, incessantly delivered.

White-throated Crake (Polluela Gorgiblanca)   *Laterallus albigularis*
6 in (16 cm). Combination of **rufous breast** and **barred flanks and undertail coverts** is diagnostic. Common throughout Caribbean lowlands and foothills; to 1,300 ft (400 m). Mainly forages on plant material in freshwater marshes and in areas of dense grass on the edges of ponds, rivers, and forests. Delivers a gradually descending low-pitched trill, practically identical to that of Ruddy Crake, but more likely to end abruptly. Calls with a rich *teap!*

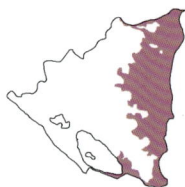

Gray-breasted Crake (Polluela Pechigrís)   *Laterallus exilis*
6 in (16 cm). Distinct **rufous nape** is diagnostic. Combination of **gray breast** and **barred flanks and undertail coverts** further distinguishes it. Very rare in Caribbean lowlands; to 700 ft (200 m). Forages on the ground at freshwater marshes, wet agricultural fields, and river and forest edge. Repeats a sharp *teap tip-tip-tip-tip* phrase (*tip* note sometimes repeated as many as 10 times), primarily monotone but may slightly descend in pitch; could be mistaken for a frog song. Call is a rich *chep!*

Yellow-breasted Crake (Polluela Pechiamarilla)   *Hapalocrex flaviventer*
5 in (13 cm). Noticeably smaller than immature Sora (p. 98). Bold **white superciliary bordered by black eye line and crown** is diagnostic. Very rare and local, with only 2 records (San Carlos vicinity, May 1917; San Juan del Norte, March 2017); most likely occurs throughout Guatuzo Plains and in other areas with suitable habitat. Forages with quick and jerky movements on floating vegetation of freshwater marshes. Delivers a high-pitched, 4-noted *kreer-ur kre-yí*, sometimes followed by a short, reedy trill. Call is a short, weak, scratchy *chiá* or *chío*.

Ruddy Crake

White-throated
Crake

Gray-breasted Crake

Yellow-breasted
Crake

Mangrove Rail (Rascón de Manglar)   *Rallus longirostris*
14 in (35 cm). The only large Nicaraguan rail with a **long bill** and **barred underparts**.
Rare and local in Pacific lowlands; known from Estero Real NR, Isla Juan Venado NR, and
Salinas Grandes salt ponds. Secretively forages in mangrove mudflats, where its scratchy
and agitated clucks and grunts are often the best clues to its presence; dueting utterances
of male and female can be raucous.

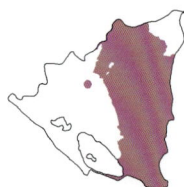

Uniform Crake (Rascón Café)   *Amaurolimnas concolor*
8 in (21 cm). Only Nicaraguan rail or crake with **uniform rufous plumage** (slightly paler
on underparts). Very rare throughout Caribbean lowlands and foothills, with few records
(known from Pearl Lagoon and San Juan River). One accidental record on the eastern slope
of Northern Highlands (Peñas Blancas NR, June 1909). Walks erect on the ground, similar to
wood-rails. Found in humid lowland forest, where it forages in leaf litter and mud. Whistles a
clear, repeating *feeép* (1 whistle per second), each whistle slightly upslurred.

Sora (Polluela Norteña)   *Porzana carolina*
9 in (23 cm). Adult has **vibrant yellow bill** that contrasts with **black mask and throat**.
Immature shows small, faint, black mask at base of bill; it is larger than the similar
Yellow-breasted Crake (p. 96). Winter resident (Oct to March). Locally uncommon at
Playitas-Moyúa-Tecomapa lagoons, and very rare in Pacific lowlands and foothills. Rare to
uncommon on western slopes of Northern Highlands; to 4,300 ft (1,300 m). Uncommon
in extreme southwestern Caribbean lowlands (along San Juan River, as far east as Bartola
RSP). Visits open marsh habitat along rivers, lagoons, and lakes. Persistently pumps tail
while foraging for aquatic prey and seeds. Gives a rolling whinny that starts sharp and
fast before descending in pitch and slowing down (lasts 2–3 seconds). Also whistles a
plaintive *wurá*, with a noticeable uptick in pitch at the end, and a high-pitched, squeaky
wi-wi-wi… series.

Spotted Rail (Rascón Moteado)   *Pardirallus maculatus*
10 in (26 cm). Dark overall, with **extensive barring and speckling** over most of body; also
note conspicuous **red spot** at base of large, tapered, yellow bill. Very rare and local, with
only 2 records (Punta El Menco, March 2000; Guatuzos WR, Aug 2013). Delivers a deep,
continuous, gruntlike *buhn-buhn-buhn*…; also makes a raspy, yelplike *kirEARH*.

Mangrove Rail

Uniform Crake

adult

Sora

immature

Spotted Rail

Purple Gallinule (Calamón Americano)   *Porphyrio martinicus*
13 in (33 cm). Adult is unmistakable, with **tricolored bill and frontal shield** and **purple iridescent feathers** on most of head, neck, and underparts. Immature is buffy brown overall; white undertail coverts are evident. Common throughout lowlands, Playitas-Moyúa-Tecomapa lagoons, and Sébaco Valley. Common on Sierra Isabelia and Sierra Dariense of Northern Highlands; to 4,500 ft (1,350 m). Prefers aquatic habitats. Forages in emergent or floating vegetation. Gives a series of hollow, mellow clucks that crescendo in intensity toward the middle. Also produces a squeaky toylike call that can turn into continuous chatter, and a nasal and plaintive *wháu* call.

Common Gallinule (Polla de Agua)   *Gallinula galeata*
13.5 in (34 cm). Adult is unmistakable, with **brilliant red frontal shield** and **yellow-tipped bill**. Nonbreeding adult and immature have duller bills, but are easily recognized by **white, jagged streaks on flanks**. Locally common breeding resident in Pacific lowlands, Playitas-Moyuá-Tecomapa lagoons, and Sébaco Valley, and on western slopes of the Northern Highlands; to 4,700 ft (1,450 m). Population numbers increase with arrival of winter residents (mid-Nov to March). Uncommon winter resident in Caribbean lowlands; present on Corn Islands. Prefers aquatic habitats containing emergent vegetation at water's edge, where it plucks food from the water's surface while swimming. Delivers a rattling chortle that rises in intensity before trailing off. Also forcefully projects a toy-horn-like *YIP!*

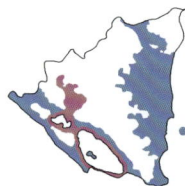

American Coot (Focha Americana)   *Fulica americana*
14.5 in (37 cm). Somewhat ducklike in appearance and behavior. Combination of **slaty body, red iris, and white bill** is unique; immature is paler overall and lacks red iris. Note white trailing edge of wing in flight. Locally uncommon breeding resident in the lakes region, Playitas-Moyúa-Tecomapa lagoons, and Sierra Isabelia and Sierra Dariense of the Northern Highlands. With the arrival of winter residents (Oct to April), becomes common throughout Pacific lowlands and in its range in the Northern Highlands; to 4,200 ft (1,300 m). Uncommon winter resident in Caribbean lowlands; present on Corn Islands. Often gregarious; forages by dabbling or diving for submerged vegetation in a variety of calm aquatic habitats. Produces a breathy, high-pitched *krah*, and a guttural *ka-ka-ka…*, often repeated at length. Also makes short clacking noises.

immature

Purple Gallinule

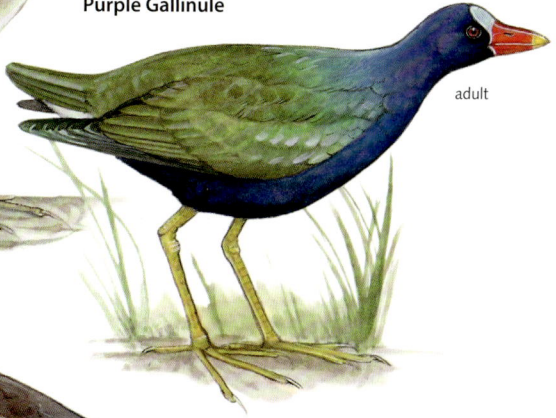

adult

adult

**Common
Gallinule**

immature

immature

American Coot

adult

THICK-KNEES Burhinidae

This small family (only 9 species) has a single representative in Nicaragua. Thick-knees are generally medium- to large-sized birds. The common name comes from bulbous ankle joints that resemble "knees." Their nocturnal hunting habits are supported by large eyes for enhanced low-light vision. Cryptic plumage allows them to blend in artfully with their arid surroundings.

Double-striped Thick-knee (Alcaraván)   *Burhinus bistriatus*
20 in (50 cm). Unmistakable. Stands tall on **long, yellow legs**. Note large, yellow eyes and **broad, white superciliaries bordered above by black lateral crown stripes**; also note **short, stout bill with black tip**. Locally common to uncommon throughout Pacific lowlands and foothills, rare in dry intermontane valleys on western slopes of Northern Highlands; to 2,300 ft (700 m). Terrestrial; inhabits arid grasslands with scattered trees, rice fields, and margins of dry marshes (e.g., Laguna de Tisma NR and Guayabo Wetlands). Primarily nocturnal and crepuscular forager; in the dark, it can be located by its red eye shine. During the day, stands motionless under shade of trees and shrubs, skillfully camouflaged within vegetation. Occurs alone or in groups of up to 20 individuals. Makes a loud, very fast, nasal honking: *pip-pip-pip*…, with an occasional discordant *piperip* added. When several individuals chorus at length, the resulting noise can have a frantic quality. Very vocal after sunset.

STILTS and AVOCETS Recurvirostridae

Stilts inhabit aquatic habitats. When foraging and breeding, they can form large, raucous groups. Long legs, long slender bills, and striking plumage typically distinguish them from other shorebirds. They gracefully move within shallow and muddy waters in search of insects, crustaceans, and other aquatic invertebrates.

Black-necked Stilt (Cigüeñuela Cuellinegra)   *Himantopus mexicanus*
14 in (36 cm). Elegant and petite. Combination of **black and white plumage**, **very long, red legs**, and **thin, long bill** makes it unmistakable. In flight, the legs extend conspicuously behind the body. Female is slightly brown on back; immature is brown rather than black. Abundant throughout Pacific lowlands and the western foothills of Northern Highlands; to 2,800 ft (850 m). Common in Guatuzo Plains and San Juan River. All populations increase from influx of NA migrants (Nov to April). Meanders through shallow, calm water and mudflats, picking prey from the surface; occurs in large numbers in a variety of aquatic habitats, including salt ponds and flooded fields. Incessantly barks a *yí-yí-yí*…, with a slight raspy quality.

American Avocet (Avoceta Americana)   *Recurvirostra america*
18 in (46 cm). Black and white plumage on back and upperwings is predominant year-round; in breeding plumage, head and neck take on an extravagant cinnamon tone (gray in nonbreeding plumage). **Long, upturned bill** is unique; male's bill is slightly straighter. In flight, broad black scapular lines stand out from white back. Uncommon to rare winter resident in Pacific lowlands, but occasionally stays throughout the year; often found just south of Isla Juan Venado WR (Salinas Grandes). Forages in shallow sections of freshwater marshes and salt ponds by sweeping submerged bill side-to-side. Screams a sharp *WEE WEE WEE*… .

OYSTERCATCHERS Haematopodidae

There is a single species in Nicaragua. These stocky, active birds are known for using their laterally compressed bills to dislodge limpets from rocky intertidal zones and pry them open in order to extract a juicy meal.

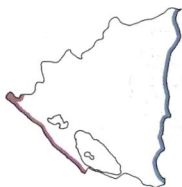

American Oystercatcher (Ostrero Americano)   *Haematopus palliatus*
18 in (46 cm). **Massive orange bill** and black head. In flight, also note on upperwing a white stripe at the base of black secondaries; white uppertail coverts stand out on black tail. Rare breeding resident on Pacific coast; when winter residents (Sept to March) boost local population; uncommon in Gulf of Fonseca and along southern Pacific coast. Rare winter resident on Caribbean coast. Forages on coastal beaches and shell bars, patrolling intertidal zones and probing the substrate. Gives a strident *PI* call, sometimes delivered with a fast, rolling, ongoing *pí-pí-pí*… . Also whistles a strained *weeur*.

Double-striped
Thick-knee

Black-necked
Stilt

nonbreeding

breeding

American
Avocet

American
Oystercatcher

Members of this family are generally distinguished from other shorebirds by their compact body and short bills, as well as their typical run-pause-run foraging strategy, observed as they search for invertebrate prey on shorelines. In breeding plumage, note plumage pattern, especially hues on upperparts, in addition to body size and structure. This shorebird family is represented in Nicaragua by 3 genera; to distinguish among the 6 species in the genus *Charadrius*, note color and size of bill and color of legs. All are fairly gregarious and can be seen in mixed-species flocks with as many as 200–300 birds. The NA migrants often have a few individuals that stay in country year-round.

Southern Lapwing (Avefría Sureña)   *Vanellus chilensis*
14 in (36 cm). Bold pattern, ornamental appearance, and relatively tall stature make it unmistakable. **Black forehead, throat, and breast** lead into otherwise white underparts. **Wispy, long black crest** projects horizontally from gray head. A recently established resident in Nicaragua (its distribution is expanding northward); there are no breeding records yet. Uncommon throughout Pacific lowlands and foothills around Playitas-Moyúa-Tecomapa lagoons and Sébaco Valley; to 1,600 ft (500 m). Uncommon along southeastern shores of Lake Nicaragua and San Juan River; accidental throughout rest of Caribbean lowlands. Gleans invertebrates from marshes, lake shores, river banks, and flooded fields; small groups of this bird often spread out to forage. Belts a loud, strained, repeated *KEEAH KEEAH KEEAH*…; in flight, gives a *eh-eh-eh*… .

Black-bellied Plover (Chorlito Gris)   *Pluvialis squatarola*
11.5 in (29 cm). **Larger size**, stocky appearance, and **thick bill** help to distinguish it from the golden-plovers. In nonbreeding plumage, note faint white superciliary on pale brownish-gray, spangled upperparts. In breeding plumage, has gray crown and blackish-gray spangled upperparts; black on underparts only reaches to lower belly (black on entire belly in American and Pacific Golden-Plover. Immature has streaked underparts, sometimes showing a buffy wash. In all plumages, note **white uppertail coverts**, **bold white wing stripes**, and **black axillaries** in flight. Common winter resident (mid-Aug to mid-May) on Pacific coast and lakes region; very rare in Sébaco Valley. Common on Caribbean coast. Passage migrants increase numbers (mid-Aug to mid-Sept and March to May) and small numbers stay year-round. Forages on sandy beaches and mudflats, and sometimes in flooded fields and mangroves. Whistles a plaintive *pweeúh*, with a momentary, but noticeable, dip in pitch.

American Golden-Plover (Chorlito Dorado Americano)   *Pluvialis dominica*
10.5 in (27 cm). Has a smaller body and a slenderer bill than does Black-bellied Plover. In nonbreeding plumage, prominent white superciliary contrasts with darker crown. In breeding plumage, note golden spangles on crown and upperparts and **entirely black underparts**; white extends only to the sides of the neck (on Pacific Golden-Plover, white extends down to undertail coverts). On female, face is not entirely black. In all plumages, note darkly barred tail, faint white wing stripes, and uniform gray underwings in flight; toes typically do not extend beyond tip of tail. Has longer primary projection than does Pacific Golden-Plover, with tertials falling short of tip of tail and primaries extending noticeably beyond it. Passage migrant in Pacific lowlands (with multiple records from Tisma NR); very rare during southern migration (mid-Aug to Nov) and rare during northern migration (March to April). Forages on sandy beaches and mudflats, individually or in small groups. Whistles a plaintive *pweelT*, with a shrill and upward inflected ending.

Pacific Golden-Plover (Chorlito Dorado del Pacífico)   *Pluvialis fulva*
10 in (25 cm). Has a smaller body and a slenderer bill than does Black-bellied Plover. In nonbreeding plumage, prominent white superciliary, often with a tawny wash, contrasts with darker crown. In breeding plumage, note golden spangles on crown and upperparts; white extends through flanks to undertail coverts on otherwise black underparts (on American Golden-Plover, white only extends to the sides of the neck). On female, face is not entirely black. In all plumages, note darkly barred tail, faint white wing stripes, and uniform gray underwings in flight; toes typically extend beyond tip of tail. Has shorter primary projection than does American Golden-Plover, with tertials barely reaching tip of tail and wings not extending noticeably beyond it. Accidental winter resident, with only 1 record that included 2 individuals at Tisma NR (April 2012). Seen occurring with American Golden-Plover.

Southern Lapwing

nonbreeding

nonbreeding

breeding

Black-bellied Plover

nonbreeding

nonbreeding

American Golden-Plover

breeding

nonbreeding

nonbreeding

Pacific Golden-Plover

breeding

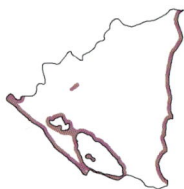

Collared Plover (Chorlitejo Collarejo)   *Charadrius collaris*

6 in (15 cm). Combination of **cinnamon wash on head** and partial white nuchal collar sets it apart from other Nicaraguan *Charadrius* plovers. Also note combination of complete black breast band, slender black bill, and flesh-colored legs. Immature has incomplete breast band. White outer rectrices reaching tip of tail are seen in flight. Common along Pacific coast and lakes region; common along Caribbean coast; and accidental at Lake Apanás. Can be found on virtually all shorelines, but typically prefers to forage in the rocky and driftwood laden zones up from the water line. Calls with a single *pip!*, but also gives a sputtering series of erratic *pip* notes.

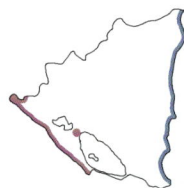

Wilson's Plover (Chorlitejo Picudo)   *Charadrius wilsonia*

8 in (20 cm). Noticeably **larger head and bill** than on other *Charadrius* plovers. Breeding male has broad, complete, black breast band; nonbreeding male and female have complete, light brown breast band; and immature has incomplete, light brown breast band. Combination of **long, thick black bill** and **flesh-colored legs** further distinguishes it; some individuals may have faint cinnamon wash on head. White outer rectrices reaching tip of tail are seen in flight. Two subspecies occur. *Beldingi* subspecies is a locally uncommon breeding resident on Pacific coast, in Tisma NR, and in Guayabo wetlands; populations grow in size with arrival of winter residents (mid-Aug to March). *Wilsonia* subspecies is a rare winter resident on the Caribbean coast. Typically breeds on sandy beaches but also forages on mudflats and salt ponds. Call is a lively and sharply rising *WHEEP!*

Semipalmated Plover (Chorlitejo Semipalmeado)   *Charadrius semipalmatus*

7 in (18 cm). In breeding plumage, **white forecrown contrasting with black mask** and **black-tipped orange bill** (short and stubby) is diagnostic among *Charadrius* plovers (bill is mostly black in nonbreeding and immature plumages). On adults, also note orange-yellow legs and complete black breast band; immature has darker legs. Has noticeably darker upperparts than does Piping Plover. Common winter resident (mid-Aug to late March) on Pacific coast and in lakes region, and uncommon in Sébaco Valley. Uncommon on Caribbean coast and Corn Islands. Some individuals persist year-round. Found at a variety of coastal and inland aquatic habitats, where it is known to use the foot-trembling method on sand or mud to force invertebrate prey into movement. Call is a double-noted *da-REEP!*, with the second note rising sharply in pitch.

Piping Plover (Chorlitejo Chiflador)   *Charadrius melodus*

7 in (18 cm). In all plumages, **yellow legs** and **short stubby bill** help distinguish it from Snowy Plover (p. 108). In breeding plumage, black midcrown and black-tipped, orange bill are sharply defined on pale face; on some individuals black breast band is complete, on others incomplete. In nonbreeding and immature plumages, breast band is pale gray and bill is black. In flight, white uppertail coverts contrast with black terminal tail band. Has noticeably paler upperparts than does Semipalmated Plover. Accidental winter resident, with 1 record on Caribbean coast (Bluefields Bay, Feb 2000). Seen on sandy beaches, foraging with other plovers. As the names implies, makes a pure, piping series of notes: *pip pip pip…* . **NT**

Collared Plover

immature

adult

Wilson's Plover

breeding
male

female/
nonbreeding
male

Semipalmated
Plover

breeding

nonbreeding

nonbreeding

Piping Plover

breeding

Snowy Plover (Chorlitejo Patinegro)   *Charadrius nivosus*

6 in (15 cm). The only *Charadrius* plover that shows a **broad black postocular stripe** (in breeding plumage). In breeding plumage, also note incomplete black breast band (stands out against pale upperparts). Slender black bill and dark legs help to distinguish it in nonbreeding plumage from nonbreeding Piping Plover (p. 106). White outer tail rectrices reaching tip of tail are seen in flight. Rare winter resident (Aug to March) along Pacific coast and at Tisma NR; half of Nicaraguan records have occurred during NA breeding season and almost exclusively at Salinas Grandes salt ponds. Call is a double-noted *puur-WEET*, with the first note ascending to an explosive second note, which is sometimes followed by a burry *chi chi*. **NT**

Killdeer (Chorlitejo Tildío)   *Charadrius vociferus*

10.5 in (27 cm). Set apart from the other *Charadrius* plovers by **black double breast band**. In flight, **rufous rump** and white wing stripe are conspicuous. Rare and local breeding resident populations occur at Chiltepe NR, Tisma NR, and Guayabo Wetlands. Countrywide winter resident (Nov to March); common in Pacific and uncommon in Northern Highlands. Uncommon in Caribbean. Although NA migrants may be found at coastal habitat, they prefer inland freshwater marshes and rice fields. Delivers a strained, onomatopoeic *kil-dee* or *kil-de-dee*.

JACANAS   Jacanidae

Jacanas patrol aquatic habitats with floating vegetation, using their extremely long toes and nails to distribute their weight. Nests are often located on floating vegetation. Females exhibit rare polyandrous behavior; they defend a territory containing up to 3 nests, with 3 males incubating and caring for their young.

Northern Jacana (Jacana Centroamericana)   *Jacana spinosa*

9 in (23 cm). **Yellow frontal shield** on black head makes it unmistakable. In flight, yellow flight feathers flash vibrantly. Immature has white superciliary, black postocular stripe, and brown upperparts; a key to identification is body shape, which is the same as that of adults. Abundant in Pacific lowlands and common in foothills. Common in Northern Highlands; to 4,300 ft (1,300 m). Common in Caribbean lowlands and foothills. Found in a variety freshwater and brackish aquatic habitats, but it is highly associated with bodies of water that have floating vegetation. Makes hoarse, erratic vocalizations, including *chura chura chét*; also gives a rapid, piercing, repeating *chét-chét-chét… .*

Wattled Jacana (Jacana Negra)   *Jacana jacana*

9 in (23 cm). **Red frontal shield and wattles** and black body make it unmistakable. Yellow flight feathers show vibrantly in flight. A SA vagrant with only 2 Nicaraguan records: on Lake Apanás (Nov 2004) and on the San Juan River, between Sábalos and El Castillo (Jan 2006). Closest breeding population occurs in western Panama.

Snowy Plover

nonbreeding

breeding

Killdeer

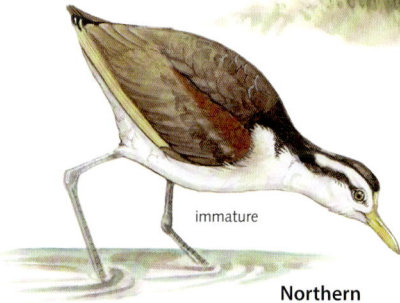

immature

Northern Jacana

adult

Wattled Jacana

Members of this family found in Nicaragua occur only as NA winter residents or passage migrants. The nondescript, drab hues of the nonbreeding plumage, which is what is most often seen in Nicaragua, make identification a challenge. Coming upon a mudflat, salt pond, or flooded field with hundreds or thousands of birds—of a variety of species—is an exhilarating and overwhelming experience. Successful identifications require patience and practice. Rather than focusing on plumage color or pattern, pay attention to a combination of subtle characteristics, including body size and shape, bill length, bill thickness, and bill curvature. Also note foraging strategy and the substrate on which the bird forages. Other important characteristics to consider are relative bill length and plumage patterns on upperwing, underwing, and tail. Immature, nonbreeding, and breeding birds of the same species may look quite different, and, in some species, body size and bill size differ between males and females, further complicating the matter.

Upland Sandpiper (Pradero)   *Bartramia longicauda*
12 in (30 cm). Combination of **thin neck, small head, plump body**, and long, yellow legs is unique. Black-tipped, yellow bill and large black eyes also aid in identification. In flight, dark rump and upperwing primaries and coverts contrast with paler, mottled plumage on upperparts and inner section of upperwing. Rare passage migrant in Pacific (Sept to Oct and mid-March to April). Accidental in Northern Highlands and Caribbean lowlands (Miraflor NR, Sept 2015; Puerto Cabezas, Nov 1966; Escondido River, Nov 1892). Forages in grasslands and open areas, with jerky run-and-stop movements. Flight call is a rich *wee-dí-dí*, made mostly at night by migrating groups.

Whimbrel (Zarapito Trinador)   *Numenius phaeopus*
17.5 in (45 cm). Combination of **long, decurved bill** and **dark brown lateral crown stripe and eye line** is unique. Immature Long-billed Curlew sometimes has bill of similar length but lacks the distinct crown pattern. In flight, looks patternless, with uniform-brown upperparts and upperwings, although extensive fine streaking shows at close range. Winter resident (Aug to March); large numbers of passage migrants arrive from mid-Aug to mid-Sept. Abundant on Pacific coast; rare at Chiltepe NR, Tisma NR, and Guayabo Wetlands; and very rare on Caribbean coast. Some individuals stay year-round. Prefers sandy beaches and mudflats, but also uses a variety of coastal aquatic habitats, as well as nearby flooded fields. Delivers a fast-paced series of squeaky, agitated, whistles: *wi-wi-wi…* .

Long-billed Curlew (Zarapito Piquilargo)   *Numenius americanus*
23 in (58 cm). **Extremely long, decurved bill** distinguishes it; also note **cinnamon color**. Shorter bill of immature may give it the appearance of a Whimbrel, but note plain crown. Cinnamon wing linings are diagnostic in flight. Rare winter resident (July to early April) on Pacific coast. Probes and picks for prey on sandy beaches, mudflats, and salt ponds. Makes high-pitched whistles that sound something like the screams of gulls; sometimes intersperses whistles with throaty tones.

Marbled Godwit (Picopando Canelo)   *Limosa fedoa*
18 in (46 cm). **Bill is long, thick, slightly upturned, and bicolored**. Cinnamon plumage and black legs help to distinguish it. In flight, note entirely cinnamon underwings. Plumage pattern and coloration distinguish it from Hudsonian Godwit (yet to be recorded in Nicaragua). Uncommon winter resident (July to March) on Pacific coast; accidental at inland locations. Rapidly and deeply probes for prey on mudflats and, occasionally, sandy beaches. Vocalizes with a rough, nasally chuckle, and gives a hoarse *kreeáh* alarm call.

Upland
Sandpiper

Whimbrel

Long-billed
Curlew

Marbled
Godwit

Ruddy Turnstone (Vuelvepiedras Rojizo) *Arenaria interpres*
9.5 in (24 cm). **Intricate plumage pattern** makes it fairly unmistakable. Short, **orange legs** and short, **wedge-shaped, black bill** also distinguish it. Nonbreeding plumage lacks the eye-catching rufous coloration, but muted head and breast pattern is still noticeable. In flight, shows white back stripe, white wing bars, and black-and-white tail pattern. A winter resident (Aug to April), but some individuals reside your-round. Common on Pacific coast; accidental at inland locations; generally very rare on Caribbean coast, but uncommon on Corn Islands. Forages on beaches and mudflats, where it uses its slightly upturned bill to turn over rocks, pebbles, and debris in search of food. Calls with a short, raspy, rattle: *pet-pet-pet*, typically given in a sequence of threes, but sometimes in greater numbers.

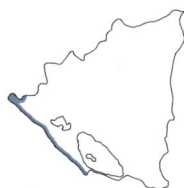

Red Knot (Correlimos Gordo) *Calidris canutus*
10.5 in (27 cm). In breeding plumage, note distinct rufous on head and underparts. **Rotund breast** and **short, thick legs** make body look stocky; **blunt-tipped bill (with pale tip)** is stouter than on Baird's Sandpiper (p. 114) and shorter than on Dunlin (p. 114). **Light barring on flanks** is also helpful. Shorter bill length and darker leg color help distinguish it from dowitchers. In flight, pale uppertail coverts contrast with darker upperparts and tail. Rare winter resident on Pacific coast (mid-July to mid-April). Picks and probes for prey on sandy beaches and mudflats; often found in groups of 10 or more. Calls include an abbreviated, hoarse *chut!*

Surfbird (Playero de Rompiente) *Calidris virgata*
10 in (25 cm). Combination of **stocky body** and **short, stout bicolored bill** separates it from other sandpipers. In nonbreeding plumage, upperparts are **uniform dark gray** and white lower underparts show spots. In flight, note bold white wing stripe and, on tail, contrast between black and white. Uncommon winter resident (Aug to early April) on Pacific coast and accidental at inland locations; populations are boosted by passage migrants (Sept and April). Scrambles along intertidal zones of rocky coastlines, often in groups. Gives a variety of squeaky *pip* utterances.

Sanderling (Correlimos Arenero) *Calidris alba*
8 in (20 cm). Most often seen in very pale nonbreeding plumage, against which the black legs, black bill (with blunt tip), and black eyes all stand out. **Black leading edge of wing** creates a shoulder look on folded wing. In breeding plumage, combination of rufous on head and breast and white underparts is unique. In flight, note bold white wing stripe and white tail with dark central rectrices. Winter resident (Aug to April). Uncommon on Pacific coast; rare at Chiltepe NR, Tisma NR, and Guayabo wetlands; accidental at other inland waters. Rare on Caribbean coast and Corn Islands. Prefers sandy beaches, where small flocks run after retreating waves to pick off exposed prey, then flee the returning waves toward dry sand. Gives multiple soft *cheep* notes while foraging; makes a weak *pit-pwít* in flight.

Ruddy Turnstone

breeding

nonbreeding

Red Knot

breeding

nonbreeding

Surfbird

breeding

nonbreeding

Sanderling

breeding

nonbreeding

Sanderlings, Ruddy Turnstones, and a Black-bellied Plover (p. 105) congregate on a beach.

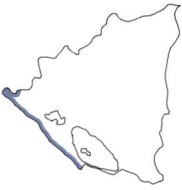

Dunlin (Correlimos Común)   *Calidris alpina*
8.5 in (22 cm). In breeding plumage, black belly is diagnostic. **Slightly decurved bill is longer** than on Red Knot (p. 112), Baird's Sandpiper, and White-rumped Sandpiper (p. 116). **Short stocky build** and **black legs** separate it from Stilt Sandpiper (p. 118). In flight, note white wing stripe and dark central rectrices. Very rare winter resident (Sept to March) on Pacific coast. Most likely seen with other foraging sandpipers. Alarm call is a screeching whistle: *PREE.*

Buff-breasted Sandpiper (Praderito Pechianteado)   *Calidris subruficollis*
8.5 in (21cm). Crisply defined markings on head, neck, and upperparts and rich coloration on underparts give it a unique look. **Large black eye** boldly stands out on **plain buffy face**. Uniform buff coloration on face and most of underparts distinguishes it from all other sandpipers. White underwings contrast with the buff underparts in flight. Very rare passage migrant in Pacific lowlands, Lake Apanás, and possibly into foothills (late Aug to early Oct and April to May). Propensity to stop in short grasslands, including flooded fields and agricultural lands, where it forages with a steady walk and bobbing head. Gives a singular, rough *prít* call. **NT**

Pectoral Sandpiper (Correlimos Pechirrayado)   *Calidris melanotos*
9 in (23 cm). Tall stature, long neck, and **heavily streaked breast that is sharply delineated from white belly** distinguish it from other *Calidris* sandpipers. Slightly decurved bill is bicolored. Male is noticeably larger than female. In flight, dark central stripe on tail contrasts with white uppertail coverts. Passage migrant (July to Nov and late Feb to May). Uncommon in Pacific lowlands and foothills (Salinas Grandes salt ponds, Tisma NR, and Guayabo Wetlands are common stopover locations), rare in Northern Highlands; to 3,300 ft (1,000 m). Rare on southeastern shores of Lake Nicaragua and San Juan River headwaters, otherwise accidental in Caribbean lowlands. More common during northward migration, when occurrence in Northern Highlands and Caribbean lowlands is likely. Prefers freshwater marshes and flooded fields, but can be seen on beaches and mudflats. Whistle call is a very short, low, tremulous *trí.*

Spotted Sandpiper (Andarríos Maculado)   *Actitis macularius*
7.5 in (19 cm). Most commonly encountered sandpiper. Often recognized by habit of incessantly pumping its tail as it forages for prey. Could be confused with larger Solitary Sandpiper (p. 120), but has shorter legs and stands tilted forward. In all plumages, note white superciliary (varying degrees of intensity), black eye line, and **brown sides of breast**. Inconspicuous white of outer rectrices and prominent white wing bar are seen in flight. Countrywide winter resident (late-July to mid-May). Common in lowlands, uncommon in foothills, and rare in highlands; to 4,100 ft (1,250 m). Adaptable to almost any type of aquatic habitat; this is the only sandpiper that is regularly seen along interior forest rivers. Usually solitary. Call is a sweet series of whistles: *twí twí twí…* (sometimes uttered as a single note).

Baird's Sandpiper (Correlimos Pasajero)   *Calidris bairdii*
7.5 in (19 cm). In all plumages, note **brown streaking on buffy breast** and white lower underparts. Slightly decurved bill is slenderer than on Red Knot and shorter than on Dunlin. **Wingtips extend beyond the tip of tail** (they do not on similar *Calidris*, except for White-rumped Sandpiper). In flight, note **dark central uppertail coverts and rectrices on gray tail** (absent on White-rumped Sandpiper, p. 116). Very rare passage migrant (Aug to Oct and Jan to mid-May). On Pacific, usually found inland, at low elevations. Also possible at a variety of elevations in Northern Highlands. Prefers to forage on the dry fringes of mudflats, pastures, and other freshwater aquatic habitats; seldom frequents beaches and rarely wades. Gives short, raspy notes, with a cricket-like quality, that sometimes transform into trills.

nonbreeding

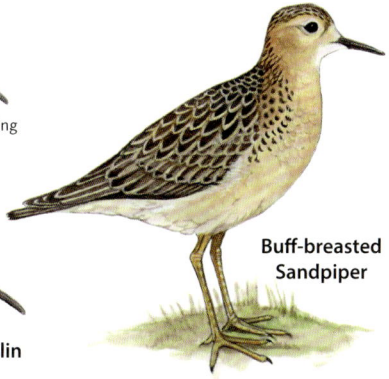

Buff-breasted Sandpiper

Dunlin

breeding

breeding

Spotted Sandpiper

Pectoral Sandpiper

nonbreeding

nonbreeding

breeding

Baird's Sandpiper

[Small *Calidris* sandpipers are commonly referred to as "peeps" and can be difficult to distinguish. Study length of folded wings in relation to the tail length, leg color, and bill shape and length in order to make accurate identifications.]

White-rumped Sandpiper (Correlimos Lomiblanco)   *Calidris fuscicollis*
7.5 in (19 cm). **Orange base of lower mandible** is diagnostic. Light to heavy streaking on breast and flanks helps distinguish it from Dunlin (p. 114) and Baird's Sandpiper (p. 114). **Wingtips extending beyond the tip of tail** help to distinguish it from all similar *Calidris* except for Baird's Sandpiper; in flight, note obvious **white uppertail coverts** (Baird's has dark center on uppertail coverts). Very rare passage migrant on Pacific coast and lakes region; accidental in the Caribbean. In Nicaragua, has been found at salt ponds and freshwater marshes; might also occur at tidal mudflats and on sandy beaches. Calls with a thin, slightly metallic warble.

Least Sandpiper (Correlimos Menudo)   *Calidris minutilla*
6 in (15 cm). The smallest peep. This is the only small *Calidris* sandpiper with **yellow legs** (though mud-covered legs can look dark). In all plumages, looks darker and browner than the Western Sandpiper and Semipalmated Sandpiper; also note brown breast with prominent streaking. **Short, blunt-tipped bill** is slightly decurved. In flight, darker than other peeps. Winter resident (mid-July to early May). Abundant in western Pacific, rare throughout eastern Pacific, Lake Apanás, San Juan River, and Caribbean coast; to 3,300 ft (1,000 m). Prefers freshwater aquatic habitats, particularly mudflats, but may also be found on sandy beaches. Call is a reedy, rising, slightly tremulous *pwee*.

Semipalmated Sandpiper (Correlimos Semipalmeado)   *Calidris pusilla*
6.5 in (16 cm). Black legs distinguish it from Least Sandpiper; **short, straight, blunt-tipped bill** helps distinguishes it from Western Sandpiper, though bill lengths sometimes overlap. Often has darker crown and ear coverts, smaller head, and stockier appearance than does Western Sandpiper. Uncommon winter resident (mid-July to early May) on Pacific coast and lakes region and accidental in Sébaco Valley and Caribbean. Populations grow when passage migrants arrive (Aug–Sept); small numbers remain year-round. Walks steadily along tidal mudflats, sandy beaches, salt ponds, and freshwater marshes, where it mainly picks prey off surface of substrate. Quickly repeats (4–6 times) a high-pitched, slightly raspy *keh* call. Flight call is similar, but often consists of a single note or of multiple notes with more time between them. **NT**

Western Sandpiper (Correlimos Occidental)   *Calidris mauri*
6.5 in (17 cm). Black legs distinguish it from Least Sandpiper. **Long, decurved, fine-tipped bill** helps distinguish it from Semipalmated Sandpiper, though bill lengths sometimes overlap. Often has paler crown and ear coverts, bigger head, and lankier appearance than does Semipalmated Sandpiper. Also note rufous on crown, ear coverts, and scapulars. Common winter resident (mid-July to mid-May) on Pacific coast and in lakes region; rare in Sébaco Valley. Populations swell when passage migrants arrive (Aug–Sept); small numbers remain year-round. Walks steadily along tidal mudflats, sandy beaches, salt ponds, and freshwater marshes, where it picks and probes for prey. Call is a clear, thin, abrupt *whí*. Also, gives a reedy trill.

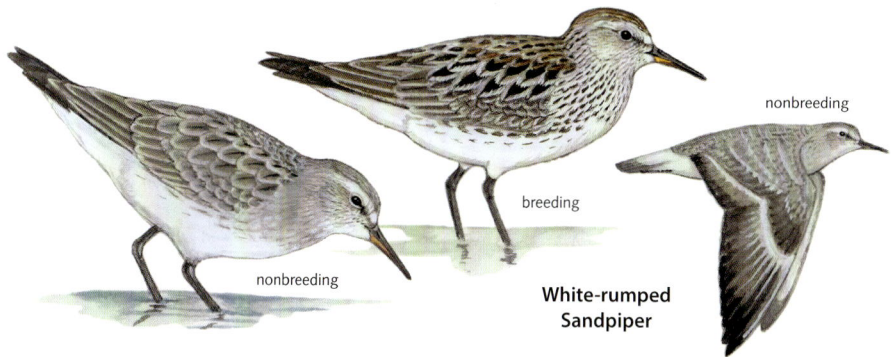

nonbreeding

breeding

nonbreeding

White-rumped Sandpiper

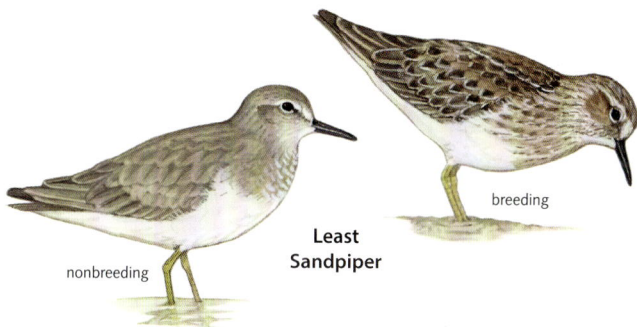

nonbreeding

Least Sandpiper

breeding

breeding

nonbreeding

Semipalmated Sandpiper

breeding

nonbreeding

Western Sandpiper

Stilt Sandpiper (Correlimos Patilargo)   *Calidris himantopus*

8.5 in (22 cm). In nonbreeding plumage, can be a challenge to identify as it resembles several other sandpipers. **Tall stature** and **dull yellow legs** separate it from Dunlin (p. 114); **black bill** is shorter than on dowitchers; and the **slightly decurved** shape of the bill distinguishes it from yellowlegs. Breeding plumage is distinct. Common winter resident (late-July to mid-May) on Pacific coast and at Chiltepe NR, Tisma NR, and Guayabo Wetlands; accidental at other inland locations. Abundant when passage migrants move through (mid-Aug to early Nov and mid-March to mid-April). Often forages in a typical dowitcher-like stitching fashion, with short and rapid probes of mud or water, in both fresh and saltwater habitats; typically not found on beaches. Call is a soft *peep*, sometimes followed by an equally soft guttural squeak. Flight call is a more grating whistle: *krear*.

Short-billed Dowitcher (Agujeta Común)   *Limnodromus griseus*

11 in (28 cm). Distinguished from very similar Long-billed Dowitcher by only subtle characteristics: base of bill is thicker; bill is generally shorter, lower mandible is often slightly decurved; when feeding, back looks flatter than on Long-billed; wing tips usually reach, or slightly extend beyond, tip of tail; tail is usually paler (visible in flight); nonbreeding Short-billed more frequently shows patterns on its face and breast than does nonbreeding Long-billed. Long, straight bill separates it from Red Knot (p. 112) and Stilt Sandpiper. Common winter resident (Aug to early April) on Pacific coast and Chiltepe NR, Tisma NR, and Guayabo Wetlands; accidental in foothills. Short-billed more common than Long-billed. Found in a variety of salt- and freshwater aquatic habitats, but prefers mudflats, where it uses its bill to probe deep and rapidly for prey. Call, a rapid *tu-tu-tu* whistle, is perhaps the best means of distinguishing it from Long-billed.

Long-billed Dowitcher (Agujeta Piquilarga)   *Limnodromus scolopaceus*

11.5 in (29 cm). Distinguished from the more common but very similar Short-billed Dowitcher by only subtle characteristics: base of bill is thinner; bill is generally longer; lower mandible is usually straight; when feeding back looks more rounded; wing tips usually do not reach tip of tail; tail is usually darker (visible in flight); nonbreeding Long-billed usually lacks patterns on face and breast. Belly is always completely cinnamon in breeding plumage. Long, straight bill separates it from Red Knot (p. 112) and Stilt Sandpiper. Uncommon winter resident (Sept to early April) on Pacific coast and lakes region, and accidental in foothills. Less common than Short-billed. Probes deep and rapidly for prey, mainly at inland freshwater aquatic habitats. Call is a single, sharp, arcing *keek* whistle, sometimes given in repetition; perhaps the best way to distinguish it from Short-billed.

Wilson's Snipe (Agachadiza Común)   *Gallinago delicata*

10.5 in (27 cm). Note boldly striped head and mantle and barred flanks. Combination of **short, stocky stature** and **long, straight bill** is unique. In flight, note white belly contrasting with dark underparts and orange tail. Rare winter resident (Oct to April) in Pacific and western slopes of Northern Highlands; accidental in Caribbean; to 4,100 ft (1,250 m). Often furtively crouches low to the ground, only to be seen when flushed. Mainly in freshwater marshes and flooded grassland, forages by probing deeply into mud and water. Calls with a raspy *rhet!* note.

nonbreeding

**Stilt
Sandpiper**

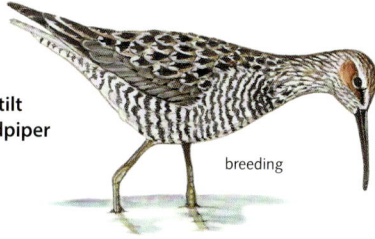

breeding

nonbreeding

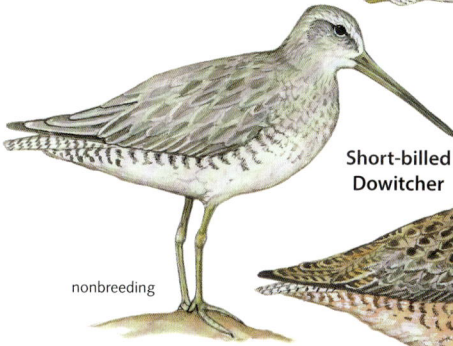

**Short-billed
Dowitcher**

nonbreeding

breeding

nonbreeding

**Long-billed
Dowitcher**

nonbreeding

**Wilson's
Snipe**

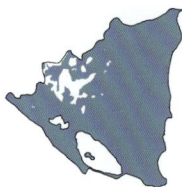

Solitary Sandpiper (Andarríos Solitario)   *Tringa solitaria*
8.5 in (22 cm). Combination of white spectacles (created by faint lore and **bold eye ring**) and **white spots on dark upperparts** characterizes this species. Note greenish legs. Has longer legs and a more erect posture than smaller Spotted Sandpiper (p. 114). In flight, note dark rump and central rectrices, conspicuous white outer tail rectrices, and uniformly dark upperwing flight feathers. Countrywide winter resident (mid-July to mid-April). Uncommon in lowlands and rare in foothills and highlands; to 3,400 ft (1,050 m). Although it frequents most aquatic habitats, prefers freshwater ponds, lakes, rivers, and streams. Picks small prey from shorelines; forages with a distinctive head bob and tail pump. Calls with a sharp *peek* while foraging and a double- or triple-noted *pwee pwee* in flight, with each note rising in pitch.

Wandering Tattler (Correlimos Vagabundo)   *Tringa incana*
11 in (28 cm). Often has a plump appearance. Most often seen in nonbreeding plumage, when it is **uniformly gray**, except for white belly and **black lores**. In breeding plumage, note heavily barred underparts. In flight, combination of uniformly gray upperwings, rump, and tail is unique. Rare winter resident (late Aug to March) on Pacific coast, and accidental at inland locations. Intermittently pumps tail as it forages alone in the intertidal zone of rocky coastlines. Whistles a rapid series of short, high-pitched notes (from 4 to 20 in number), with only slight oscillation in pitch and sometimes a slight rise in intensity.

Greater Yellowlegs (Andarríos Patiamarillo Grande)   *Tringa melanoleuca*
14 in (36 cm). Tall, slender, and elegant, with **long, bright yellow legs**. Much larger than Lesser Yellowlegs; **stout and slightly upturned bill** is one and a half times the length of the head (on Lesser, length of bill matches length of head). Most often seen in nonbreeding plumage, when it is pale overall; in breeding plumage, showcases darker barring and streaking. White rump and uniform wing color seen in flight are similar to that of Lesser Yellowlegs, Stilt Sandpiper (p. 118), and Wilson's Phalarope (p. 122). Countrywide winter resident (late July to April). Common in Pacific and uncommon in Northern Highlands; to 3,300 ft (1,000 m). Rare in Caribbean lowlands. Some individuals stay year-round at Salinas Grandes salt ponds, Tisma NR, and Guayabo wetlands. Frequents a variety of aquatic habitats but prefers shallow waters or mudflats; actively walks to find prey, which it picks from surface or subsurface of water. Call is a loud, slightly raspy series of notes: *TEW TEW TEW…*; in flight or when flushed, usually gives 3 or more notes.

Willet (Playero Aliblanco)   *Tringa semipalmata*
15 in (38 cm). Despite inconspicuous gray plumage, robust build and **long, straight, stout bill** make this shorebird relatively easy to identify. Also note bluish-gray legs. Any potential confusion is resolved when **black-and-white wing pattern** is seen in flight. Most sightings are of *inornatus* subspecies, but *semipalmatus* subspecies, which is smaller and has brown tones, also occurs in Nicaragua. Abundant winter resident (mid-July to early May) on Pacific coast and western shorelines of Lake Managua and Lake Nicaragua; rare along the San Juan River and the Caribbean coast. A few individuals stay year-round along the Pacific coast. Picks and probes in surf and calm waters of sandy beaches, but also frequents a variety of other aquatic habitats. Often gives a muffled chatter punctuated by a rich double-noted *ki-dít* or a piercing and slightly downslurred *WEEIT!*

Lesser Yellowlegs (Andarríos Patiamarillo Chico)   *Tringa flavipes*
10.5 in (27 cm). Tall, slender, and elegant, with **long, bright yellow legs**. Much smaller than Greater Yellowlegs; length of **slender, straight bill** matches length of head (on Greater, bill is one and a half times the length of the head). Most often seen in nonbreeding plumage, when it is pale overall; in breeding plumage, shows darker barring and streaking . In flight, white rump and uniform wing color are similar to that of Greater Yellowlegs, Stilt Sandpiper (p. 118), and Wilson's Phalarope (p. 122). Winter resident (mid-July to mid-May), with high numbers of passage migrants (Oct and Nov). Abundant in Pacific lowlands and foothills; to 1,600 ft (500 m). Accidental in Northern Highlands and rare in Caribbean lowlands. Often found in flocks that actively forage by picking at surface of water at estuaries, freshwater marshes, and flooded fields (less often on riverbanks). While foraging, calls with a repeated series of downslurred, clear whistles: *tew-tew-tew…* . In flight or when flushed, typically only makes 2 notes.

**Solitary
Sandpiper**

nonbreeding

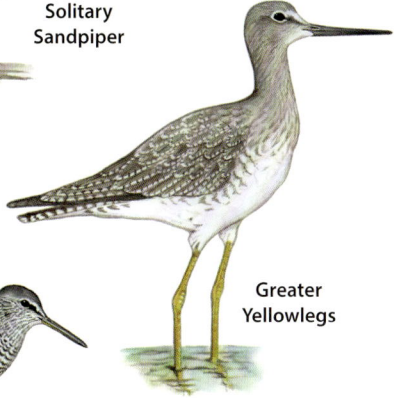

**Greater
Yellowlegs**

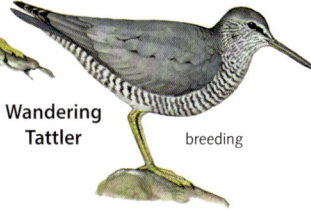

**Wandering
Tattler**

breeding

nonbreeding

nonbreeding
inornatus ssp.

nonbreeding
semipalmatus ssp.

Willet

breeding

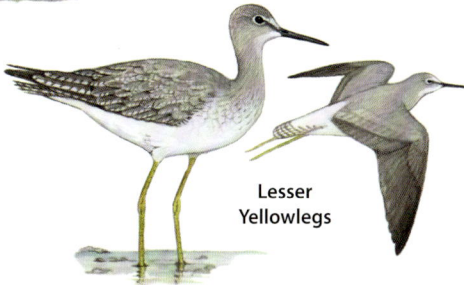

**Lesser
Yellowlegs**

Wilson's Phalarope (Falaropo Tricolor)   *Phalaropus tricolor*
9 in (23 cm). In breeding plumage, male and female are both unmistakable. Nonbreeding birds can be confused with *Tringa* sandpipers but are distinguished by **thin, straight, fine-tipped bill** and contrast between white underparts and gray upperparts. Lack of black patch behind the eye separates it from the other phalaropes. White rump and uniform wing color seen in flight is similar to that of Greater Yellowlegs (p. 120), Lesser Yellowlegs (p. 120), and Stilt Sandpiper (p. 118). Mainly a passage migrant (mid-Aug to mid-Sept and April to May), with higher numbers during southward migration; small numbers stay as winter residents. Locally common at Salinas Grandes salt ponds and rare in rest of Pacific lowlands. This is the most terrestrial of the phalaropes. It searches with outstretched neck on mudflats and erratically jabs bill to pick prey from ground or air; also swims and wades in shallow freshwater sources. Call is a soft, throaty *whek*.

Red-necked Phalarope (Falaropo Cuellirrojo)   *Phalaropus lobatus*
8 in (20 cm). In breeding plumage, distinctive. **Black patch behind the eye** distinguishes it from Wilson's Phalarope; **thin bill** and **faint white stripes on gray back** distinguish it from Red Phalarope. White wing stripe and dark central rectrices are seen in flight. Mainly a passage migrant (mid-July to Dec and in April), with higher number of stopover birds during southward migration. Common on Pacific Ocean pelagic waters and rare on Pacific coast (several records at Salinas Grandes salt ponds). Various Jan records in Gulf of Fonseca indicate that some individuals winter in Nicaraguan waters as well. Large flocks gather to feed at tidelines, where they often spin up plankton by rapidly twirling on the water's surface. Call is a low, harsh *kheck*.

Red Phalarope (Falaropo Rojo)   *Phalaropus fulicarius*
8.5 in (22 cm). Unmistakable if seen in unique breeding plumage. **Black patch behind the eye** distinguishes it from Wilson's Phalarope; **short, thick bill** and **uniform gray back** distinguish it from Red-necked Phalarope. **Yellow base of bill** (variable in extent) is diagnostic if seen. In flight, note white wing stripe and dark central rectrices. Passage migrant (Aug to Jan and late March to April); common on Pacific Ocean pelagic waters and very rare on Pacific coastal waters; accidental at inland locations. Small flocks forage on water's surface, rapidly twirling to spin up plankton.

Wilson's Phalarope

breeding female

breeding male

nonbreeding

nonbreeding

The customary spinning motion of a phalarope on the water (Wilson's shown).

Red-necked Phalarope

nonbreeding

nonbreeding

breeding female

nonbreeding

Red Phalarope

breeding female

nonbreeding

nonbreeding

These gull-like birds breed in polar regions and migrate to tropical waters for the boreal winter. To attain food, they practice kleptoparasitism, in which they steal prey captured by other birds. They are mostly seen in pelagic waters, but sometimes wander into coastal waters. Members of this family have hooked bills and, usually, a white patch on dark primaries. The unmistakable central rectrices of breeding birds are not normally seen in Nicaragua, making identification very difficult at times, given an array of color morphs on both adults and immature plumages. All pale-morph adults have black caps, whereas dark-morph adults are almost completely black. In most cases, body size, shape, and flight pattern are important keys to identification.

Pomarine Jaeger (Págalo Pomarino)   *Stercorarius pomarinus*
18.5 in (47 cm); WS 52 in (132 cm). Has a larger body and head, broader wings, and stouter bill than Parasitic Jaeger. Breeding adult distinguished by **long, twisted central rectrices** with rounded tips. Both pale and dark morphs have a **bicolored bill**. Pale morph has a **black half-hood**; on dark morph, head is entirely dark. All plumages show a **second pale underwing patch** at base of primaries (on wing lining). Winter resident (Sept to April). On Pacific, common on pelagic waters and very rare on coastal waters. In Caribbean, uncommon on pelagic waters and very rare on coastal waters. In flight, has stiff, heavy wingbeats.

Parasitic Jaeger (Págalo Parásito)   *Stercorarius parasiticus*
16.5 in (42 cm); WS 46 in (117 cm). Has a smaller body and head, narrower wings, and thinner bill than Pomarine Jaeger. Breeding adult distinguished by the **short, pointed central tail rectrices**. Both pale and dark morph adults have **solid black bills**; all immature morphs have bicolored bill. Although plumage color varies greatly on immatures, most show **some level of cinnamon coloration**. Winter resident (Sept to April). On Pacific, common on pelagic waters and very rare on coastal waters. In Caribbean, uncommon on pelagic waters and very rare on coastal waters. Note fast, falcon-like flight as it persistently harasses other seabirds.

Long-tailed Jaeger (Págalo Colilargo)   *Stercorarius longicaudus*
15 in (38 cm); WS 43 in (109 cm). The palest and most petite of the jaegers. Distinguished by **short, thick bill**. On breeding plumage, **long, pointed central rectrices** are unmistakable; **gray mantle contrasts with dark primaries**. There is a variety of immature morphs, ranging from pale to dark; all show blunt-tipped central rectrices. Rare winter resident on pelagic waters of Pacific Ocean; very rare on the Caribbean Sea; only accidental in coastal waters. Flight is more graceful and ternlike than that of other jaegers. This species is less aggressively kleptoparasitic than other jaegers.

Pomarine
Jaeger

breeding
pale morph

immature
pale morph

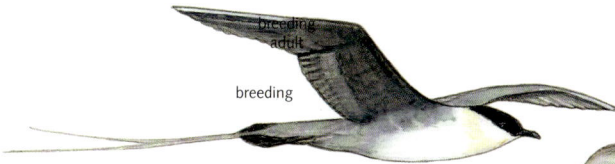

breeding
pale morph

immature
pale morph

Parasitic
Jaeger

breeding
adult

breeding

nonbreeding

Long-tailed
Jaeger

immature
pale morph

The members of this cosmopolitan family occur almost exclusively in aquatic habitats. Gulls are typically larger bodied, with broad wings, stout bills, and squared tails; terns typically have smaller bodies, with narrow, pointed wings, slender bills, and forked tails. Gulls are opportunistic feeders, with a variety of foraging strategies, while terns mainly plunge-dive into water or hawk insects from the air. With both gulls and terns, identification is best made by relying on several features; a photo for later study is sometimes necessary for the most difficult cases. There is only 1 skimmer in Nicaragua; its unique bill makes it relatively easy to identify.

Swallow-tailed Gull (Gaviotín Trasnochador)   *Creagrus furcatus*
22.5 in (57 cm). In all plumages, note **large eyes**; on adults, note **vibrant red orbital ring**. In flight, also note **forked tail** and distinct pattern of black, gray, and white on upperwing (only smaller Sabine's Gull has the same pattern). Accidental pelagic migrant on Pacific Ocean; both adults and immature have been recorded. Highly pelagic; picks prey off the surface of the sea at night and rests on the water during the day.

Sabine's Gull (Gaviotín Ahorquillado)   *Xema sabini*
13.5 in (34 cm). On adult, combination of relatively short, **yellow-tipped, black bill** and **slate hood with black border** is unique. In nonbreeding plumage, remnant of hood shows only as a nuchal collar. Immature has dusky crown and nape. In flight, note slightly **forked tail** and distinct pattern of black, gray, and white on upperwing (only larger Swallow-tailed Gull has the same pattern). Common passage migrant (late July to Nov and April to May) on pelagic waters of Pacific Ocean and rare on coastal waters; consistently seen on Gulf of Fonseca, where small numbers may be winter residents. Forages on the surface of pelagic waters; also frequents the protected waters of gulfs and bays. Calls with a throaty rattle or a single raspy note.

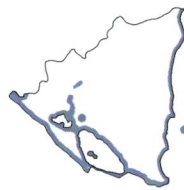

Laughing Gull (Gaviota Reidora)   *Leucophaeus atricilla*
16.5 in (42 cm). Bill is longer and stouter than on smaller Franklin's Gull; **broken white orbital ring is narrower**. In nonbreeding plumage, remnant of black hood shows as gray streaking on hindcrown. First and second year adults typically have gray on the nape and sides of breast, while immature is dusky brown overall. In breeding and nonbreeding plumages, note black wingtips and white tail in flight; in first winter plumage, note broad black terminal band on tail and black primaries. Winter resident (Nov to May); sizable numbers remain in the lowlands year-round. Abundant on Pacific coast and lakes region, and rare at Lake Canoas, Playitas-Moyúa-Tecomapa lagoons, Sébaco Valley, and Lake Apanás. Common on Caribbean coast and on large rivers of Caribbean lowlands. Found at a wide variety of both coastal and inland aquatic habitats. Calls have a squeaky, whining quality, similar to that of Franklin's Gull but lower-pitched; calls often change into a cackle: *ha ha ha… .*

Franklin's Gull (Gaviota de Franklin)   *Leucophaeus pipixcan*
14.5 in (37 cm). Bill is shorter and slenderer than on larger Laughing Gull; **broken white orbital ring is broader**. In breeding plumage, note black hood; in all other plumages, note half-hood. First and second year adults typically have clean white nape and breast, while immature has dusky brown upperparts and white underparts. In flight, black-and-white wingtips distinguish breeding and nonbreeding plumages from Laughing Gull; first winter plumage has incomplete black terminal band on tail and black wingtips. Passage migrant (Nov to Dec and April to May); winter residents occasionally occur, with small numbers staying year-round. Uncommon on Pacific pelagic and coastal waters and in lakes region; accidental on Caribbean coast. Frequents a variety of both coastal and inland aquatic habitats, as well as pelagic waters. Calls have a squeaky, whining quality, similar to those of Laughing Gull but higher-pitched.

Swallow-tailed Gull

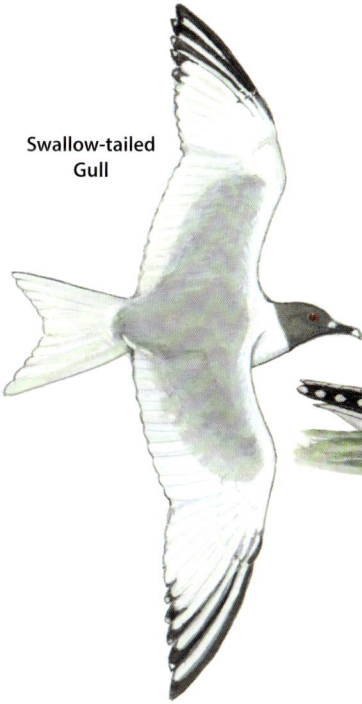

nonbreeding

Sabine's Gull

breeding

immature

immature

nonbreeding

breeding

first year

breeding

Laughing Gull

nonbreeding

breeding

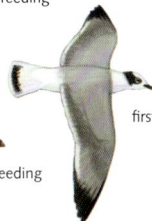

first year

breeding

Franklin's Gull

[Although Herring Gull is the only *Larus* gull that occurs with regularity, there is always a chance that a rarer species is hiding in plain sight among other gulls. Members of this genus are notoriously difficult to identify because of the number of plumage changes between immature and adult plumages. Among the 5 *Larus* gulls recorded in Nicaragua, a daunting 34 plumages are possible. General posture, body size, bill shape and size, and grayness of mantle are all important keys to identification.]

Ring-billed Gull (Gaviota de Pico Anillado)   *Larus delawarensis*
17.5 in (44 cm). Smallest of the white-headed gulls. Has a **short, stout bill**. The Ring-billed and the larger Herring Gull have the palest gray mantle among *Larus* in Nicaragua. On second year and older adults, combination of **black ring (towards tip of yellow bill)**, pale iris, and yellow legs is diagnostic. First year adults with black-tipped, flesh-colored bills and flesh-colored legs are similar to first year California Gull and first and second year Herring Gull, but have much lighter underparts; immature is primarily dusky brown, with a black bill that has some amount of pale color at the base. Very rare winter resident (Nov to early May) on both coastlines; accidental at inland locations with appropriate habitat. Most likely to be found amid Herring Gull flocks.

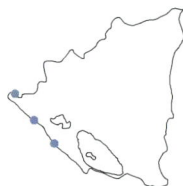

California Gull (Gaviota Californiana)   *Larus californicus*
21 in (53 cm). A medium-sized gull with a white head and **long, relatively slender bill**. Irises are dark at all ages. Gray mantle is darker than that on smaller Ring-billed Gull and larger Herring Gull, but paler than that of Lesser Black-backed Gull (p. 130) and larger Kelp Gull (p. 130). On second year and older adults, combination of dark iris and **black and red markings towards tip of yellow bill** is diagnostic. As second year adult sometimes resembles many other *Larus* plumages, it is important to note body size and bill shape. Immature has variable brown coloration, often with a wash of cinnamon, and a black bill that sometimes shows a pale base. Accidental winter resident (Jan to June) on Pacific coast. Most likely to be found amid Herring Gull flocks.

Herring Gull (Gaviota Argéntea)   *Larus argentatus*
25 in (64 cm). Largest white-headed gull. Has a **long and relatively stout bill** that shows a bulge towards the tip of the lower mandible. Herring Gull and smaller Ring-billed Gull have the palest gray mantle among *Larus* in Nicaragua. On mature adults, combination of pale iris, flesh-colored legs, and **red spot on lower mandible of yellow bill** is diagnostic. The mantle color of the various plumages between second and third year adults is key to distinguishing it from other *Larus*; during these plumage stages the bill typically has a black tip. First year adult resembles immature, but with flesh color on base of bill; immature is uniformly dark brown with a black bill. Winter resident (Nov to early May); uncommon on Pacific coast and rare throughout lakes region and Caribbean coast. Scavenges on beaches, marshes, mudflats, and harbors.

nonbreeding

nonbreeding

first year

first winter

breeding

Ring-billed Gull

immature

California Gull

nonbreeding

breeding

nonbreeding

breeding

Herring Gull

first/second year

nonbreeding

nonbreeding

Lesser Black-backed Gull (Gaviota Sombría)   *Larus fuscus*
21 in (53 cm). Medium-sized gull with white head and relatively **short, slender bill**.
Among *Larus* in Nicaragua, dark gray mantle is the darkest except for larger Kelp Gull.
On mature adult, combination of pale iris, bright yellow legs, and **red spot on lower
mandible of yellow bill** is diagnostic. The mantle color of the various plumages
between second and third year adults is key to distinguishing it from other *Larus*. First
year adult resembles immature but begins to show yellow at the base of lower mandible;
immature has black bill and darker upperparts and wings that slightly contrast with
paler head and underparts. Vagrant from Europe (Sept to April). Most likely to be found
amid flocks of Herring Gulls; only 3 records (Isla Juan Venado WR, Jan 2013; Bluefields
Bay, March 2013; Corn Islands, Dec 2016).

Kelp Gull (Gaviota Cocinera)   *Larus dominicanus*
23.5 in (60 cm). Medium-sized gull with white head and relatively **long, stout bill** that
has a bulge towards tip of lower mandible. Slate-gray mantle is the darkest among *Larus*
in Nicaragua. On mature adult, combination of dull yellow legs and **red spot on lower
mandible of yellow bill** is diagnostic. Note that eye color on adults ranges from dark to
pale. The mantle color of the various plumages between second and third year adults is
key to distinguishing it from other *Larus*; first year adult resembles immature, which has a
black bill and is typically darker overall than immature Lesser Black-backed Gull. Vagrant
from South America; there are multiple records of one individual (Guayabo Wetlands,
2013). Most likely to be found amid flocks of Herring Gulls.

Brown Noddy (Tiñosa Común)   *Anous stolidus*
15.5 in (40 cm). The **silvery white crown** on **dark brown body** could only be confused
with the yet to be recorded Black Noddy (p. 416); note length and thickness of bill.
Immature lacks most of the white coloration on head. **Notched, long, graduated tail**
gives it a distinct profile in flight. Uncommon pelagic migrant on pelagic waters of both
Pacific Ocean (*pileatus* subspecies*)* and Caribbean Sea (*stolidus* subspecies). Rarely seen
on coastlines. Forages on surface of water by swooping and hovering.

immature/first
year

breeding

nonbreeding

**Lesser
Black-backed Gull**

breeding

breeding

immature/first
year

Kelp Gull

second year

Brown Noddy

Sooty Tern (Charrán Sombrio)   *Onychoprion fuscatus*

16 in (41 cm). **Deeply forked, black tail and white outer rectrices** seen in flight distinguish it from all other black-and-white terns. Distinguished from Bridled Tern by white forehead that extends to eye, black crown that is continuous with **black back**; and in flight, note contrast on the underwing between the black primaries and white wing linings. Immature resembles breeding Black Tern but is larger and has a deeply forked tail. Accidental pelagic migrant on both Pacific Ocean and Caribbean Sea; possible at inland aquatic habitats. Typically only seen over pelagic waters, where it picks food from surface of water.

Bridled Tern (Charrán Embridado)   *Onychoprion anaethetus*

15 in (38 cm). **Deeply forked, gray tail and white outer rectrices** seen in flight separate it from all other black-and-white terns. Distinguished from Sooty Tern by white forehead that transitions into a thin superciliary and often extends beyond the eye; **pale nuchal collar** that separates **dark gray back** and black crown; and, in flight, only a slight contrast on the underwing between the black-tipped primaries and the white underwing. Immature and first year adults are similar to other immature tern species but distinguished by tail structure and color. There is a breeding colony on the Farallones Islets (Gulf of Fonseca), from May to Oct; otherwise it is an uncommon pelagic migrant over both Pacific Ocean and Caribbean Sea. Although usually seen over pelagic waters picking food from surface, it is more likely to be seen from the coastline than Sooty Tern.

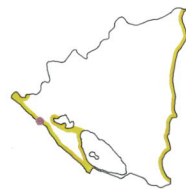

Least Tern (Charrán Menudo)   *Sternula antillarum*

9 in (23 cm). Smallest of the terns. In breeding plumage, combination of **black-tipped, yellow bill, black crown and nape**, and **white forehead** is diagnostic. In nonbreeding and immature plumages, small size should distinguish it from all similar birds except for Black Tern. Distinguished from Black Tern by yellow legs, head pattern, and lighter upperparts and wings. Black outermost primaries on pale wings are helpful to see in flight. Locally abundant at Salinas Grandes salt ponds, where there is a breeding population (June to Aug); uncommon as passage migrant (Aug to Sept and March to May) on Pacific coast and western shoreline of Lake Managua and Lake Nicaragua; rare passage migrant on Caribbean coast. Prefers coastal waters and calm waters of gulfs, estuaries, salt ponds, and lake shores, where it is seen bouncing around in flight before plunge-diving for fish. Calls with a combination of high-pitched squeals and raspy shrieks.

Black Tern (Fumarel Negro)   *Chlidonias niger*

10 in (25 cm). In breeding plumage, has **black head and body**; no other gull or tern— in any plumage—has both the black head and body. In nonbreeding plumage, black only shows on midcrown and on black ear patch. In flight (all plumages), note **gray underwings** and short, notched tail; in flight (nonbreeding plumage), shows a **dark bar extending from the front base of the wings onto the breast**. Common winter resident (July–May) in pelagic waters of Pacific Ocean and on Pacific coast, lakes regions, and San Juan River; first year birds persist in large flocks year-round at Salinas Grandes salt ponds and Guayabo Wetlands. Rare passage migrant (Aug to early Nov and late April to May) on pelagic waters of Caribbean Sea and on Caribbean coast. Feeding flocks of 15 to 300 typically occur over calm pelagic waters, salt ponds, and marshes but also can be seen resting on sand or mudflats. Gives a sibilant, high-pitched *tee* call and screams a harsh *KEAH!*

immature

adult

Sooty Tern

adult

immature

breeding

Bridled Tern

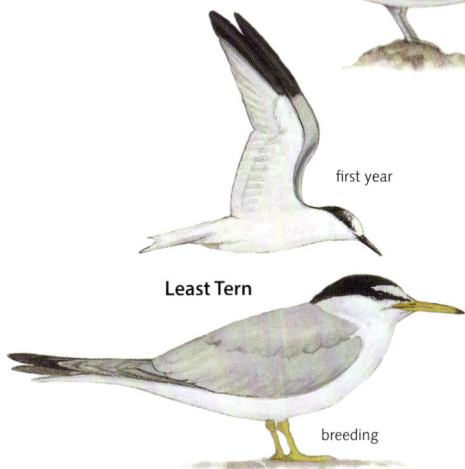

first year

Least Tern

breeding

nonbreeding

Black Tern

breeding

Roseate Tern (Charrán Rosado)   *Sterna dougallii*

15.5 in (39 cm). The palest of the *Sterna* terns. In breeding plumage, **black bill with a red base** is diagnostic. In nonbreeding plumage, black only shows on midcrown. In flight, note deeply forked **white tail**. Accidental pelagic migrant on pelagic water of Caribbean Sea and on Caribbean coast; only 1 record (Corn Islands, 1955). Calls with a harsh and raspy *kri-áh!*

Common Tern (Charrán Común)   *Sterna hirundo*

14.5 in (37 cm). Among the 4 *Sterna* terns, Common Tern is the only one whose **tail does not extend past the wingtips** (see when perched); compared to similar terns, the short legs are only longer than the very short legs of Artic Tern. In breeding plumage, bill is red with a black tip. In nonbreeding plumage, black crown is reduced to midcrown and **dark bar is seen on shoulder** of folded wing. On perched and flying birds, note **extensive black on outer primaries**; in flight, also note black outer rectrices on deeply forked white tail. Winter resident (Nov to May), but sizable numbers remain year-round. Common on Pacific coast and lakes region, uncommon on pelagic waters of Pacific Ocean, and rare on Caribbean coast. Actively hovers and swoops over bodies of water before plunge-diving for small fish. Calls include a raspy *keat keat keat…*, and a descending, scratchy *kee-at*.

Arctic Tern (Charrán Ártico)   *Sterna paradisaea*

15.5 in (39 cm). When perched, distinguished from other *Sterna* terns by **very short legs**, round head, and short neck. In breeding plumage, bill is entirely red. In nonbreeding plumage, black crown is reduced to midcrown and dark bar is not evident on shoulder of folded wing. Outer primaries are only black-tipped, so folded wings are a uniform gray; black outer rectrices on deeply forked and long, white tail are seen in flight. Very rare passage migrant on coastal and pelagic waters of Pacific (Aug to Nov and April to May). Gives harsh, grating *kri-yáh* grunts, intermixed with high-pitch chipping notes.

Forster's Tern (Charrán Cejinegro)   *Sterna forsteri*

14.5 in (37 cm). With longer legs, Forster's stands taller than Common Tern; it is also paler overall. In breeding plumage, bill is orange with a black tip. In nonbreeding plumage, black crown is reduced to a mask. In flight, **note white outer rectrices** on deeply forked tail, and **white rump that contrasts with gray tail and back**. Rare winter resident (Nov to April) on Pacific coast and lakes region; locally uncommon at Guayabo Wetlands, where small numbers stay year-round. Accidental on Caribbean coast. Found at a variety of coastal and other aquatic habitats. Uses a variety of typical tern foraging techniques—hawking, surface-seizing, and plunge-diving. Repeats a fast, throaty *ark-ark-ark…*, and a single, descending *aaarrk*.

Large-billed Tern (Charrán Picudo)   *Phaetusa simplex*

15 in (38 cm). **Very large yellow bill** contrasts with black crown. In flight, **striking upperwing pattern** is created by black primaries, white secondaries, and gray mantle. Vagrant from South America, with 3 independent records (Guayabo Wetlands, Aug 2013; Guatuzos WR, Aug 2013; San Juan del Norte, Dec 2012).

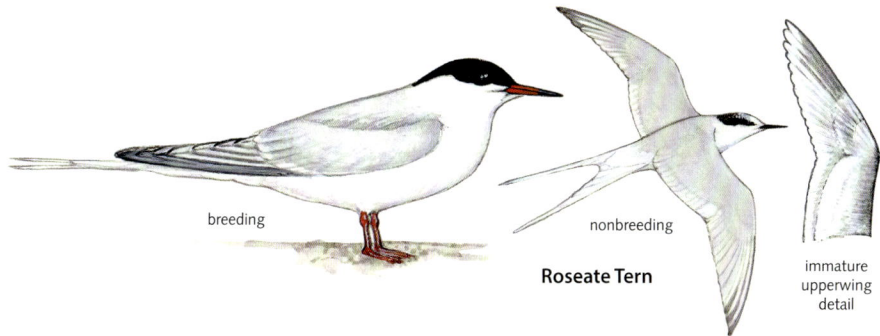

breeding

nonbreeding

Roseate Tern

immature
upperwing
detail

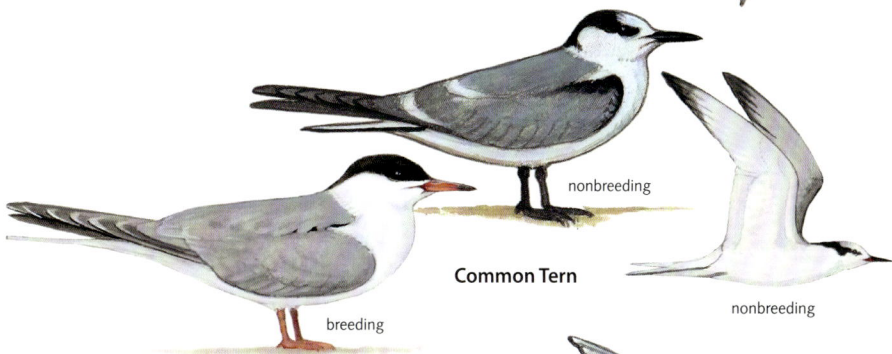

nonbreeding

Common Tern

breeding

nonbreeding

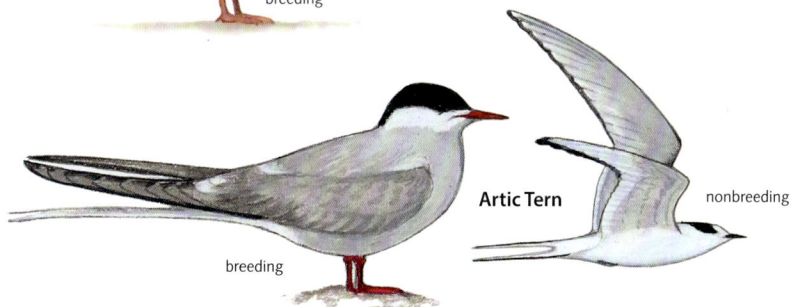

Artic Tern

nonbreeding

breeding

nonbreeding

nonbreeding

Large-billed Tern

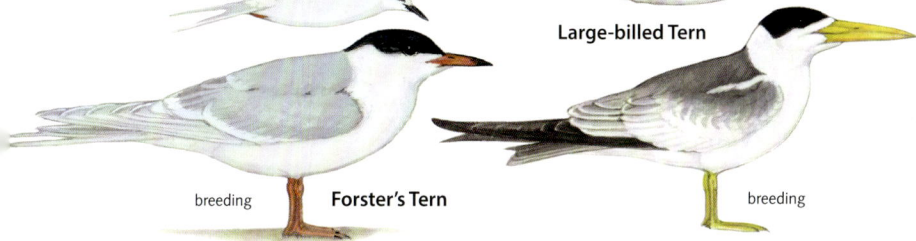

breeding

Forster's Tern

breeding

Gull-billed Tern (Charrán Piquinegro) — *Gelochelidon nilotica*

14 in (36 cm). **Thick, short bill** and **long legs** give it a gull-like appearance. In breeding plumage, note black bill and black crown and nape. In nonbreeding and immature birds, bill is thicker than on other similar black-and-white terns; also note whiter upperparts and **black legs**. Short, forked tail is seen in flight. Uncommon winter resident on Pacific coast and lakes region; noticeable numbers stay year-round; rare on Caribbean coast. Forages on surface of water or in aerial fashion over estuaries, mudflats, salt ponds, and freshwater marshes. Makes a variety of nasal calls, including a hollow *oawh!* and a strained *eh-eh-eh*.

Sandwich Tern (Pagaza Puntiamarilla) — *Thalasseus sandvicensis*

15 in (38 cm). **Black bill with a yellow tip** is unique among terns. Bill is long and slender. Of the terns that have a black bill, the Sandwich Tern is the only one that has a black crest. In nonbreeding plumage, black crown is reduced to midcrown. In flight, white underwings are black tipped on the primaries and short tail is deeply forked. Uncommon winter resident (Sept to early May) in pelagic waters of both coasts, along coastlines, and in lakes region; individuals and small groups persist year-round. Forages along the coast, in a variety of aquatic habitats; also forages on the shore of bodies of fresh water. Often plunge-dives for food. Gives a throaty *kreáh!*

Caspian Tern (Pagaza Piquirroja) — *Hydroprogne caspia*

21 in (53 cm). If Royal Tern or other similar terns are nearby, note **bulky body** and **very heavy red bill** on Caspian. If close, also note dark tip on bill. Immature and nonbreeding Caspian Terns are the only terns with a black crown and a brightly colored bill that lack a white forehead. In flight, extensive black primaries on underwings contrast with white wings; short tail is only slightly forked. Winter resident (Nov to May), but individuals persist year-round. Uncommon on Pacific coast and lakes region, and rare on San Juan River headwaters and Caribbean coast. Forages at estuaries, mudflats, and freshwater shorelines and marshes; often vigorously plunge-dives for food. Gives shrill, slightly hoarse whistles.

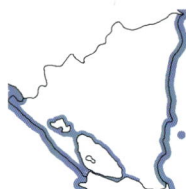

Royal Tern (Pagaza Real) — *Thalasseus maximus*

20 in (51 cm). Backwards **protruding crest** creates a flat-headed appearance. On stout bill, note **slightly upturned lower mandible**; color varies from yellow to orange; bill is thinner than that of Caspian Tern, and shorter than that of Elegant Tern. In nonbreeding plumage, black crown is reduced to midcrown. In flight, white underwings are black-tipped on the primaries and tail is forked. Abundant winter resident (Sept to mid-March) on both coasts, lakes region, San Juan River headwaters, and Corn Islands; individuals and groups persist year-round. Forages at estuaries, salt ponds, mudflats, shorelines, and marshes, where it plunge-dives for food; often seen resting in large groups on sandy beaches and mudflats. Gives a throaty, crackling *kreáh!*

Elegant Tern (Pagaza Elegante) — *Thalasseus elegans*

17 in (43 cm). **Long, shaggy crest** often extends down the nape. **Bill thin and slightly decurved**; varies from yellow to orange; it is longer than that of Royal Tern bill. In nonbreeding plumage, black crown is reduced to midcrown. In flight, white underwings are black tipped on the primaries and tail is forked. Uncommon winter resident (Sept to April) on Pacific coast and in lakes region; individuals and groups persist year-round. Accidental on Carribean coast. Forages at estuaries, salt ponds, mudflats, shorelines, and marshes, where it plunge-dives for food; often mixes with groups of Royal and Sandwich Terns. Gives a rough, rattling *kree-yáh*. **NT**

nonbreeding

Gull-billed Tern

breeding

nonbreeding

Sandwich Tern

breeding

breeding

Caspian Tern

nonbreeding

nonbreeding

immature

Royal Tern

breeding

Elegant Tern

nonbreeding

breeding

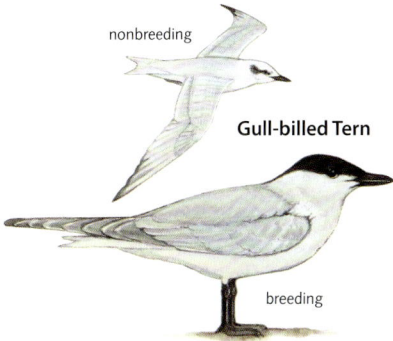

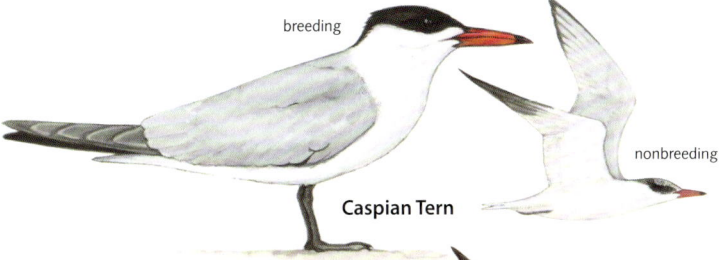

A mixed group of Royal and smaller Elegant terns on a beach.

Black Skimmer (Picotijera Americano)   *Rynchops niger*
18 in (46 cm). **Bill is long and laterally compressed** and **lower mandible extends well beyond the length of the upper mandible**. Uncommon winter resident (mid-Aug to mid-March) on Pacific coast, lakes region, and San Juan River headwaters; locally abundant at Salinas Grandes salt ponds, Isla Juan Venado WR, and Guayabo Wetlands, where groups stay year-round. Rare on Caribbean coast. Gracefully flies low over calm waters, with lower mandible dropped into water feeling for fish; rests during the day on mudflats and sandbars.

SHEARWATERS and PETRELS   Procellariidae

Found on marine waters throughout the world, members of this family are generally pelagic and have very large distributions, except during breeding months, when they nest on oceanic islands. They feed on fish, krill, and squid. Sightings generally occur from vessels on offshore waters, where fleeting views of birds in flight are the norm and identification can be quite difficult. It is important to note overall plumage pattern, shape and size of body, and relative wing and tail proportions. Flight pattern is also a distinguishing quality. Undoubtedly, more species will be confirmed in Nicaraguan waters as a result of future efforts to document birdlife on the seldom visited pelagic waters.

Black-capped Petrel (Petrel Diablotín)   *Pterodroma hasitata*
16 in (41 cm); WS 38.5 in (98 cm). Striking black-and-white pattern on upperparts (**white rump and nape** contrast with **dark cap**) is diagnostic among pelagics. In worn plumage, nape sometimes looks dusky. A pelagic vagrant on Caribbean Sea. **EN**

Parkinson's Petrel (Pardela de Parkinson)   *Procellaria parkinsoni*
17.5 in (45 cm); WS 46 in (116 cm). **Stout, pale bill (with a black tip)** contrasts with **all-dark plumage**. Has a longer wingspan and a shorter tail than does the similar, dark morph Wedge-tailed Shearwater. An uncommon pelagic migrant on Pacific Ocean pelagic waters; accidental on coastal waters of both coasts. Sometimes associated with mixed-species flocks or pods of dolphins and small whales. Scavenges for scraps. **VU**

Wedge-tailed Shearwater (Pardela Colicuña)   *Ardenna pacifica*
18 in (46 cm); WS 40 in (101 cm). Dark, pale, and intermediate morphs are possible; all have a **slender, gray bill with a black tip** and a **long, pointed tail** (when not fanned out). A smaller wingspan helps to separate the dark morph from the similar Parkinson's Petrel and the pale morph from the similar Pink-footed Shearwater (p. 140). A common Pacific Ocean pelagic migrant (common Sept to May; rare June to Aug; can be seen year-round); accidental on coastal waters. Feeds by contact-dipping in pelagic waters. Sometimes occurs in flocks of up to 2,500 individuals, but can also be seen in small numbers on coastal waters.

Black Skimmer

Black-capped Petrel

Parkinson's Petrel

pale morph

dark morph

Wedge-tailed Shearwater

Pink-footed Shearwater (Pardela Patirrosada)   *Ardenna creatopus*
18.5 in (47 cm); WS 45 in (114 cm). Combination of mainly white underparts and **pale, black-tipped bill** distinguishes it from other Pacific shearwaters. **Pink feet** are diagnostic but are often not visible. Common Pacific Ocean pelagic passage migrant (mid-March to June and Sept to Dec), but occurs year-round. Can be seen from shore but generally forages in pelagic waters, where it plunge-dives for prey. Sometimes in very large flocks of up to 1,200; generally associated with mixed-species flocks and pods of dolphins and small whales. **VU**

Galapagos Shearwater (Pardela de Galápagos)   *Puffinus subalaris*
12 in (30 cm); WS 25.5 in (65 cm). Sharp contrast between **black upperparts and white underparts** distinguishes it from other similar shearwaters. It is further distinguished by **small size**. Pelagic migrant on Pacific Ocean; common on pelagic waters and rare on coastal waters.

Black-vented Shearwater (Pardela Culinegra)   *Puffinus opisthomelas*
14.5 in (36 cm); WS 31.5 in (80 cm). **Small size** distinguishes it from other Pacific shearwaters with similar plumage; border between dark upperparts and white underparts is poorly defined. **Bill is slender and gray**. Note dark undertail coverts. Very rare pelagic migrant on Pacific Ocean (Oct to March); accidental on coastal waters. Flies low to the water with quick, choppy wingbeats.

Pink-footed
Shearwater

Galapagos
Shearwater

Black-vented
Shearwater

Covering the oceans worldwide, the members of this family are the smallest of the pelagic birds. Their two habits of hovering low over the water and pattering their feet on the water's surface while feeding (mainly zooplankton), make it easy to identify them to the family level, but identifying a given species is a challenge. Species in Nicaragua can be organized into pale-rumped and dark-rumped groups, which is a helpful starting point. It is important to observe tail pattern and, among the pale-rumped species, the characteristics of the rump patch. All storm-petrels in Nicaragua have a pale band on the upperwings, though this band changes subtly from species to species.

Leach's Storm-Petrel (Paíño Añapero)   *Oceanodroma leucorhoa*
8 in (20 cm); WS 18 in (45 cm). Both white- and dark-rumped morphs have a **forked tail, long, pointed wings**, and a pale wing band that reaches leading edge of wing. **White rump patch** on the white-rumped morph (*leucorhoa* subspecies) is narrower than that on the Wedge-rumped Storm-Petrel, and is often divided by a dusky stripe. Besides the size difference, more extensive upperwing bar, shorter legs, shallower fork in the tail, and thinner bill help distinguish the dark-rumped morph (*chapmani* subspecies) from the Black Storm-Petrel. A year-round Pacific Ocean pelagic migrant (uncommon Aug to Sept and April to May; rare at other times of the year). Accidental on coastal waters. May also occur on the Caribbean Sea. Usually found over pelagic waters, in small numbers; flies erratically, with deep wingbeats.

Wedge-rumped Storm-Petrel (Paíño Danzarín)   *Oceanodroma tethys*
7 in (17 cm); WS 14.5 in (37 cm). Noticeably smaller than white-rumped morph of Leach's Storm-Petrel, though it is difficult to note size difference if they are not seen together; **fork in tail is shallower** than on Leach's and **white rump patch is broader** (extending onto lower flanks). Pale upperwing band does not reach leading edge of wing. Common pelagic migrant on Pacific Ocean (late April to Nov). In search of prey, practices surface-dipping and surface-seizing in pelagic waters; rarely seen from shore.

Black Storm-Petrel (Paíño Negro)   *Oceanodroma melania*
9 in (23 cm); WS 20 in (51 cm). **Largest storm-petrel**; further distinguished from dark-rumped morph of Leach's Storm-Petrel by **deeply forked tail** and relatively **rounded head**. **Long legs** often dangle in flight. Pale upperwing band is usually less extensive than on Leach's. A pelagic migrant on Pacific Ocean that can be seen year-round. Patterns of abundance are not well understood. Forages mainly over coastal and shelf waters, with characteristic slow, deep wingbeats that are almost ternlike.

Least Storm-Petrel (Paíño Menudo)   *Oceanodroma microsoma*
6 in (14 cm); WS 13.5 in (34 cm). **Smallest storm-petrel**. The **wedged-tail** looks short; the pale upperwing bar is usually diffuse and does not reach the leading edge of the wing. A year-round Pacific Ocean pelagic migrant (uncommon July to Jan and March to May; rare at other times of the year). Prefers shelf waters, where it forages with a batlike, weak, fluttering flight; rarely seen from shore.

dark-rumped morph
chapmani ssp.

white-rumped morph
leucorhoa ssp.

**Leach's
Storm-Petrel**

**Wedge-rumped
Storm-Petrel**

**Black
Storm-Petrel**

**Least
Storm-Petrel**

SUNBITTERN Eurypygidae

The Sunbittern is the sole member of its family. It ranges from Guatemala to northern South America. Although it resembles a heron in body shape and in behavior, its closest relative is the Kagu of New Caledonia, the sole member of the only other family in the Order Eurypgiformes. Sunbitterns are generally cryptic. They are strictly associated with watercourses and aquatic habitats adjacent to dense humid forests, where they feed on a variety of invertebrates and small vertebrates.

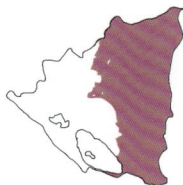

Sunbittern (Ave Sol) *Eurypyga helias*
19 in (48 cm). Combination of large, horizontal body, long neck, and **black head** with **white superciliary** and **malar stripe** is distinctive. Also note orange bill and legs. Rare to locally uncommon in Caribbean (in Saslaya NP); to 3,900 ft (1,200 m). Found in humid lowland forest along rivers and streams (frequently standing on boulders), and within flooded forests, where it forages on the ground in a stalk-and-strike manner; sometimes forages on river sandbars. Although generally cryptic in coloration and behavior, the flashy sunburst markings on the primaries are visible in flight or display, a behavior most likely intended to alarm and confuse potential predators. Song is a mellow and monotone, low-pitched *wiiiiiup* whistle (1 second in length). Calls with a short, pleasant trill.

STORKS Ciconiidae

These large wading birds primarily occur in the Old World. There are only three New World species, all found in the Neotropics. Storks have a long neck, long legs, and a stout, long bill. On broad wings and a short tail, they often soar for long periods of time, a behavior not shared by herons and egrets. Both Nicaraguan species are mostly white, with a bare head and neck. Conspicuous in every way, they fly between shallow aquatic habitats with slow wingbeats and with the neck and legs fully extended, in order to find fish, amphibians, reptiles, and insects. Lacking the ability to vocalize, they communicate by clapping their mandibles.

Jabiru (Jabirú) *Jabiru mycteria*
53 in (135 cm); WS 90 in (229 cm). Huge; this is the largest bird in Nicaragua. Note black, featherless head, black neck with a bright **red collar** at the base (visible only at close range), and all-white body; **massive bill is slightly upturned**. Immature is mottled with brown and gray, and readily identified by large size. In flight, all-white underwings and undertail distinguish it from smaller Wood Stork. Rare throughout most of Pacific lowlands and Playitas-Moyuá-Tecomapa lagoons, though it is somewhat reliably found in Apacunca GRR; also rare in Caribbean lowlands. Frequents a variety of aquatic habitats, including freshwater marshes, grasslands, flooded fields, lake and river shorelines, and swamps within lowland pine savanna. Uses its large bill to catch fish, mollusks, amphibians, snakes, and birds. Undertakes seasonal movements depending on water levels, seeking shallow waters where prey concentrate. Found alone or in pairs, and sometimes with Wood Storks.

Wood Stork (Cigüeña Americana) *Mycteria americana*
38.5 in (98 cm); WS 61 in (155 cm). **Bare, steel-gray head and neck** and **heavy, decurved bill**. In flight, **black flight feathers** separate it from larger Jabiru and smaller White Ibis (p. 164). Locally common to abundant in Pacific lowlands, including the lakes region and San Juan River headwaters. Uncommon at Lake Apanás; can be seen flying between aquatic habitats on western slopes of Sierra Isabelia and Sierra Dariense in the Northern Highlands; to 4,300 ft (1,300 m). Generally uncommon throughout Caribbean lowlands. Found alone or in flocks, with large numbers gathering at seasonally dry wetlands, where fish, crustaceans, amphibians, and other prey items concentrate.

An adult Jabiru at its nest.

Sunbittern

Jabiru

adult

adult

immature

Wood Stork

TROPICBIRDS   Phaethontidae

These aerial seabirds range throughout tropical regions of the world. The 3 species in this family have ternlike bodies, black masks, and long, streaming central tail rectrices. They nest on small islands and wander great distances when not breeding.

Red-billed Tropicbird (Rabijunco Piquirrojo)   *Phaethon aethereus*
20 in (51 cm), not including long central rectrices; WS 44 in (112 cm). **Long central rectrices** and heavy **red bill** make the adult unmistakable. Immature lacks long tail rectrices but can be distinguished from other white, ternlike birds by the **bold black barring** on upperparts and **black-tipped rectrices** on a tapered tail. Uncommon pelagic migrant on pelagic waters of the Pacific Ocean and very rare on Pacific coastal waters; occasionally seen on Farallones Islets in the Gulf of Fonseca. Status and abundance is unknown in Caribbean Sea, but most likely an uncommon pelagic migrant. Generally travels at sea alone or in pairs. Hunts for fish and squid by plunge-diving into the water from considerable heights.

BOOBIES   Sulidae

These masterful plunge-divers occur in tropical marine waters around the world, where they feed on fish and squid. They have a streamlined body shape, with a long pointed bill, narrow and slightly angled wings, and a long graduated tail. Females are larger than males; immatures often do not acquire adult plumage until three to five years of age. Identification is often a challenge, but note the color of bill and feet and the size of the body. Nicaragua hosts very few breeding colonies; most individuals that occur in the country have likely traveled thousands of kilometers from colonial nests on small distant islands.

Masked Booby (Piquero Blanco)   *Sula dactylatra*
32 in (81 cm). Combination of **large, white body** and **black mask, flight feathers, and tail** distinguishes it from the smaller boobies. **Greenish-yellow bill** distinguishes it from almost identical Nazca Booby. Note clean, broad, well-defined white nuchal collar on most immatures. A pelagic migrant; common over Pacific Ocean and rare over Caribbean Sea. Accidental on coastal waters of both coasts. Attempts shallow vertical plunges to catch fish (often flying fish). Sometimes associates with shearwater flocks.

Nazca Booby (Piquero Piquinaranja)   *Sula granti*
34.5 in (88 cm). Combination of **large, white body** and **black mask, flight feathers, and tail** distinguishes it from the smaller boobies. **Reddish-orange bill** distinguishes it from almost identical Masked Booby. Most immatures lack white nuchal collar; if present, it is smudged, narrow, and poorly defined. Uncommon pelagic migrant on pelagic waters of Pacific Ocean. Accidental on coastal waters of Pacific. Catches fish by plunge diving up to 3 or 4 meters into the water.

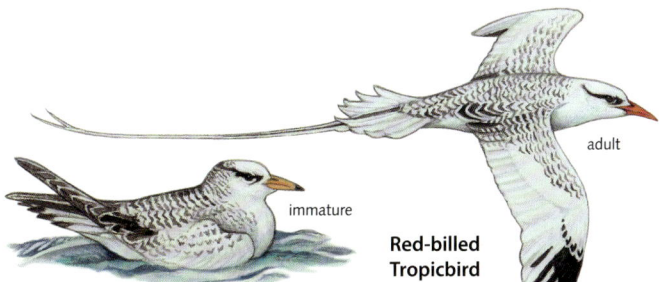

immature

adult

Red-billed Tropicbird

adult

immature

Masked Booby

adult

immature

Nazca Booby

Blue-footed Booby (Piquero Patiazul)   *Sula nebouxii*
32 in (81 cm). **Blue feet** and **streaked head** are unique. On immature, feet are duller blue; also note brown head and throat. Combination of foot color, dark bill, and lack of nuchal collar help to distinguish it from other similar boobies. A very rare pelagic migrant that occurs over coastal and pelagic waters of the Pacific Ocean; note, however, that there is a breeding colony on the Farallones Islets in the Gulf of Fonseca. Often forages for fish in warmer, clearer inshore waters, where it can be seen from mainland.

Brown Booby (Piquero Pardo)   *Sula leucogaster*
30 in (76 cm). This is the most commonly seen booby in Nicaragua. All adults have brown upperparts. On most adults, note brown on head and lower breast that contrasts with white underparts and underwing linings; on adult male of *brewsteri* subspecies, note pale color of head. Immature is entirely brown; contrast in hue between breast and belly helps distinguish it from both immature and Pacific dark-morph Red-footed Booby. Abundant on Gulf of Fonseca and pelagic waters of Pacific Ocean (*brewsteri* subspecies), but uncommon to coastal waters and Caribbean Sea. Fishes by angled plunge-dives from low heights.

Red-footed Booby (Piquero Patirrojo)   *Sula sula*
28 in (71 cm). Four color morphs occur in Nicaragua: Pacific white and dark morphs and Caribbean white and dark morphs. Adults of all 4 have diagnostic **red feet** and **blue bill with pink base**. If red feet are difficult to see in flight, note small body size and angled wings. Also note variation in tail patterns between the color morphs. All-brown immature can be confused with immature Brown Booby, but faint, dark breast band on immature Red-footed helps distinguish them. A pelagic migrant in pelagic waters; uncommon over Pacific Ocean and rare over Caribbean Sea. Potentially accidental to coastal waters. Skims gracefully over water. Often feeds at night.

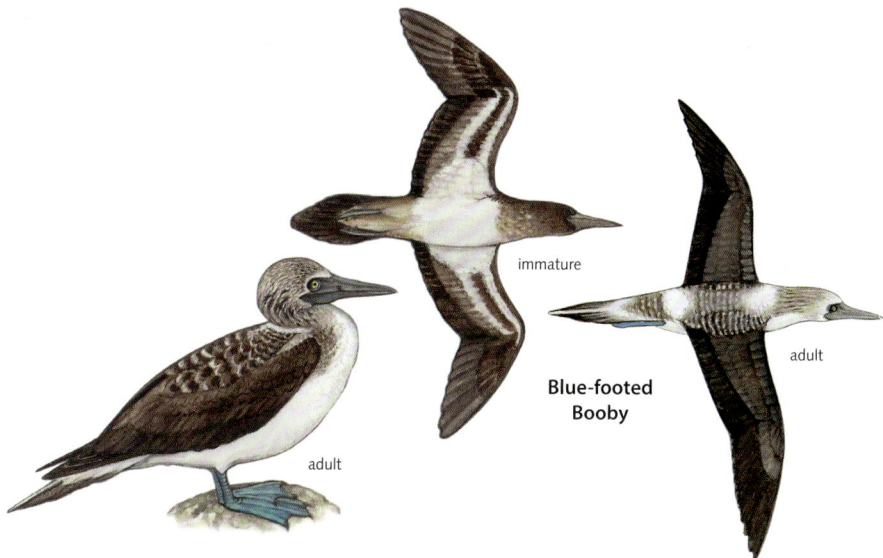

immature

Blue-footed Booby

adult

adult

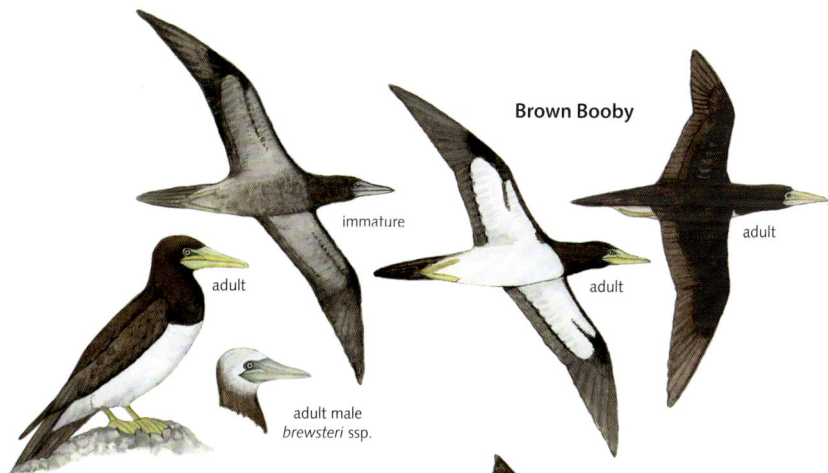

Brown Booby

immature

adult

adult

adult

adult male
brewsteri ssp.

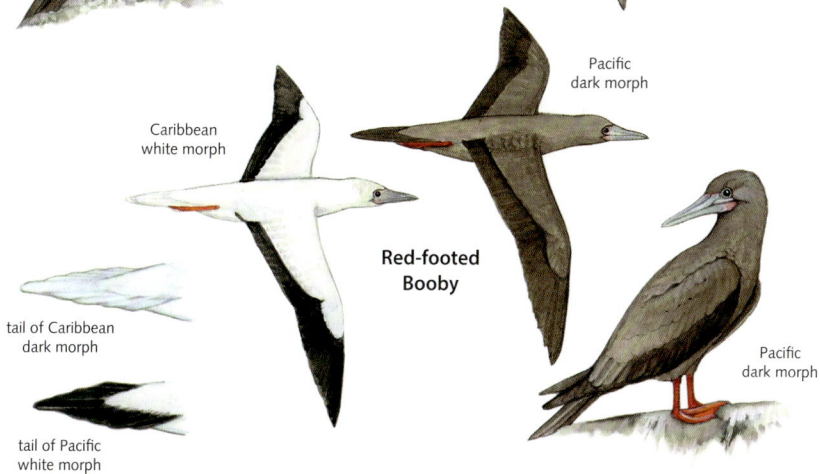

Caribbean
white morph

Pacific
dark morph

Red-footed Booby

tail of Caribbean
dark morph

Pacific
dark morph

tail of Pacific
white morph

CORMORANTS Phalacrocoracidae

These medium to large waterbirds are generally glossy black as adults; in breeding plumage, colorful skin features or ornamental plumage stand out clearly against the glossy black. Although cormorants spend their lives in and around water, their plumage is not entirely waterproof; they are often seen perched in sunny spots with their wings spread wide to dry.

Neotropic Cormorant (Cormorán Neotropical)   *Phalacrocorax brasilianus*
26 in (66 cm). Similar to Anhinga, but **stocky, hooked bill** and mainly **glossy black plumage** distinguish it; immature has dusky underparts and head and whitish throat. Abundant in lowlands, as well as in Playitas-Moyúa-Tecomapa lagoons, Sébaco Valley, and Lake Apanás; uncommon on Corn Islands. Encountered in practically all aquatic habitats, both coastal and inland, where it dives in the water for prey.

DARTERS Anhingidae

This pantropical family is represented by a single species in the Neotropics. When not in the water in search of fish, they often perch on vegetation with wings spread open in order to dry out their feathers.

Anhinga (Aninga)   *Anhinga anhinga*
35 in (89 cm). **Long, thin neck** and **long, sharp bill** distinguish it from the Neotropic Cormorant. The silvery-white **streaked and spotted upperwing coverts** are conspicuous on its black body. Female has buffy head and neck. Abundant throughout lowlands. Prefers freshwater and brackish aquatic habitats (in close proximity to forest). Swims stealthily, often with only its snakelike head above the water's surface. Spears prey with its sharp bill.

adult

adult

immature

Neotropic Cormorant

Anhinga

female

male

FRIGATEBIRDS   Fregatidae

Members of this small pantropical family are often observed in large numbers soaring effortlessly high above coastal waters. Well known for their kleptoparasitic behavior towards gulls, terns, boobies, and other seabirds, they engage in aerial piracy by pestering their targets until the catch is spewed out—and retrieved by the frigatebird. But more commonly, they acquire their own food, by plucking fish, jellyfish, and crustaceans from the sea surface. They nest in impressively large colonies on the rocky outcroppings and mangrove forests of islands, where the male's inflated red gular sac can be seen in display.

Magnificent Frigatebird (Rabihorcado Magno)   *Fregata magnificens*
37 in (95 cm). Has an almost prehistoric appearance: It is **large**, with a **black body, long pointed wings**, and a long, deeply forked tail that is often held closed. Male has brilliant red gular sac; female has large white band across breast; immature exhibits varying amounts of whitish plumage on breast and head. Abundant on Pacific Ocean and coastline and reaches as far inland as Lake Managua, Lake Nicaragua, and the San Juan River headwaters; common on Caribbean Sea and coastline, sometimes occurring inland along large rivers. There is one known breeding colony, on the Fallarones Islets in the Gulf of Fonseca. Soars high over coastal and pelagic waters, but also frequently over lakes and large rivers fairly close to the coast.

PELICANS   Pelecanidae

Pelicans have large bodies and massive bills. Their voluminous, extendable gular sacs aid them in the capture of fish. The 2 species that occur in Nicaragua fly with their heads retracted back toward their body.

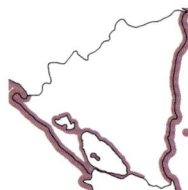

Brown Pelican (Pelícano Pardo)   *Pelecanus occidentalis*
50 in (126 cm). In breeding and nonbreeding plumage, adult has **silvery-gray body**. Immature has brownish upperparts and whitish underparts. Abundant on Pacific coastal waters, with a breeding colony on the Gulf of Fonseca; reaches lakes region and San Juan River headwaters. Common on Caribbean coastal waters and inland, following large rivers; also common on Corn Islands. Performs spectacular plunge-dives into water to catch fish. Can often be seen riding updrafts from waves while maintaining tight, linear flocks.

American White Pelican   *Pelecanus erythrorhynchos*
(Pelícano Blanco Americano)
62 in (157 cm). A large, white pelican; in flight, **black primaries and outer secondaries** are diagnostic. Soaring birds can be confused with Wood Stork (p. 144), but in flight note retracted position of head and neck (Wood Stork flies with neck extended). Uncommon winter resident (late Aug to early June) on Pacific coast and in lowlands; although uncommon, sometimes seen in very large numbers (1,500+). Found on freshwater aquatic habitats and coastal bays and estuaries, where they swim on the water's surface in search of fish.

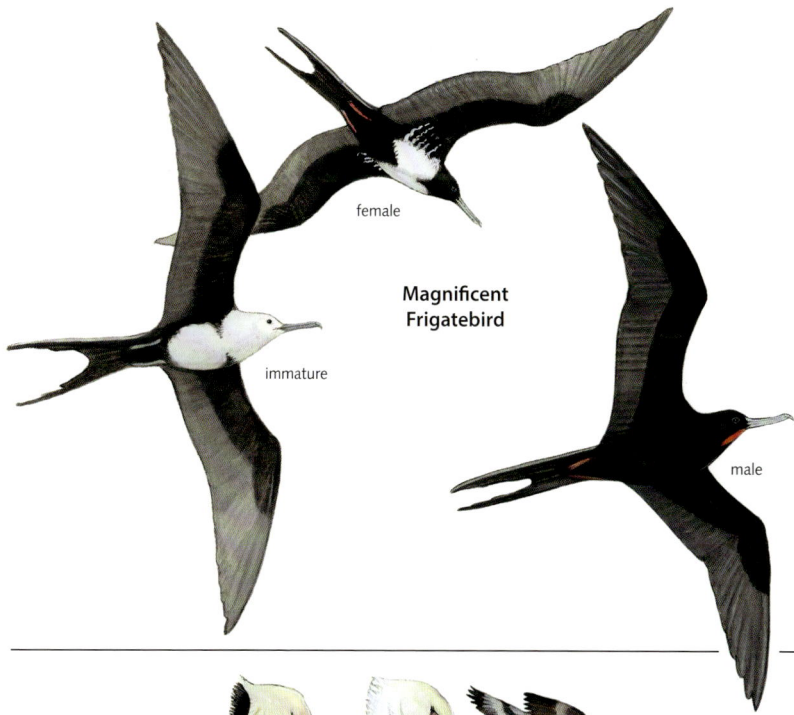

**Magnificent
Frigatebird**

female

immature

male

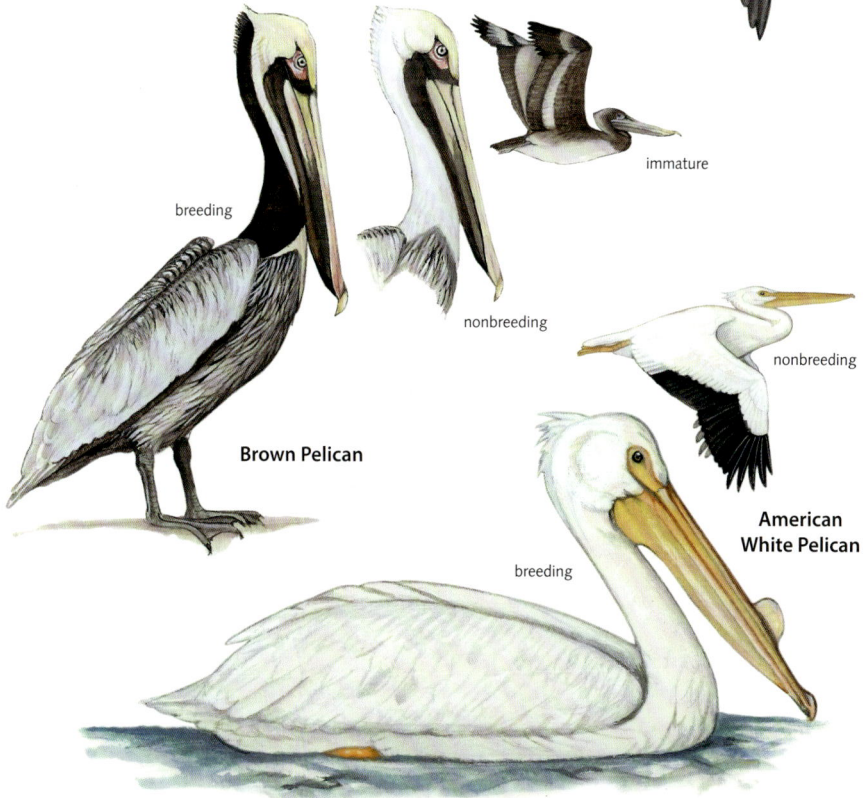

breeding

nonbreeding

immature

Brown Pelican

nonbreeding

**American
White Pelican**

breeding

This is a cosmopolitan family of wading birds that are easily observed in Nicaragua. Their long necks and legs and short tails make them well-suited for life in shallow aquatic habitats. Long, pointed bills equip them to spear fish, crustaceans, and small vertebrates, and pick mollusks and aquatic insects from the water. They fly with necks retracted and legs extended, on broad, rounded wings. Sexes are similar but adults and immatures often have different plumages. Of the species that breed within Nicaragua (many colonially), some have populations that are added to by NA migrants.

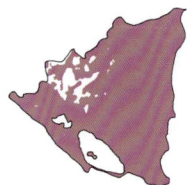

Pinnated Bittern (Avetoro Neotropical)   *Botaurus pinnatus*

24 in (62 cm). **Buffy plumage, thin, white superciliary** (seen at close range), and white throat are distinctive. Combination of **fine, dark brown barring on hindneck** and lack of black malar stripe distinguishes it from American Bittern; streaking on foreneck and breast further distinguishes it from immature tiger-herons. Rare countrywide, but locally common at Playitas-Moyúa-Tecomapa lagoons, Lake Apanás, San Miguelito Wetlands, Guatuzo Plains, and along San Juan River; to 3,300 ft (1,000 m). Found in freshwater marshes dominated by reeds and, often, water hyacinth. Inconspicuous and solitary, the Pinnated Bittern is easily overlooked; when alarmed, it freezes with its neck and head extended upward, allowing for perfect camouflage within reedy vegetation. During breeding season, vocalizes with a strange and hollow *pop pop* that is followed by a low and resonant *boonk*. When alarmed, calls with an agitated *EH! EH! EH!… .*

American Bittern (Avetoro Norteña)   *Botaurus lentiginosus*

28 in (71 cm). **Black malar stripe** (descending the neck) and **unstreaked buffy-brown hindneck** distinguish it from Pinnated Bittern. Note boldly streaked foreneck and breast; further distinguished from immature night-herons by slender body and bill and lack of white spots on wing coverts. Accidental winter resident; there are only 2 records in Nicaragua (Lake Apanás, Dec 2010; Indio Maíz BR, Nov 2016). Expected in freshwater and reedy marshes. Very shy, inconspicuous, and cryptically patterned—and easily overlooked. When alarmed, calls with a gruff *auk!*

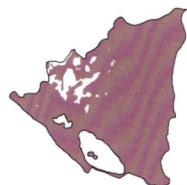

Least Bittern (Avetorillo Pantanero)   *Ixobrychus exilis*

12 in (30 cm). Distinctly smaller than other members of this family. **Buffy overall**, with **white scapular stripes on mantle** and **buffy patch on wing coverts** (very noticeable in flight). On male, note bold blue-black crown and back (duller on female and immature). Uncommon breeding resident countrywide (known at Tisma NR, Guayabo Wetlands, Lake Apanás, Guatuzo Plains, and San Juan River, but also look for it at Apacunca GRR); population increases with arrival of NA migrants (Oct to March); to 3,300 ft (1,000 m). Solitary and secretive, the Least Bittern is easily overlooked. Prefers freshwater marshes with tall emergent vegetation, where it skillfully clings to reeds and perches above water to survey for prey. When alarmed, flies short distances, with dangling legs, before landing and posing with bill pointing up to resemble surrounding vegetation. When breeding, makes a soft, throaty, somewhat froglike *guah*. Also makes a scratchy *ek ek ek…* (5–7 notes in total).

Pinnated
Bittern

American
Bittern

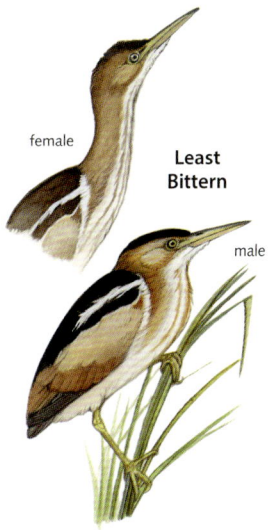

female

Least
Bittern

male

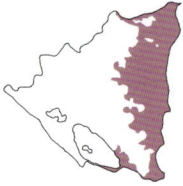

Rufescent Tiger-Heron (Garza Tigre Colorada)   *Tigrisoma lineatum*
27 in (68 cm). On adult, note **rufous head and neck**. On immature, feathered white throat and shorter bill distinguish it from immature Bare-throated Tiger-Heron; immature distinguished from other immature tiger-herons by head and neck that show more rufous coloration (only noticeable with sufficient experience). Beware of confusion with much smaller Green Heron (note differences in size, posture, predation behavior, and habitat). Rare to uncommon in Caribbean lowlands. Wades in calm waters of forested streams, slow-moving rivers, and swamps (Fasciated Tiger-Heron found in fast-moving rivers), before ambushing prey from a motionless, ready-to-strike stance. Solitary and inconspicuous, it is easily overlooked in its low-light habitats; more active at dawn and dusk. Emits a guttural, haunting *huuá huuá huuá*, of varying speed and length.

Bare-throated Tiger-Heron (Garza Tigre Gorgilisa)   *Tigrisoma mexicanum*
31 in (79 cm). **Bare (featherless) yellow throat** in all plumages sets it apart from all other *Tigrisoma* and *Botaurus* herons. Uncommon throughout lowlands and foothills, locally common on Lake Nicaragua shorelines, banks of San Juan River, Isla Juan Venado WR, and Caribbean swamps; generally to 1,500 ft (450 m) but sometimes up to 2,300 ft (700 m). Usually solitary; wades in fresh and brackish waters, preferably bordered by forest (often seen resting on tree branches). Makes a deep, hoarse growl that is repeated at length: *woah woah woah…*. Also barks a quick *ruoaw!*

Fasciated Tiger-Heron (Garza Tigre de Río)   *Tigrisoma fasciatum*
25 in (63 cm). **A fringe of white feathers on center of throat** is diagnostic (on Bare-throated Tiger-Heron, throat is featherless and there is more yellow at base of bill); body is smaller and **bill is blacker, shorter, and thicker** than on other tiger-herons. Immature is almost identical to immature Rufescent Tiger-Heron, but averages less rufous on head and neck (only noticeable with sufficient experience). Very rare in Indio Maíz BR (first country report at Bartola River headwaters, 2012); based on records from Honduras and Costa Rica, potentially also found in Bosawas Biosphere Reserve and Guatuzo Plains. Almost always found on fast-moving, rock-strewn rivers and streams, where it perches on boulders in a motionless stance before striking at prey. Solitary, shy, and skulking—and easily overlooked. Utters a guttural bleat, sometimes repeated, with last bleat louder and drawn out.

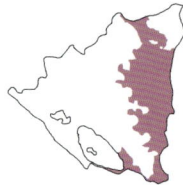

Agami Heron (Garza Pechicastaña)   *Agamia agami*
28.5 in (73 cm). Unmistakable. **Very long and slender bill,** glossy **blue-green upperparts, rich chestnut underparts**, and **silvery feathers on side of neck**. Uncommon and local in Caribbean lowlands and along southern shoreline of Lake Nicaragua (consistent records near Bluefields, Papaturro River in Guatuzos WR, and San Juan River tributaries). Prefers swamps and slow-moving streams bordered by dense vegetation. Solitary and secretive; waits motionless or walks very slowly in search of prey. Frequently overlooked because of behavior and habitat. Produces a deep, gravelly purr; also makes a croaking *erh erh erh…*, often given in flight. **VU**

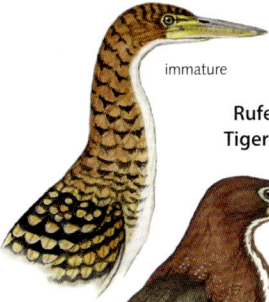

immature

**Rufescent
Tiger-Heron**

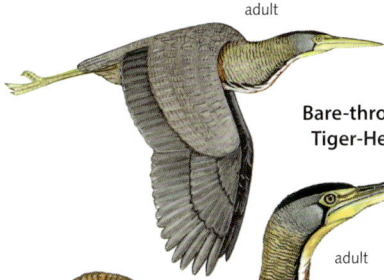

adult

adult

**Bare-throated
Tiger-Heron**

immature

adult

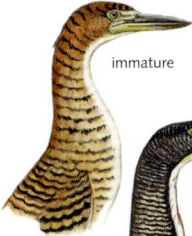

immature

**Fasciated
Tiger-Heron**

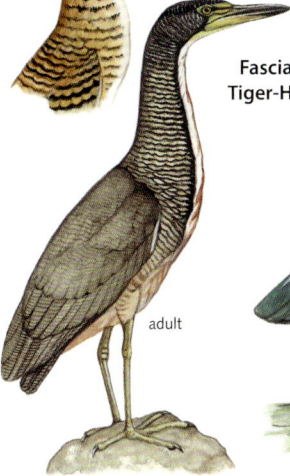

adult

Agami Heron

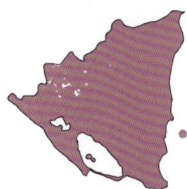

Great Egret (Garzón Grande)   *Ardea alba*

39 in (99 cm). **Largest white heron** in Nicaragua. Very **long neck**, **all-yellow bill**, and **black legs** distinguish it from other white, large-bodied herons and egrets. Common breeding resident countrywide, population increases with the presence of NA migrants (Sept to April); to 4,600 ft (1,400 m). Found in both fresh- and saltwater marshes, where it feeds in shallow waters; feeds alone but nests in colonies. Utters a variety of deep bleats and growls.

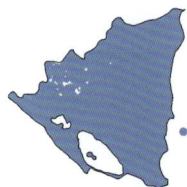

Snowy Egret (Garzeta Patiamarilla)   *Egretta thula*

24 in (61 cm). Elegant and svelte, with **long, slender black bill** and noticeable **yellow lores**. Combination of **black legs** and **bright yellow feet** is diagnostic (back of legs are dull yellow on immature). Common winter resident countrywide (Sept to April), with a small subset of the population remaining throughout boreal summer (most likely breeding); to 4,600 ft (1,400 m). Frequents fresh and brackish waters, especially along lakeshores, river mouths, estuaries, and ponds. Feeds singly or in groups, searching for fish, crustaceans, amphibians, and small reptiles. Utters a variety of strained honks and thin croaks.

Cattle Egret (Garcilla Bueyera)   *Bubulcus ibis*

20 in (51 cm). Smaller and **stockier** than any other white heron and egret. **Short, yellow legs and bill** and short neck further distinguish it from other white egrets. Abundant countrywide; to 4,600 ft (1,400 m). Seen in flocks, roaming grasslands and agricultural fields to catch insects flushed by grazing cattle. Roosts and nests colonially in trees near ponds or lakeshores. Native to Africa; began colonization of the Americas by 1877. Produces an abbreviated and muffled *aht aht aht…*, with occasional squabbles.

Little Blue Heron (Garceta Azul)   *Egretta caerulea*

24 in (61 cm). Combination of **black-tipped bill** and **dull green legs** (in all plumages) is unique among herons and egrets. Immature is further distinguished from white egrets by dusky primary wing tips, seen in flight. Common winter resident countrywide (Sept to April), with many individuals (mostly immature) remaining in country throughout boreal summer; to 3,300 ft (1,000 m), rare to uncommon at higher elevations. Found in fresh- and saltwater habitats with shallow, calm waters. Usually alone, standing quietly as it waits for prey. Gives a raspy and somewhat feisty *raowh!*

Reddish Egret (Garceta Rojiza)   *Egretta rufescens*

30 in (76 cm). On both dark and white morphs, **shaggy, feathered neck**, **white iris**, **black-tipped bill with pink base** (breeding), and **gray legs** distinguish it from other white herons and egrets (nonbreeding and immature bills are entirely or mostly dark). Lack of white belly distinguishes it from smaller Tricolored Heron (p. 160). Winter resident on both coasts (Oct to Feb) with some immatures remaining throughout boreal summer; uncommon on Pacific and rare on Caribbean. Visits shallow saltwater habitats such as beaches and salt ponds. When capturing fish, raises wings as it runs and jumps. Gives a weak, guttural croak or bleat. **NT**

Great Egret

Snowy Egret

Cattle Egrets often forage for food flushed by grazing livestock.

breeding

nonbreeding

nonbreeding

Cattle Egret

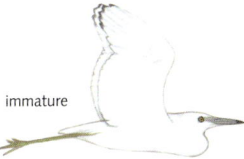

immature

adult

Little Blue Heron

immature molting into adult plumage

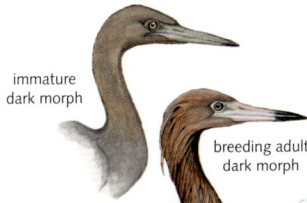

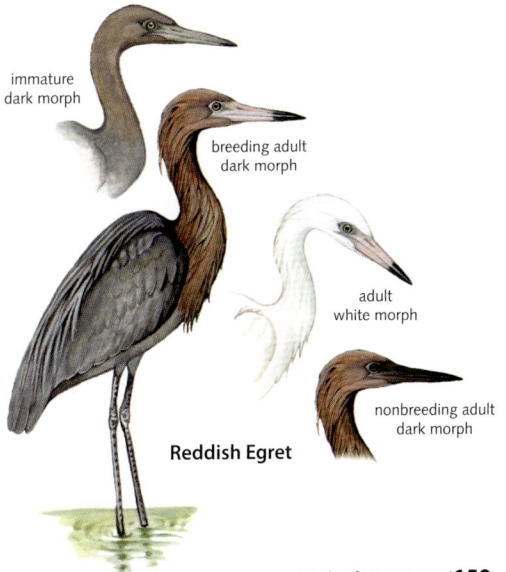

immature dark morph

breeding adult dark morph

adult white morph

nonbreeding adult dark morph

Reddish Egret

Great Blue Heron (Garzón Azul)   *Ardea Herodias*

47 in (120 cm). Largest heron in Nicaragua, with very **long neck and legs**. Has a **steely, blue-gray body**, blackish shoulders, **rufous thighs**, and **black-and-white striped crown**. Common winter resident countrywide (Sept to April), with some individuals remaining throughout boreal summer; to 4,600 ft (1,400 m). Found at a variety of fresh- and saltwater habitats. Waits quietly for prey in shallow waters. Gives throaty croaks and agitated bleats.

Striated Heron (Garcilla Estriada)   *Butorides striata*

16.5 in (42 cm). Could easily be misidentified as a Green Heron (with which it was formerly considered conspecific), but black crown and hindneck contrast with **gray or buffy cheeks and neck**. Vagrant with only 1 country record (San Juan River, March 2009); range may be expanding northward. Utters a strained *cha cha cha*.

Green Heron (Garcilla Capiverde)   *Butorides virescens*

17 in (44 cm). Stocky. Note **rich, chestnut head, neck, and breast** and **greenish-gray upperparts**; in breeding season, yellow bill and legs turn orange. On Striated Heron, note gray or buffy cheeks and neck. Abundant breeding resident, countrywide; population increases with the arrival of NA migrants (Sept to March); to 4,200 ft (1,300 m). Found at a variety of fresh- and saltwater habitats, including marshes, lakeshores, riverbanks, ponds, and mangroves. Powerfully strikes at fish and amphibians from perch at water's edge. Commonly heard vocalizing with a short, choppy, strained *ka ka ka ka*, often given in flight. In breeding season, gives a deep, muffled *koaw*.

Tricolored Heron (Garceta Tricolor)   *Egretta tricolor*

24 in (60 cm). Only heron or egret on which **white belly contrasts with dark upperparts**. Breeding residents are joined by NA migrants (Sept to May). Common throughout Pacific lowlands, Playitas-Moyúa-Tecomapa lagoons, and Sébaco Valley; and uncommon at Lake Apanás and in Caribbean lowlands, but locally common around southeastern shores of Lake Nicaragua and San Juan River headwaters. Accidental in Sierra Isabelia and Sierra Dariense of the Northern Highlands. Found at fresh- and saltwater marshes, ponds, riverbanks and lakeshores, as well as mangroves and estuaries. Usually solitary; with wings raised, captures prey. Barks a throaty scream when alarmed.

adult

immature

Striated
Heron

adult

Great Blue
Heron

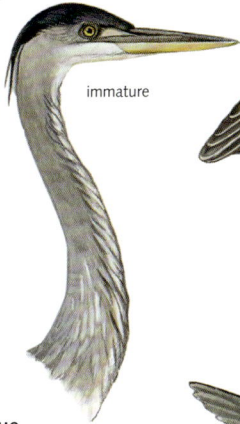

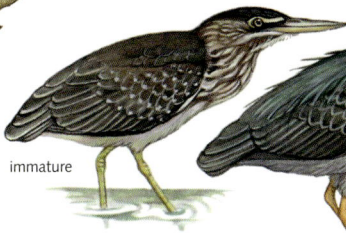

adult

immature

adult

Green
Heron

immature

adult

adult

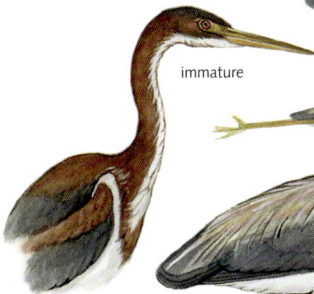

Tricolored
Heron

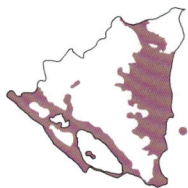

Black-crowned Night-Heron (Martinete Capinegro)   *Nycticorax nycticorax*
25 in (64 cm). **Black crown, nape, and back** lead into **gray wings** and contrast with **white underparts**; plumage pattern similar to that of Boat-billed Heron, but compare bill shape and size. Immature is easily confused with immature Yellow-crowned Night-Heron, but note paler plumage, more yellow on slender bill, shorter legs and neck, and larger white spots on wing coverts. Common and local breeding resident throughout lowlands, Playitas-Moyúa-Tecomapa lagoons, and Sébaco Valley; rare at Lake Apanás; population increases with arrival of NA migrants (Oct to April). Forages at night in aquatic habitats, where it patiently waits for prey; during the day, sometimes perches in a tree at water's edge. When disturbed, gives a raspy *aught*, similar to that of Yellow-crowned Night-Heron but slightly lower-pitched.

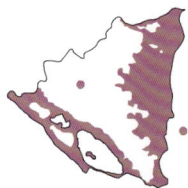

Yellow-crowned Night-Heron (Martinete Cangrejero)   *Nyctanassa violacea*
24 in (61 cm). **Black face** punctuated by **white ear patch** and **pale yellow crown** is unique. Immature is easily confused with immature Black-crowned Night-Heron, but note overall darker coloration, more black on stouter bill, longer legs and neck, and smaller white spots on wing coverts. Uncommon breeding resident in lowlands, and accidental in Northern Highlands (Santa María de Ostuma, Matagalpa Dept., March 1983); population increases with arrival of NA migrants (Oct to April). Waits motionless or walks slowly in pursuit of crabs at beaches, mangroves, ponds, and riverbanks. Less nocturnal than Black-crowned Night-Heron. Gives a deep and curt *augh* when disturbed, similar to that of Black-crowned Night-Heron but slightly higher-pitched.

Boat-billed Heron (Pico Cuchara)   *Cochlearius cochlearius*
19.5 in (50 cm). **Huge bill**, with upper mandible resembling the hull of a boat, makes it unmistakable (although plumage pattern is similar to that of Black-crowned Night-Heron). Uncommon throughout lowlands and foothills and uncommon to locally common at Isla Juan Venado WR, Guatuzo Plains, and along San Juan River and tributaries; to 1,600 ft (500 m). Strictly nocturnal. Forages alone by wading slowly in shallow, muddy water. Roosts by day in trees with very dense foliage, making it easy to overlook; colonial nester. Barks a nasally, scratchy *rah! rah! rah!... .*

immature

**Black-crowned
Night-Heron**

adult

immature

adult

**Yellow-crowned
Night-Heron**

immature

**Boat-billed
Heron**

adult

A riverside rookery of Boat-billed Herons.

These large-bodied wading birds are found nearly worldwide. Typically they have long legs and bills, which are de-curved on ibises and spatulate on spoonbills. The attention-grabbing bills are used to probe in mud for crustaceans, fish, amphibians, and small invertebrates. Sexes are similar, but in some species nonbreeding and immature birds show different plumages. All are strong flyers showcasing long, broad wings; the necks are extended in flight. Members of the family are quite conspicuous given their daytime activity, group foraging, and colonial roosting and nesting in trees near water.

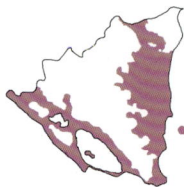

White Ibis (Ibis Blanco)   *Eudocimus albus*
25 in (64 cm). **Red face, bill, and legs** and **white plumage** make adult unmistakable; all other Nicaraguan ibises have dark plumage. In flight, black primary wingtips set it apart from white storks, herons, and egrets. On immature, combination of white underparts and brown upperparts distinguishes it from the other ibises. Locally common in Pacific lowlands and foothills; to 1,500 ft (450 m). Rare to uncommon in Caribbean lowlands. Found at fresh- and saltwater habitats; prefers estuaries, mangroves, salt ponds, swamps, and lake and river shorelines.

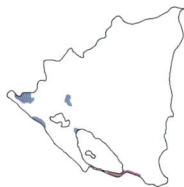

Glossy Ibis (Ibis Brillante)   *Plegadis falcinellus*
23 in (58 cm). **Brown iris** and **gray facial skin with thin pale bluish borders** separate it from almost identical White-faced Ibis; immature Glossy Ibis is indistinguishable from immature White-faced Ibis (p. 418). From a distance, can also be confused with Green Ibis, but it lacks the stocky silhouette. Uncommon breeding resident in Guatuzo Plains and San Juan River headwaters; uncommon and local winter resident (mid-Aug to March) in the Pacific; to 1,500 ft (450 m). Prefers freshwater marshes, but also found in mangroves and coastal lagoons. Associates loosely with White Ibis and herons while feeding in shallow waters.

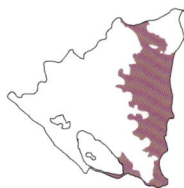

Green Ibis (Ibis Verde)   *Mesembrinibis cayennensis*
31 in (80 cm). **Dark, glossy green plumage** appears black in poor light. Stockier than all other ibises. Uncommon in Caribbean lowlands. Found in swamps and gallery forests; feeds at water's edge by probing the mud with its bill for worms and insects. When perched, gives a low-pitched, soft, guttural *kuuáh*; also makes a loud, resonant *kluh-kluh-kluh*…, repeated at length and often when in flight.

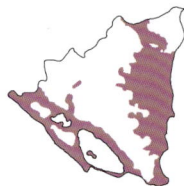

Roseate Spoonbill (Espátula Rosada)   *Platalea ajaja*
31 in (80 cm). **Only pink bird** in Nicaragua; **spatulate bill** is unique. Locally common in Pacific lowlands; uncommon throughout Caribbean lowlands; locally uncommon in Playitas-Moyúa-Tecomapa lagoons, Sébaco Valley, and Lake Apanás. Possible to see as it moves between wetlands, in search of muddy areas at marshes. Probes in mud for crustaceans, aquatic insects, amphibians, and small fish.

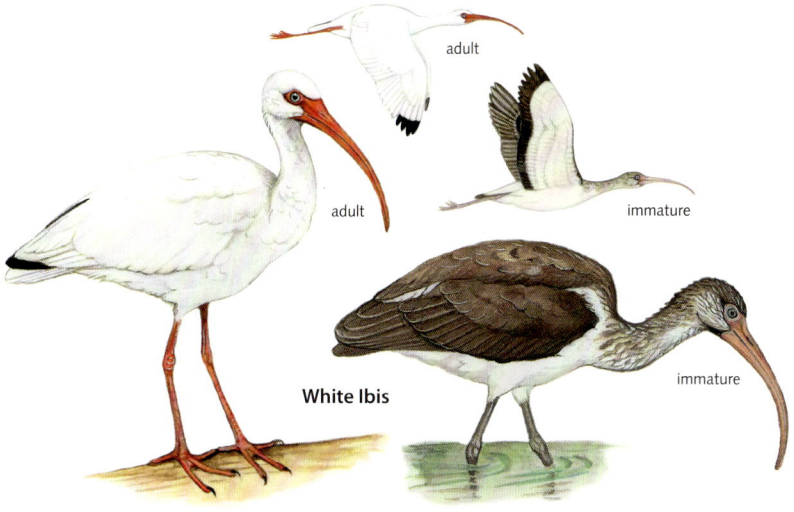

adult

adult

immature

immature

White Ibis

nonbreeding/
immature

Glossy Ibis

Green Ibis

breeding

breeding
adult

breeding
adult

**Roseate
Spoonbill**

immature

Birds of Nicaragua / **165**

An exclusively New World family of raptors that are known for eating carrion. Their featherless heads minimize surfaces for growth of pathogens picked up from bacteria-laden food. Hooked bills assist them in picking flesh from decomposing carcasses. They are most often seen soaring high on long, broad wings as they search for food. Although they share the name, these vultures are not closely related to the Old World Vultures of Europe, Asia, and Africa. They look similar due to convergent evolution.

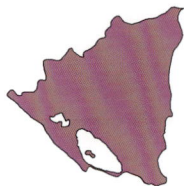

Turkey Vulture (Zopilote Cabecirrojo)   *Cathartes aura*
29 in (74 cm); WS 69 in (174 cm). Only vulture with a **red head**; resident population has a white band on the nape. Immature has a black head, similar to that of Black Vulture, but is distinguished by **long wings and tail** and the **two-toned underwing pattern** seen in flight, a characteristic shared with Lesser Yellow-headed Vulture and Zone-tailed Hawk (p. 184). In flight, wings are held in a dihedral position. A breeding resident; abundant countrywide at all elevations. Breeding residents are joined by winter residents (Sept to May). Distinctively tilts side-to-side while flying—often relatively low to the ground—over open areas in search of carrion; can also find food in densely vegetated areas due to a well-developed sense of smell. Winter residents and passage migrants join kettles of other species of migrating raptors as they move on thermal updrafts, mainly in the Caribbean lowlands; numbers in kettles can rise to tens of thousands in a short period of time.

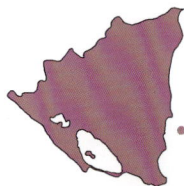

Black Vulture (Zopilote Negro)   *Coragyps atratus*
24.5 in (62 cm); WS 58 in (150 cm). **Black head and neck** separate it from all other vultures except for immature Turkey Vulture. **Broad wings**, **white primaries** on underwings, and a **short tail** are readily noticed in flight. Wings are held horizontally when soaring. Abundant countrywide, at all elevations. Can be seen in all terrestrial habitats but prefers open areas in close proximity to human activity. When not perched in groups beside a food source, black vultures often soar high, looking for carrion.

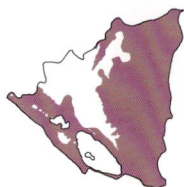

Lesser Yellow-headed Vulture   *Cathartes burrovianus*
(Zopilote Cabecigualdo)
24 in (61 cm); WS 61.5 in (156 cm). From a distance **multiple colors on head** may simply look orange-yellow. In flight, can easily be mistaken for a Turkey Vulture, but it is smaller. Immature has a gray head. Rare throughout lowlands and foothills, but locally uncommon on shores of Lake Managua and Lake Nicaragua, the Mosquitia, and San Juan River headwaters; to 1,300 ft (400 m). Usually seen alone, flying low to ground over grasslands, lowland pine savanna, and a variety of aquatic habitats; prefers to eat fish and reptiles.

King Vulture (Zopilote Rey)   *Sarcoramphus papa*
31 in (79 cm); WS 72.5 in (184 cm). The largest vulture; contrasting **black-and-white** plumage and gaudy, **vibrantly colored head** are unmistakable. Immature is black with white mottling on wing linings. Rare in Pacific and Northern Highlands and uncommon in Caribbean; to 6,200 ft (1,900 m). Although it can be seen soaring high over an impressive variety of habitats, it prefers primary forests. Perches on tallest trees within forest, with wings stretched out.

Zone-tailed Hawk

adult

immature

Black Vulture

Turkey Vulture

adult

adult

King Vulture

adult

immature

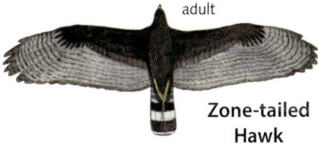

Lesser Yellow-headed Vulture

A Lesser Yellow-headed Vulture flying low over marshland.

OSPREY   Pandionidae

A truly cosmopolitan species that is the sole member of its family. Other commonly used names around the world (fish eagle, river hawk, sea hawk, and fish hawk) are a clue to its dietary preferences. It boasts long, curved claws and heavily scaled legs that aid in getting a good grip on slippery fish.

Osprey (Águila Pescadora)   *Pandion haliaetus*
23 in (59 cm); WS 64 in (163 cm). Note dark brown upperparts, mainly white underparts, and a **white head** with a **broad, dark eye line**. Further distinguished in flight by long, narrow wings that are cocked back. Locally common winter resident (Sept to April); to 4,300 ft (1,300 m). A population of non-breeding individuals stays year-round. A fishing specialist that hunts over all aquatic habitats that have water deep enough to support fish and to sustain plunge-diving from great heights. Calls with a shrill whistle: *weeAT!* Also gives casual, short, high-pitched whistles.

HAWKS, KITES, and EAGLES   Accipitridae

A diverse family of birds that occur throughout the world. These diurnal raptors are known for their strong hooked bill and powerful talons used for killing and ripping. In Nicaragua alone there are 21 genera; species vary in size (from the diminutive Tiny Hawk to the gigantic Harpy Eagle), morphology, preferred habitat, and hunting strategies. Although females typically look like males, they are usually larger. Several factors make identifications a challenge. Most species show distinct differences between adult and immature plumages and a few are polymorphic. And, a number of immatures have brown upperparts and white underparts streaked with brown. Additionally, some Forest-Falcons (family Falconidae) resemble members of this family. For these reasons, it is important to note wing and tail morphology and to compare the sizes of birds soaring in close proximity to each other. It takes a lot of practice to note fine distinctions and even the most experienced of birders cannot always make a definitive identification.

Snail Kite (Elanio Caracolero)   *Rostrhamus sociabilis*
17.5 in (44 cm). Note **slender, deeply hooked bill** and **white band at base of tail**. Male is slaty-black overall; female is brown overall, with pale white superciliary and throat and tawny and white underparts. Immature is paler than female, and lacks red iris and orange cere and legs of adults. Locally common at Estero Real NR, Apacunca GRR, Tisma NR, and Guayabo Wetlands; uncommon along shores of Lake Nicaragua, Guatuzo Plains, and San Juan River; rare throughout Pacific lowlands and Playitas-Moyúa-Tecomapa lagoons. Almost exclusively found at aquatic habitats such as freshwater or brackish marshes, ponds, lakes, and lagoons; also found at rivers if main food source, apple snails (genus *Pomocea*), is available. Produces a dry, ratcheting, series of notes: *eht-eht-eht… .*

Northern Harrier (Aguilucho Rastrero)   *Circus hudsonius*
20.5 in (52 cm). In all plumages, note distinctive **white uppertail coverts** and **owl-like facial disc**. Male has gray head; female has brown head and white superciliary, white line below the eye, and white on throat; immature is similar to female but with cinnamon underparts and less streaking. Long, slender tail and long wings stand out in flight. Uncommon winter resident (Oct to mid-April). Occurs throughout Pacific, Northern Highlands, and extreme southwestern Caribbean lowlands; country record for elevation is 4,300 ft (1,300 m), but it may occur higher. Flies low to ground, with slow wingbeats and, often, with a teetering motion, in search of prey in open areas (marshes and other open aquatic habitats, grassland, agricultural fields, scrub, and thorn forest). When migrating, flies high and in a direct path.

Osprey

adult female

adult male

immature

Snail Kite

adult male

adult female

adult male

immature

adult female

Northern Harrier

Gray-headed Kite (Elanio Cabecigrís)   *Leptodon cayanensis*
20 in (51 cm). Note contrasting **black upperparts**, **gray head**, and **white underparts** on perched individuals. In flight, also note overall shape of broad, rounded wings; long, rounded tail; and black wing linings contrasting with black-and-white barred flight feathers. Black tail has 3 broad, white bands (third usually not visible) and thin, white terminal band. Unlike adult, immature has yellow orbital skin, cere, and legs. Immatures vary in plumage, with dark, pale, and intermediate morphs; immature pale morph resembles larger Black-and-white Hawk-Eagle (p. 188). Uncommon and local in lowlands and foothills. Hunts small prey items within dry forest, gallery forest, lowland humid forest, and mangroves. Makes a moderately paced, series of repeated chuckling notes: *erh!-erh!-erh!…* (5 notes per second); series, of variable length, slowly rises in intensity until a slight drop-off at end. Also calls with a nasal, meow-like *err-AWH!*, with the second syllable distinctively inflected upward.

Hook-billed Kite (Elanio Picoganchudo)   *Chondrohierax uncinatus*
16.5 in (42 cm). **Heavy hooked bill**, **yellow supralorals**, and **pale eyes** obvious in all plumages (including black morph). Male has gray, barred underparts; female has orange, barred underparts; immature has mainly white underparts. Size of bill can vary. In flight, note wings pinched in at the base, which creates a rounded, broad-winged appearance. Rare countrywide, but locally uncommon on Sierra Isabelia and Sierra Dariense of Northern Highlands, Guatuzo Plains, and along San Juan River; to 4,600 ft (1,400 m). Found in virtually all habitats; perches for long periods of time while scanning for terrestrial or aquatic snails. Delivers a rapid, squeaky series of notes: *rhi-rhi-rhi…* (complete call 1–2 seconds in length), with a slight arc in intensity. Sound suggests calls made by woodpecker or woodcreeper.

Double-toothed Kite (Elanio Gorgirrayado)   *Harpagus bidentatus*
13.5 in (34 cm). **Black center stripe on white throat** is diagnostic. Adult has variable gray and rufous barring on belly, flanks, and thighs. On male, barring continues to breast, with rufous patches on the sides; breast is solid rufous on female. Immature is easily confused with immatures of other species that have brown upperparts and brown streaks on white underparts; note light streaking and center throat stripe. Wings extend forward in flight. Rare countrywide; to 4,600 ft (1,400 m). When hunting, often perches in forested habitat at mid-canopy or higher; regularly soars. Whistles a high-pitched, rhythmic, 5-note combination: *feeeu-pí píp feeeu-píp*. Call is a thin whistle: *seeúp*. In flight, gives a sharp, high-pitched *chip* or descending *cheeu*.

immature
pale morph

immature
intermediate morph

adult

adult

**Gray-headed
Kite**

adult male

immature

adult female

**Hook-billed
Kite**

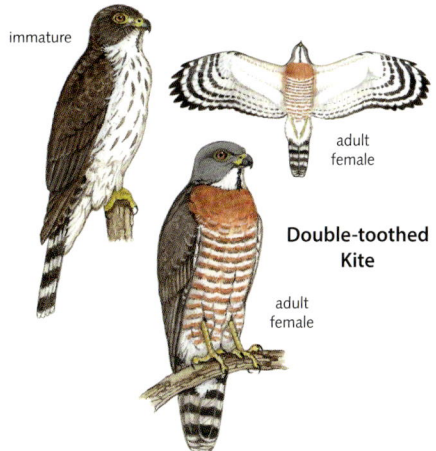

adult
black morph

adult
female

immature

adult
female

**Double-toothed
Kite**

adult
female

A Double-toothed Kite accompanying
a group of foraging Capuchin Monkeys
through the forest canopy.

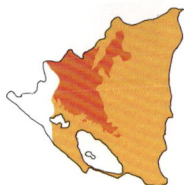

Swallow-tailed Kite (Elanio Tijereta)   *Elanoides forficatus*
23 in (58 cm). Virtually unmistakable, with **long, deeply forked tail** and stark contrast between black and white plumage. Common breeding migrant (mid-Jan to early Sept), and some individuals remain year-round. Occurs countrywide. Typically breeds in foothills and highlands; SA passage migrants occur throughout most of lowlands; to 6,200 ft (1,900 m). Present in most habitats but generally absent from arid habitats. Hunts small prey above the forest canopy. In migration, flocks can number up to 250; sometimes mixes with Plumbeous Kite or Mississippi Kite to forage or migrate. Gives a fast succession of sharp, shrieking *SI-WEET!* notes.

Pearl Kite (Elanio Carigualdo)   *Gampsonyx swainsonii*
8.5 in (22 cm). Very small raptor with richly colored head. Note **small black mask**, buffy cheeks and forehead, and blue-gray half-hood. In flight, **rufous flanks and thighs** stand out from white underparts and underwings. Common in Pacific; rare to uncommon on western slopes of North Highlands. Prefers open areas and forest edge, mainly in arid habitats such as dry forest, thorn forest, and scrub. Often perches in the open; dives from perch to pounce on prey; sometimes hovers before capture. Repeats a continuous *sí-sí-sí…*, with short, piercing notes.

White-tailed Kite (Elanio Azul)   *Elanus leucurus*
15.5 in (40 cm). Mostly white, with **black upperwing coverts**, gray upperwings and upperparts, and **black carpal patch** on pale wing linings. Common in lowlands and foothills, uncommon in highlands; to 4,400 ft (1,350 m). Influx of NA migrants increases local population (Dec to March). Mainly occurs in open areas, where it flies relatively low to ground in search of prey. Frequently hovers over prey before gracefully dropping to ground with distinctive wings elevated and tail dropped. Whistles a high-pitched, up-slurred *feep* and a downslurred *feeur*. Also delivers a weak, raspy scream.

Mississippi Kite (Elanio Colinegro)   *Ictinia mississippiensis*
14 in (35 cm). A small, sleek, pale raptor. Note **white secondaries** on dark upperwings and slightly notched, square **black tail**. Primaries show hints of rufous. May be confused with Plumbeous Kite (both species occur April to mid-May); note color of primaries, completely black tail, and wing tips just reaching the tip of the tail on perched birds. Immature shows heavy, rufous streaking on underparts, and narrow white barring on tail. Passage migrant (Sept to Oct and April to mid-May). Uncommon in Pacific, common in Northern Highlands; to 4,600 ft (1,400 m). Rare in western section of Caribbean. Most often seen migrating in kettles and riding thermals (up to 500 individuals).

Plumbeous Kite (Elanio Plomizo)   *Ictinia plumbea*
14 in (35 cm). A small, sleek raptor. Most readily identified by **bold rufous primaries** and **narrow, white bands on square, black tail**. May be confused with Mississippi Kite (both species occur April to mid-May); note color of primaries, barring on the tail, and wing tips noticeably extending beyond the tip of the tail on perched birds. Immature shows heavy, brown-gray streaking on underparts, and broad white barring on tail. Breeding migrant (Feb to early Aug). Uncommon on eastern slopes of Northern Highlands, rare in Pacific; to 4,600 ft (1,400 m). Found in forest habitats; demonstrates aerial prowess above canopy and at forest edge as it snatches insects on the wing or gleans prey from tops of vegetation. Whistles a high-pitched *TEE-heeur*; also calls with a descending sequence of notes: *ti-ti-tu*.

Swallow-tailed Kite

Pearl Kite

immature

adult

adult

**White-tailed
Kite**

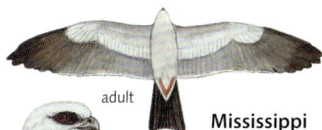

adult

**Mississippi
Kite**

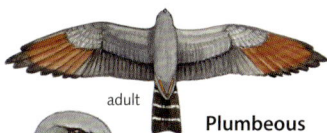

adult

**Plumbeous
Kite**

adult

immature

adult

immature

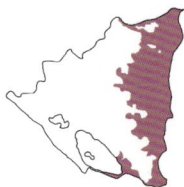

Black-collared Hawk (Gavilán Collarejo)   *Busarellus nigricollis*
20 in (51 cm). Combination of rufous body, pale head, and **broad, black upper breast patch** is distinctive. In flight, wings are long and broad; short broad tail barely projects beyond trailing edge of wings. Immature is paler, with black streaking on breast and brown barring on rest of underparts, underwings, and undertail. Locally common in San Miguelito Wetlands, Guatuzo Plains, and San Juan River headwaters, but possible along large rivers and other appropriate habitat in Caribbean lowlands. Exclusively found at aquatic habitats with stagnant or slow moving fresh or brackish water, where it captures fish from the water's surface. Produces 2 unique screams: one is a low-pitched, bleating scream with a dry and scratchy quality; the other is a high-pitched, weak, raspy *weeurweáh*.

Savanna Hawk (Gavilán Sabanero)   *Buteogallus meridionalis*
23 in (58 cm). **Black-tipped flight feathers** and **white band on black tail** stand out against pale cinnamon body. Immature also has the black-tipped flight feathers but is generally duller. A vagrant from Panama, with only 2 records in Nicaragua; both were adults seen in the Pacific lowlands (Feb 2012 and March 2014). Within their normal distribution, they typically hunt by walking on the ground or ambushing prey from a low perch in open areas. Screams a rich, downslurred *YI-eaaah*, which trails off in intensity and terminates with a hoarse tone.

Harris's Hawk (Gavilán Charreteado)   *Parabuteo unicinctus*
20.5 in (52 cm). **Rufous scapulars and thighs** contrast with **dark brown body**. In flight note black and white pattern on tail; also note contrast between rufous wing linings and black flight feathers. Immature has dark brown blotches on cinnamon breast and belly. Uncommon in Pacific lowlands, southeastern shore of Lake Nicaragua, and San Juan River headwaters; accidental in Northern Highlands (San Rafael del Norte, March 1892). Prefers thorn forest, scrub, and freshwater marshes. Groups of birds often cooperate when pursuing prey. Screams a very dry, raspy *erhhhh*.

adult

immature

adult

**Black-collared
Hawk**

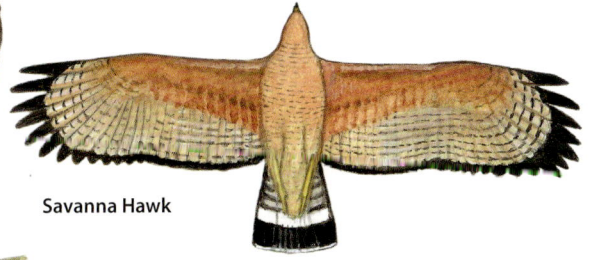

Savanna Hawk

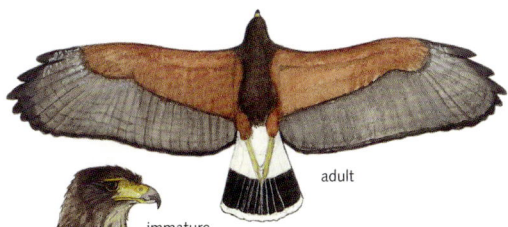

adult

adult

immature

Harris's Hawk

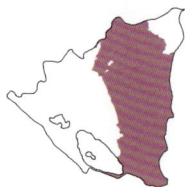

Tiny Hawk (Gavilán Chico)   *Accipiter superciliosus*
M 8 in (21 cm); F 10.5 in (26 cm). **Small size and dark, fine barring on underparts** distinguish it from other raptors. **Black crown** is prominent on gray head. Immature has similar pattern as adult but with brown or rufous coloration. Very rare in Caribbean lowlands and foothills. Inhabits humid lowland forest, second growth forest, and forest edge. From secretive canopy perch, waits for prey (known to specialize on hummingbirds). Makes a repeated series of sharp whistle notes: *whi whi whi… .*

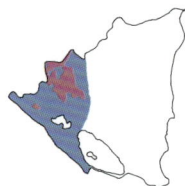

Sharp-shinned Hawk (Gavilán Pajarero)   *Accipiter striatus*
M 11 in (27 cm); F 14 in (35 cm). Two distinct sub-species occur in Nicaragua. *Chinogaster* subspecies (White-breasted Hawk): **Dark upperparts** (slaty-black on male, slightly browner on female) contrast sharply with **white underparts**. Immature has faint white superciliary and light streaking on throat and breast. Uncommon in highland pine and pine-oak, and cloud forests of Northern Highlands; above 3,000 ft (900 m). Small disjunct population occurs on Casita Volcano. *Velox* subspecies: Distinguished from very similar larger Cooper's Hawk by **uniform blue-gray upperparts** (female slightly browner) and **smaller, rounded head**. Immature is brown overall, with heavy brown streaking on white breast and belly. In flight, on both subspecies, note long, squared-tail and small head that barely extends beyond leading edge of wing. Winter resident (Oct to April); rare in Pacific and uncommon in Northern Highlands, primarily in foothills and highlands. Found in a diversity of open habitats, but more likely in forests. Slowly repeats a sharply whistled *wheep!, wheep!, wheep!* (1 call per second) and delivers a strident, scolding, rattle: *chu-chu-chu… .*

Cooper's Hawk (Gavilán Palomero)   *Accipiter cooperii*
16 in (40 cm). Similar to *velox* subspecies of smaller Sharp-shinned Hawk but note that blue-gray upperparts (female slightly browner) contrast with **paler nape and darker crown; head is larger and squarish**. Immature is brown overall, with light brown streaking on white breast and belly. In flight, note long, rounded-tail and large head that extends noticeably beyond leading edge of wing. Rare winter resident (Oct to April) in Pacific and Northern Highlands, primarily in foothills and highlands. Found hunting birds and small mammals at forest edge, in semi-open areas (e.g., coffee farms and roadsides), and at freshwater marshes. Call is a varying repetition of *eh eh eh…*, each note nasal and abbreviated.

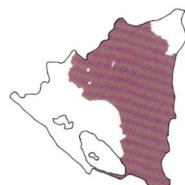

Bicolored Hawk (Gavilán Bicolor)   *Accipiter bicolor*
M 14 in (36 cm); F 18 in (45 cm). **Rufous thighs** on long legs are diagnostic (though not always visible). Also note **amber iris** contrasting with **dark head**. Underparts on immature range from white to buffy; partial to complete nuchal collar is also white to buffy. Compare light morph immature to Collared Forest-Falcon and immature Barred Forest Falcon. Rare throughout eastern slopes of Northern Highlands and Caribbean; to 4,300 ft (1,300 m). Ambush hunts from mid-canopy or canopy perch, within forest or on forest edge. Gives an erratic, nasal, and squeaky honking, as well as a quickly repeated, barking: *keh keh keh… .*

immature   adult

Tiny Hawk

adult
chinogaster ssp.

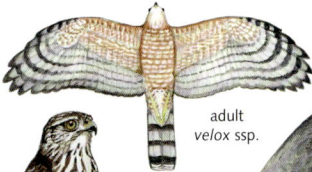

adult
velox ssp.

immature
velox ssp.

adult
velox ssp.

immature
chinogaster ssp.

adult
chinogaster ssp.

**Sharp-shinned
Hawk**

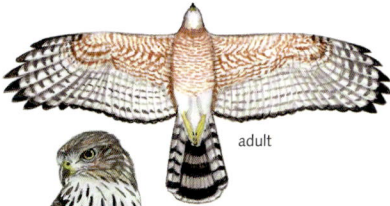

adult

adult

**Cooper's
Hawk**

immature

**Bicolored
Hawk**

adult

immature

immature

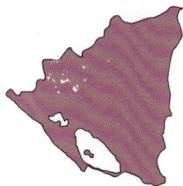

Common Black Hawk (Gavilán Cangrejero)   *Buteogallus anthracinus*

20.5 in (52 cm); WS 46.5 in (118 cm). Despite size difference, can be confused with Great Black Hawk or Solitary Eagle, and should be carefully compared. Up close, note **yellow lores** and **black thighs**. On perched birds, wingtips reach tip of tail or just short of it. In flight, note one **white tail band**, **protruding secondaries** that create broad-winged look, **short tail** slightly extending beyond trailing edge of wing, and extended feet only reaching white tail band. Immature has dark malar stripe, often broad and messy, wavy bands on tail, and barred thighs. Locally common in lowlands and rare in foothills and highlands; to 4,600 ft (1,400 m). Found on coastlines, lakeshores, marshes, and rivers, where it searches for aquatic prey. Whistles a fast series of shrill notes that climax in pitch and intensity before falling. Also calls with a shrill, downslurred *seeea* whistle.

Great Black Hawk (Gavilán Negro)   *Buteogallus urubitinga*

24 in (61 cm); WS 50.5 in (128 cm). Despite size difference, can be confused with Common Black Hawk or Solitary Eagle, and should be carefully compared. Up close, note **gray to pale yellow lores** and, usually, **white barred thighs**. On perched birds, wingtips fall noticeably short of tail tip. In flight, note **2 white tail bands** from below; **white uppertail coverts** from above; **protruding secondaries** that create broad-winged look; **long tail** extending beyond trailing edge of wing, and extended feet reaching to at least middle of second white tail band. Immature lacks malar stripe and has a finely banded tail and barred thighs. Rare to uncommon countrywide; to 5,200 ft (1,600 m) but could occur higher. Most often found within forested habitat and is known to frequent river edges and other aquatic habitats, even if in open areas. Gives a drawn-out and high-pitched scream (similar to that of Solitary Eagle) that is mostly monotone but diminishes in intensity at the end.

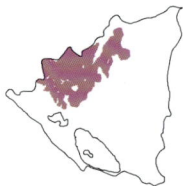

Solitary Eagle (Águila Solitaria)   *Buteogallus solitarius*

29 in (74 cm); WS 67 in (170 cm). Despite size difference, can be confused with Common Black Hawk or Great Black Hawk, and should be carefully compared. Up close, note **pale yellow lores, black thighs**, massive bill, and thick legs. On perched birds, wingtips reach tail tip or slightly beyond it. In flight, note 1 **white tail band**, **very short tail** barely projecting beyond trailing edge of wing, and extended feet reaching beyond white tail band, almost to the tail tip. Immature lacks malar stripe and has dark markings on breast; pale tail has no obvious bands except for dark subterminal band (on older birds); also note brown thighs. Very rare and local in Northern Highlands and Bosawas Biosphere Reserve; from 1,300 to 4,300 ft (400 to 1,300 m), but may disperse to adjacent lowlands on occasion. Inhabits highland pine and pine-oak forest, and humid lowland forests, particularly in mountainous terrain, where updrafts are ideal for soaring. Gives a drawn-out and high-pitched scream (similar to that of Great Black Hawk), that is mainly monotone but sometimes shifts to a higher pitch. Also calls with a short and powerful *WHI!* **NT**

**Common
Black-Hawk**

adult

immature

adult

adult

**Great
Black-Hawk**

adult

immature

adult

Solitary Eagle

adult

immature

Barred Hawk (Gavilán Pechinegro)   *Morphnarchus princeps*
24 in (61 cm). Contrast between **black head and breast** and **finely black barred underparts** is diagnostic. In flight, underparts, underwings, and wing linings appear gray at a distance, but white band on short black tail is prominent. Very rare and local in Caribbean, with known records from Saslaya NP and Indio Maíz BR; from 1,300 to 5,400 ft (400 to 1,650 m), but accidental in lowlands in close proximity to foothills and highlands. Reaches northernmost distribution in Nicaragua. Found in hilly terrain in humid lowland forest and cloud forest, where it hunts from mid-canopy perches; often seen soaring in pairs. Gives a broken *weEEoo* whistle, with a sharp upward inflection in pitch before slowly descending. Quickly repeats a *weep weep weep…*, sometimes with an escalating intensity.

White-tailed Hawk (Gavilán Coliblanco)   *Geranoaetus albicaudatus*
23 in (57 cm). Combination of **rufous scapulars** on **slaty-gray upperparts** is unique. Rare dark morph is slaty-gray overall and lacks rufous scapulars. Very similar to Short-tailed Hawk (p. 182) in flight, but larger and has black subterminal band on white tail. Immature is mostly dark; also note pale breast patch, faint black subterminal band, and pale tail. Populations are disjunct and vary in abundance. Common in the Mosquitia; rare throughout Pacific lowlands and foothills; very rare in Guatuzo Plains and San Juan River headwaters; and locally very rare in Northern Highlands (i.e., Moropotente NR). Prefers open habitat and semi-arid habitats such as dry forest, thorn forest, scrub, lowland pine savanna, and grasslands. Soars frequently, but also hovers above or follows after grass fires in search of prey. Delivers a rhythmic and slightly scratchy *diwii di-wí di-wí di-wí di-wí di-wí*; in the twin notes, each of the two syllables stands out distinctly.

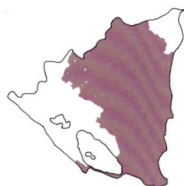

White Hawk (Gavilán Blanco)   *Pseudastur albicollis*
20 in (51 cm). Combination of **striking white plumage, black flight feathers** and **black subterminal band on tail** makes it unmistakable. Common in Caribbean (primarily in lowlands and foothills) and rare on eastern slopes of the Northern Highlands; to 3,900 ft (1,200 m). Very rare on Rivas Isthmus and on southern borders of Lake Nicaragua. Found in humid lowland forest or at forest edge. Perches in mid-canopy, from where it flies out to hunt; soars regularly. Gives a thin, dry *weEEah* scream.

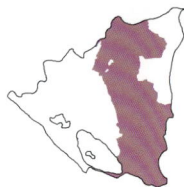

Semiplumbeous Hawk (Gavilán Dorsigrís)   *Leucopternis semiplumbeus*
15 in (38 cm). Note striking contrast between **slate-gray upperparts** and **white underparts**. Vibrant **orange cere and feet** distinguish it from similar Slaty-backed Forest-Falcon (p. 226); shorter tail further distinguishes it. Broad, short wings and narrow white subterminal band on black tail are notable in flight. Immature is very similar but with black streaks on throat and flanks. Unommon in Indio Maíz BR, otherwise rare in Caribbean lowlands and foothills. Hunts in understory and mid-canopy of humid lowland forest, dropping onto prey from a low perch; does not soar. Repeats a shrill *werEEP!*, with a distinct upward inflection in pitch.

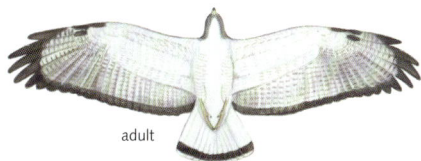

adult

immature

Barred Hawk

White-tailed Hawk

adult

White Hawk

Semiplumbeous Hawk

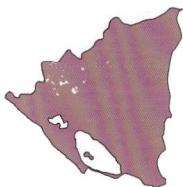

Roadside Hawk (Gavilán Chapulinero)   *Rupornis magnirostris*

15 in (38 cm). A frequently seen hawk. Slim, with **gray head** that washes onto breast, sometimes mottled with brown tones. **Heavily rufous-barred underparts** and broad black and gray bands on square tail are consistent field marks. In flight, note black-tipped rufous primaries. Immature has pale superciliary, heavy brown streaks on throat and breast, and narrow black bands on tail. Common countrywide; to 4,600 ft (1,400 m). Prefers open or disturbed habitats such as roadsides, fields, secondary forest, forest edge, lowland pine savanna, and scrub, where it perches low to wait for prey. Screams a squeaky, whinny *WEEaaa* (typically less drawn-out and stronger than than that of similar Gray Hawk), which sharply rises in pitch at beginning before slowly descending. Also calls with a short, squeaky *weá*.

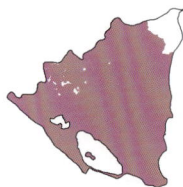

Gray Hawk (Gavilán Gris)   *Buteo plagiatus*

17 in (43 cm). **Fine gray-and-white barred underparts** on uniform gray body; up close, note black iris. In flight, black-tipped outermost primaries and broad black terminal tail band stand out from below. Immature is easily confused with immatures of other species that have brown upperparts and brown streaks on white underparts; note bold pattern on head created by brown crown, eye line, and malar stripe, and heavily streaked underparts. Common countrywide; to 4,600 ft (1,400 m). Hunts from perches in a variety of habitat, but prefers forest and forest edge and is found near water (marshes, gallery forests, and river edge). Screams a squeaky, piercing *WEEaaa* (typically more drawn-out and weaker than that of Roadside Hawk), which sharply rises in pitch at beginning before slowly descending.

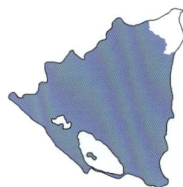

Broad-winged Hawk (Gavilán Aludo)   *Buteo platypterus*

16 in (41 cm). On breast, note heavy **rufous mottling** (rufous is more dispersed on rest of underparts). **Broad black malar stripe** borders whitish throat; head is dark. In flight, also note black trailing edge of wings and pointed wing tips. Immature is easily confused with immatures of other species that have brown upperparts and brown streaks on white underparts; note pale superciliary and malar stripe. (Mostly black dark morph occurs infrequently.) Common winter resident in Pacific and Northern Highlands; uncommon in Caribbean. Abundant passage migrant (mid-Sept to mid-Oct and mid-March to mid-April) in Pacific (mainly southward migration) and southwestern Caribbean (mainly northward migration). Perches in mid-canopy of forest or forest edge to search for prey; typically seen in flight during migration, when it often joins kettles of other species of migrating raptors that number in the thousands. Delivers a high-pitched, strained *pi-eeeee*, mainly even-pitched.

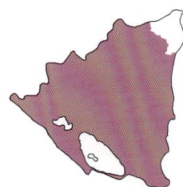

Short-tailed Hawk (Gavilán Colicorto)   *Buteo brachyurus*

16 in (41 cm). Two color morphs occur. On the pale morph, note **white forehead and throat** that contrast with **dark head**. In flight, it is very similar to White-tailed Hawk (p. 180), but it is smaller and has light gray barring and a black subterminal band. Immature is distinguished from immatures of other species that have brown upperparts and brown streaks on white underparts by minimal streaking on sides of breast. The dark morph is all black except for traces of white on forehead and white on base of flight feathers and tail. Immature is spotted white on much of underparts and on wing linings. Common throughout Pacific and Northern Highlands, and rare to uncommon in Caribbean. Most often seen soaring; inhabits a diversity of habitats. Delivers a descending, screaming *weeea* that ends abruptly and with a slight drop in pitch.

adult

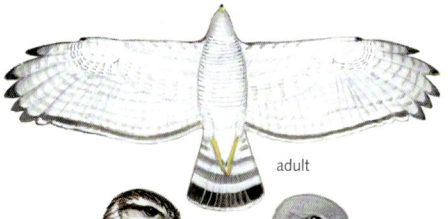
adult

adult

immature

**Roadside
Hawk**

adult

immature

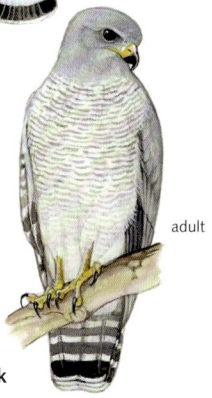
adult

Gray Hawk

adult

immature

adult

immature

**Broad-winged
Hawk**

adult

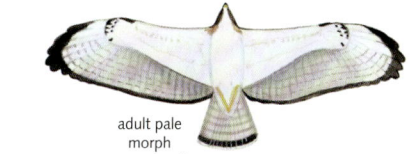
adult pale
morph

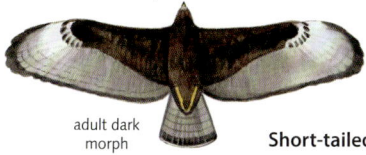
adult dark
morph

**Short-tailed
Hawk**

immature
pale morph

adult pale
morph

Swainson's Hawk (Gavilán Pechioscuro)   *Buteo swainsoni*

20 in (51 cm); WS 51.5 in (131 cm). The pale morph is the most common of the 3 color morphs. On it and intermediate morphs, **clean, white throat** contrasts with **dark breast band**. Dark morph is uniformly dark. Soars on long, pointed wings raised in dihedral position; on all 3 morphs, note **light undertail coverts** and **black subterminal tail band** in flight. Immature pale morph is easily confused with immatures of other species that have brown upperparts and brown streaks on white underparts; note bold malar stripe and heavy streaking on breast that often forms a band. Rare winter resident (mid-Sept to April) in Pacific, Guatuzo Plains, and along San Juan River; to 4,600 ft (1,400 m). Abundant passage migrant (mid-Sept to mid-Oct and mid-March to April) in Pacific, Northern Highlands, Guatuzo Plains, and along San Juan River. Upward of 5,000 individuals can be seen in migrating mixed-species raptor kettles. Gives a weak and slightly raspy *wíaaaah* scream, slowly descending in pitch and intensity.

Zone-tailed Hawk (Gavilán Impostor)   *Buteo albonotatus*

20 in (51 cm). Perhaps the more appropriate name for this bird is the Spanish common name, which means "imposter hawk," as its plumage and behavior closely resemble that of the Turkey Vulture (p. 166). In flight, 2-toned underwings, and black and white bands on long, square tail help to distinguish it from all other black raptors. Immature has sparse white spotting on breast. Rare breeding resident in Pacific and very rare in Northern Highlands; to 4,600 ft (1,400 m). With the arrival of winter residents (Oct to May), becomes slightly more abundant; uncommon in Pacific and rare in Northern Highlands and Caribbean. A habitat generalist that is often seen soaring with wings held in dihedral position and tilting body side-to-side, behavior that mimics that of Turkey Vultures. Produces a scream (*rhEEEEaa*) that rises in pitch at the beginning and falls at the end.

Red-tailed Hawk (Gavilán Colirrojo)   *Buteo jamaicensis*

22 in (56 cm). Separate breeding resident (*kemsiesi* subspecies) and winter resident (*borealis* and *calurus* subspecies) populations occur in Nicaragua; all have **rufous tail** and **dark bar on leading edge of wings** (seen in flight). *Kemsiesi* subspecies has **cinnamon underparts** except for **white breast**; may include dark morphs. On immature, white underparts show variable amounts of streaking; also note multiple narrow dark bands on pale tail. Uncommon breeding resident in Northern Highlands and San Cristóbal-Casita NR; primarily above 2,000 ft (600 m). Also found in the Mosquitia. Rare winter resident (Oct to April) throughout breeding distribution, throughout Pacific, and in the Northern Highlands; mainly consists of *borealis* subspecies but sometimes includes *calurus* subspecies. Both usually have a heavy, brown streaked band on lower belly, but intermediate and dark morphs may occur. Gives a dry, raspy, scream (*TEEaaaaah*) that begins with a burst of intensity then slowly descends in intensity and pitch; delivery is typically continuous but it sometimes is interrupted at midpoint of scream.

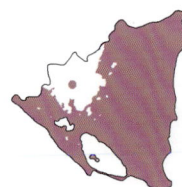

Crane Hawk (Gavilán Ranero)   *Geranospiza caerulescens*

19 in (48 cm). Only black raptor with **orange legs** and **red iris** on all **black head**. In flight, note **white, crescent-shaped bar on outer primaries** and long tail with 2 broad white bands. Immature has white plumage on forehead, superciliaries, and throat, and white barring on much of underparts and thighs, except for large white vent. Locally rare in lowlands and foothills. Accidental in Northern Highlands (Yali Volcano NR, Nov 2010). Prefers forested areas near freshwater marshes, rivers, and mangroves. Conspicuous in the canopy as it clambers around and flaps its wings while searching with its feet for prey hidden in trunk holes. Whistles a downslurred and plaintive squeal: *WEEeeeur*, similar to that of Roadside Hawk (p. 182), but less nasal.

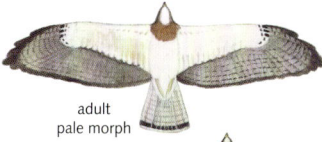

adult
pale morph

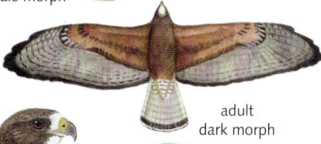

adult
dark morph

**Swainson's
Hawk**

adult
pale morph

immature
pale morph

adult

**Zone-tailed
Hawk**

adult

immature

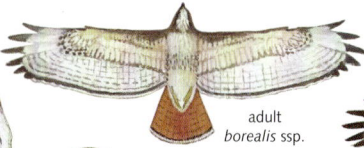

adult
borealis ssp.

**Red-tailed
Hawk**

adult
kemsiesi ssp.

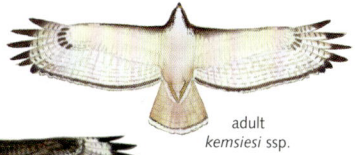

adult
kemsiesi ssp.

adult *calurus* ssp.
dark morph

immature
kemsiesi ssp.

**Crane
Hawk**

adult

immature

adult

Crested Eagle (Águila Crestada)   *Morphnus guianensis*
31.5 in (80 cm); WS 66.5 in (162 cm). **Single-pointed black crest** is iconic in both pale and dark morphs. Pale morph (more likely to be seen) has gray head and breast and pale, tawny barring on white underparts; white wing linings stand out in flight. Dark morph is mostly black, with black-and-white barring on underparts; white wing linings with black barring stand out in flight. Despite smaller size, immature can be confused with immature Harpy Eagle; note relatively longer and narrower wings and tail and slenderer tarsi. Historically, very rare throughout Caribbean lowlands, but now likely only to be found in large tracts of humid lowland forest in Bosawas Biosphere Reserve (in the north) and Punda Gorda NR and Indio Maíz BR (in the south). Perches atop the canopy and soars above it. Whistles a pure *fwiá fwí-fwih* 3-note phrase, or simply repeats a *fweeá* of a similar quality. **NT**

Harpy Eagle (Águila Harpía)   *Harpia harpyja*
38 in (97 cm); WS 80 in (203 cm). **Massive size** and **double-pointed black crest** make it unmistakable. Gray head vividly contrasts with black breast; on generally white underparts, note black bars on thighs. In flight, note black mottled wing linings. Despite larger size, immature can be confused with immature Crested Eagle; note relatively shorter and broader wings and tail and thicker tarsi. Historically, very rare throughout Caribbean lowlands, but now likely only to be found in large tracts of humid lowland forest in Bosawas Biosphere Reserve (in the north) and Punda Gorda NR and Indio Maíz BR (in the south). Two historical records suggest a larger historical distribution (San Juan del Sur, ca. 1900; Matagalpa Dept., Oct 1907). Only found in large tracts of humid lowland forest, where it hunts from perches in mid-canopy and canopy; has a surprisingly high level of maneuverability within dense canopy structure and rarely soars. Screams a piercing *WEEyaaaah*. **NT**

adult
dark morph

immature

adult
pale morph

Crested Eagle

adult
pale morph

adult

Harpy Eagle

adult

immature

Black Hawk-Eagle (Aguililla Negra)   *Spizaetus tyrannus*
27 in (68 cm); WS 55.5 in (141 cm). Predominantly black but note **white barred thighs, tarsi, and vent**; also note **short bushy crest**. In flight, recognized by striking **black-and-white barred pattern** on broad wings and long tail. Immature has black mottling and streaking on white head and breast; in flight, note heavy black mottling on white wing linings. Uncommon on eastern slopes of Northern Highlands and northern region of Sierra Chontaleña; uncommon in Saslaya NP, otherwise rare throughout Caribbean; to 6,600 ft (2,000 m). Two historical records in Pacific (Mombacho Volcano, Feb 1896; Sierras Managua, ca. 1900). Frequently seen soaring high above fragmented humid lowland forest, highland pine and pine-oak forest, and cloud forest. Whistles a *wi wi wi-wEEur* phrase, or sometimes simply an arcing *wEEur*.

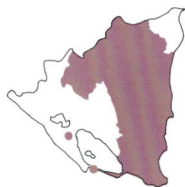

Ornate Hawk-Eagle (Aguililla Penachuda)   *Spizaetus ornatus*
25 in (63 cm); WS 51 in (130 cm). **Tawny head plumage** and **black pointed crest** make it unmistakable. When it soars (on short, broad wings), note heavily barred underparts. Black barring on white flanks and thighs help distinguish immature from similar, mainly white raptors. Rare on eastern slopes of Northern Highlands and throughout Caribbean. Two historical records from southern Pacific (Sierras Managua, ca. 1900; San Emilio, April 1905). Inhabits humid lowland forest, highland pine and pine-oak forest, and cloud forest, where it perches in mid-canopy to scan for prey; often seen soaring directly above canopy or perched on emergent snag. Screams short and piercing *WEEuw* whistles, or a combination of *weuw weh-weh* whistles. **NT**

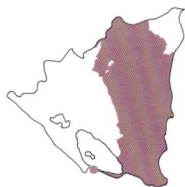

Black-and-white Hawk-Eagle   *Spizaetus melanoleucus*
(Aguililla Blanquinegra)
23 in (58 cm); WS 51 in (130 cm). Orange cere and yellow iris contrast with **small black mask** to create a menacing facial expression. **Short black crest** often only appears as a patch on white head. Resembles immature pale-morph Gray-headed Kite (p. 170). In flight, underwings are white, primaries have black tips, and wings are pointed. Rare throughout Caribbean lowlands and foothills. One historical record from Rivas Isthmus (San Emilio, April 1905). Requires undisturbed humid lowland forest; frequently seen soaring. Delivers 4–5 piping, arcing whistles: *fwEEo!*

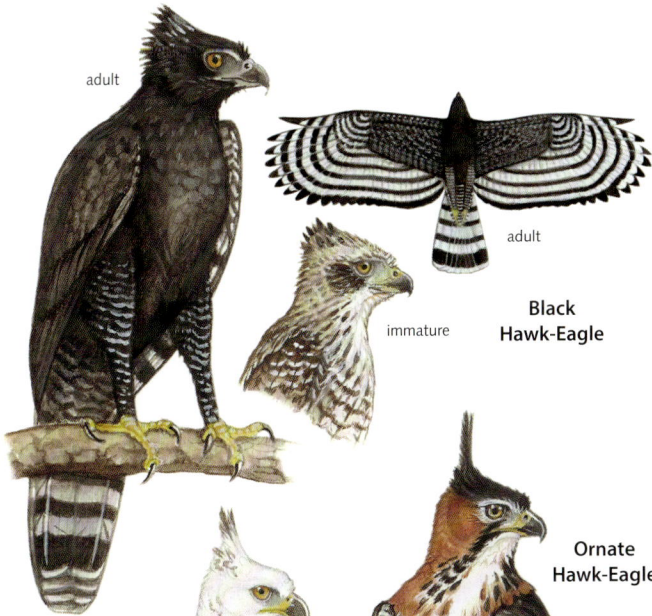

adult

adult

immature

**Black
Hawk-Eagle**

**Ornate
Hawk-Eagle**

immature

adult

adult

**Black-and-white
Hawk-Eagle**

BARN-OWLS Tytonidae

Charismatic and cosmopolitan, these nocturnal raptors are known for their habit of hunting in open areas and their ability to thrive in urban locations. Barn-Owls typically have a large head, slender body, pale coloration, and a prominent heart-shaped facial disc.

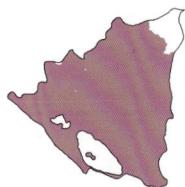

Barn Owl (Lechuza Común) *Tyto alba*

16 in (41 cm); WS 40 in (100 cm). **Pale, heart-shaped facial disc** is diagnostic; male typically has white underparts, while female typically has cinnamon underparts. In flight, from below, pale coloration is striking, even in low light situations. Locally common countrywide; to 5,200 ft (1,600 m). Flies low and effortlessly over agricultural fields, urban areas, forest edge, and marshes; roosts in dense vegetation, tree cavities, and a variety of human made structures within urban areas. Produces a dry, grating scream: *kreeEE*; also makes a dry, high-pitched, chattering rattle: *cheít cheít cheít... .*

OWLS Strigidae

This is a diverse cosmopolitan family of raptors. Not all owls are nocturnal, but most have sophisticated adaptations for hunting in low-light settings. While screech-owls (*Megascops*) and pygmy-owls (*Glaucidium*) are easily distinguished from other owls, it is often a challenge to distinguish among the species within each genus, in part due to variable color morphs and cryptic plumage patterns. While it is important to note habitat and distribution, one of the best keys to identification is vocalizations (especially important at night!). To make daytime sightings, look for large amounts of fecal droppings and regurgitated pellets on the ground below daytime roosts; also look for raucous mobbing songbirds. Some species have poorly known distributions in Nicaragua.

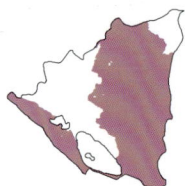

Striped Owl (Búho Listado) *Pseudoscops clamator*

14.5 in (37 cm). This owl is easily distinguished by a **white facial disc with a black rim**; has long ear-tufts. Also note heavily streaked underparts, **rufous forehead, and crown with black streaks**. Rare in Pacific lowlands and foothills. Accidental on eastern slopes of Northern Highlands; rare in Caribbean. Prefers semi-open and open areas within dry forest; also inhabits humid lowland forest edge and river edge. Flies low over open areas in search of prey or observes from an exposed perch; roosts in dense understory or even on ground. Uses 2 distinctly different vocalizations. Gives a deep and mournful *wuóh* that slightly arcs in pitch, and also delivers a high-pitched, shrill, shriek: *heeAH*.

Barn Owl

Striped
Owl

A Striped Owl perched
at a roadside at night.

Pacific Screech-Owl (Tecolotito Sabanero)

Megascops cooperi

9.5 in (24 cm). The **palest of the *Megascops***, and typically with noticeable, **long ear-tufts**. On underparts, note long, thin streaking; toes only show bristles. Common throughout Pacific lowlands and foothills, and uncommon on western slopes of Northern Highlands; to 2,300 ft (700 m). Inhabits a variety of habitats. When hunting, perches on understory branches at the edge of open areas; roosts in dense vegetation and tree cavities. Sings a rolling chuckle: *oah-oah-oah* (2 seconds in length); often increases in intensity and slows in pace.

Whiskered Screech-Owl (Tecolotito Manchado)

Megascops trichopsis

7 in (18 cm). Small. There are various color morphs, ranging from gray to rufous. Short ear-tufts are often only barely noticeable. **Short, broad streaks on underparts and feathered toes** help distinguish it from other *Megascops*. Rare on Sierra Dipilto-Jalapa, Sierra La Botija, and Tepesomoto-Pataste NR; from 2,500 to 4,900 ft (750 to 1,500 m). Accidental on Casita Volcano. Reaches southernmost distribution in Nicaragua. Prefers highland pine and pine-oak forest and forest edge, where it typically hunts in flight. Often roosts on horizontal branches of trees within dense vegetation, nestled up against the trunk. Toots an evenly paced sequence of low-pitched, montone *hoo* notes (entire sequence lasting 2 seconds). Also sings a series of slightly hoarse *hoo* notes, with erratic and choppy spacing between the notes.

Vermiculated Screech-Owl

Megascops guatemalae

(Tecolotito Vermiculado)

8.5 in (21 cm). Two subspecies occur in Nicaragua, each with its own geographical distribution. Both subspecies have gray and rufous morphs; short ear-tufts that are often not noticeable; and bare toes. *Guatemalae* subspecies: Underparts are more coarsely streaked and with sparser vermiculation than on *vermiculatus*. White eyebrows are noticeable; facial disc, with black border, is distinct. Rare on eastern slopes of Northern Highlands and in Bosawas Biosphere Reserve; to 4,600 ft (1,400 m), but possibly higher. *Vermiculatus* subspecies: Underparts are more finely streaked and with denser vermiculation than on *guatemalae*. White eyebrows are not noticeable and facial disc is indistinct because of faint black border. Rare in humid lowland forests in southern Caribbean. Both subspecies deliver a rapid series of repeated, wooden, low-pitched notes: *hu-hu-hu*… (call of *guatemalae* subspecies is even slightly lower pitched and is slower paced than that of *vermiculatus*). Call is easily confused with that of the Cane Toad (*Rhinella marina*), which is more resonating.

Pacific
Screech-Owl

rufous
morph

Whiskered
Screech-Owl

gray
morph

Vermiculated
Screech-Owl

rufous morph
vermiculatas ssp.

gray morph
vermiculatus ssp.

guatemalae ssp.

Crested Owl (Búho Penachudo)   *Lophostrix cristata*

16 in (41 cm). Both pale and dark morphs have almost comically **long, slender, white ear-tufts** that extend from the eyebrows. Underparts show very fine vermiculation that is not always possible to detect. Rare on eastern slopes of Northern Highlands and in Caribbean; to 4,300 ft (1,300 m). Found in cloud forest, humid lowland forest, and secondary forest, where it is found at forest edges and along streams. Delivers a low, vibrating growl or purr: *wh-h-hoar-r-r* (1 second in length), which terminates in a slight crescendo.

Great Horned Owl (Búho Grande)   *Bubo virginianus*

20 in (50 cm). **Bulky, very large, and with pronounced ear-tufts**. Not likely confused with other tufted owls; **heavily barred underparts** should remove any doubt. Rare in Pacific foothills and locally rare on western slopes of Northern Highlands; above 700 ft (200 m). Rare in the Mosquitia. Inhabits dry forest, highland pine and pine-oak forest, cloud forest, and lowland pine savanna. Often perches mid-canopy at forest edge or in open areas to survey for small mammal prey; roosts within forests. Sings a pure, but slightly tremulous, 4-noted *hoo-hoo hoo hoo*, with slightly more emphasis on the second note. Also sings a deep, monotone *hoo, hoo, hoo*.

Stygian Owl (Búho Oscuro)   *Asio stygius*

16 in (40 cm). Unique and unmistakable; has extremely dark plumage. Large profile highlights **long, sickle-shaped ear-tufts towards the center of the head**, which can be laid flat when relaxed. Only owl with **pale forehead patch** and **heavily streaked underparts with broad cross-bars**. Rare and local, with records only from Sierra Isabelia (possibly overlooked in appropriate habitat throughout Northern Highlands); above 3,900 ft (1,200 m). Reaches southernmost distribution in Nicaragua. Prefers cloud forest and forest edge, where it hunts from mid-canopy perches; also hunts while in flight. Consistently gives a deep, mellow, short *huó* (repeated every 5 seconds).

Burrowing Owl (Búho Llanero)   *Athene cunicularia*

9.5 in (24 cm). Unmistakable. Pale brown overall, with extensive **white spotting** and white eyebrows. **Long legs** and **round head** create a unique profile for a terrestrial bird. Accidental winter resident with only 1 record (Moropotente NR, Jan 2017). Prefers open areas such as grasslands and other arid terrain, where it perchs low to ground (on rocky outcroppings, for example) to survey for prey.

dark
morph

Crested Owl

pale
morph

Great
Horned Owl

Stygian
Owl

Burrowing
Owl

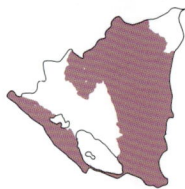

Spectacled Owl (Búho de Anteojos)   *Pulsatrix perspicillata*
18.5 in (47 cm). Easily recognized by diagnostic **white spectacles,** formed by connection of malar stripe, lore, and superciliary. Immature is completely white, with starkly contrasting black facial mask. Locally uncommon in Pacific lowlands and foothills. Locally uncommon in Caribbean and very rare on eastern slopes of Northern Highlands; to 4,900 ft (1,500 m). Found in dry forest (slightly humid), cloud forest edge, humid lowland forest and edge. Hunts from perch at mid-canopy and roosts in dense vegetation, often along streams. Delivers a muffled but resonating series of deep notes: *wúh-wúh-wúh-woo-woo-woo,* reminiscent of the sound made by shaking a sheet of tin.

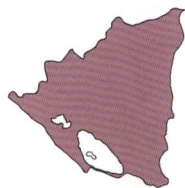

Mottled Owl (Cárabo Café)   *Ciccaba virgata*
14 in (36 cm). Combination of no ear-tufts, **heavily mottled breast**, streaked underparts, and a brown facial disc (with faint white rim) distinguishes it from owls of similar size. Occurs throughout the country; common in Pacific and Northern Highlands, and uncommon in Caribbean. Inhabits a wide range of forested and semi-forested habitats, where it hunts from mid-canopy along forest edge and in semi-open areas; moves to lower levels of canopy to roost in dense vegetation. Delivers a strong, sliding, double-noted *wuUU*, the second note louder and with a twangy quality; sometimes makes a longer, more complex sequence: *hu hu wuUU wuUU wuUU hu.*

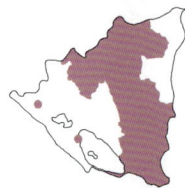

Black-and-white Owl (Cárabo Negriblanco)   *Ciccaba nigrolineata*
15 in (38 cm). Unmistakable. No other owl has **black barred underparts** or **black facial disc with white spectacled eyebrows.** Uncommon on eastern slopes of Northern Highlands and rare throughout Caribbean. Accidental in Pacific, with 2 historical records, one at San Cristóbal Volcano (June 1917), the other at Mombacho Volcano (ca. 1900). Occurs in cloud forest and humid lowland forest, but may be found in other habitats as well. Mainly hunts from a perch, but will also catch prey while in flight. Delivers a distinctive, nasal *hwua-hwua-hwua-hwua-HWUOH-HWUOH* series that builds to 1 or 2 explosive final notes. Also calls with a single, calm *hwuoh.*

Spectacled Owl

Mottled Owl

Black-and-white Owl

Northern Pygmy-Owl (Mochuelo Serrano)   *Glaucidium gnoma*
6.5 in (16 cm). There are various color morphs, ranging from gray to rufous; plumage pattern is the same on all morphs. **Faint white spots on head** distinguish it from Ferruginous Pygmy-Owl. Light streaking on underparts and **5 or 6 whitish tail bands** (only 4 visible on undertail) help to separate it from the other Pygmy-Owls; no range overlap with Central American Pygmy-Owl. Reaches southernmost distribution in Nicaragua. Rare to uncommon in Sierra Dipilto-Jalapa and Sierra Isabelia of Northern Highlands; above 3,300 ft (1,000 m). Found in highland pine and pine-oak forest, cloud forest, and forest edge. Hunts primarily at dawn and dusk but sometimes also during the day. Toots 2 or 3 hollow whistles: *hu-hu* or *hu-hu-hu*; repeated sequences often form a rhythmic pattern.

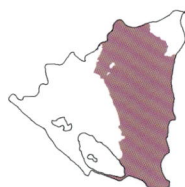

Central American Pygmy-Owl   *Glaucidium griseiceps*
(Mochuelo Centroamericano)
5.5 in (14 cm). **Bold white spots on head**, heavy streaking on underparts, and **3 to 4 broken pale tail bands** (only 2 visible from undertail) help to distinguish it from other Pygmy-Owls. Smaller and shorter-tailed than Ferruginous Pygmy-Owl; no range overlap with Northern Pygmy-Owl. Uncommon in Caribbean lowlands and foothills. Regularly hunts both during the day and at night; perches in mid-canopy of humid lowland forest, forest edge, secondary forest, and in plantations. Toots a series of 4–15 dulcet whistles (*hu*), at a rate of 3-4 per second and with consistent, equal spacing between whistles; faster-paced, and with shorter sequences, than that of Ferruginous Pygmy-Owl.

Ferruginous Pygmy-Owl (Mochuelo Herrumbroso)   *Glaucidium brasilianum*
6 in (15 cm). There are various color morphs, ranging from brown to rufous; plumage pattern is the same on all morphs. **Short, white streaks on head**, heavy streaking on underparts, and **5 to 7 pale tail bands** (only 4 or 5 visible from undertail) help to distinguish it from other Pygmy-Owls. Larger and longer-tailed than Central American Pygmy-Owl. Occurs throughout the country; common in Pacific and Northern Highlands, and rare in Caribbean. Found in almost all terrestrial habitats and in urban areas; actively hunts primarily at dawn and dusk but also during the day. Toots an incessant series of grainy, forced whistles (*hu*), at a rate of 2–3 per second and with consistent, equal spacing between whistles; slower paced than that of Ferruginous Pygmy-Owl and with longer sequences.

Northern Pygmy-Owl

rufous morph

gray morph

Central American Pygmy-Owl

Ferruginous Pygmy-Owl

false eyes on back of head

The beautifully colored members of this family occur in tropical regions of the Americas, Africa, and Asia. They are most diverse in the Neotropics. Trogons have compact bodies, long square tails, and short stout bills used to pick off arboreal fruit and to snatch large insects. Exhibiting strong sexual dimorphism, males are more boldly colored, showing metallic green or blue on upperparts and chest. Key field marks include color of bill, color and shape of orbital ring, and pattern on undertail. These forest residents perch motionless for long periods of time, and it is possible to walk directly underneath them without noticing their presence. When they do move, they fly only short distances, with an undulating flight.

Elegant Trogon (Trogón Collarejo)   *Trogon elegans*
11 in (28 cm). Male is very similar to smaller male Collared Trogon, but note **orange-red orbital ring**, **broad white tips on undertail**, and **conspicuous gray wing coverts**. On female, **white ear patch** is diagnostic, and red is restricted to lower underparts. Uncommon in Pacific (not extending east of the lakes) and in dry intermontane valleys on western slopes of Northern Highlands; to 4,400 ft (1,350 m). Inhabits dry forest, thorn forest, gallery forest, tall second growth, and edges of habitat. Forages in mid-canopy and canopy, alone or in pairs. Repeats *qwah-qwah* a varying number of times; has a hoarse, plaintive quality. Also gives a faster *oh oh oh…*, followed by a rolling *wicka-wick-werk*-werk-*werk*.

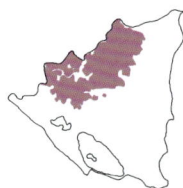

Collared Trogon (Trogón Colibarreteado)   *Trogon collaris*
10 in (25 cm). Male is very similar to larger male Elegant Trogon, but note lack of orbital ring, **narrow white tips on undertail**, and **darker wing coverts**. Female lacks ear patch, **red underparts extend to white collar,** and undertail pattern is unique (light gray with narrow, black subterminal band and white tips). Common in Northern Highlands and in highlands of the Caribbean (Quirragua NR, Musún NR, and Saslaya NP); above 2,600 ft (800 m). Rare to uncommon in Caribbean foothills, occasionally found in lowlands. Occurs in cloud forest, humid lowland forest, and forest edge. Forages from understory to mid-canopy; alone or in pairs. Whistles 2 (sometimes 3) sweet, melancholic, downslurred *feur* notes; very similar to that of Black-throated Trogon but each note is slightly shorter. Also delivers a soft, short trill.

Elegant
Trogon

male

female

bronzy uppertail on male
differs from green uppertail
of male Collared Trogon

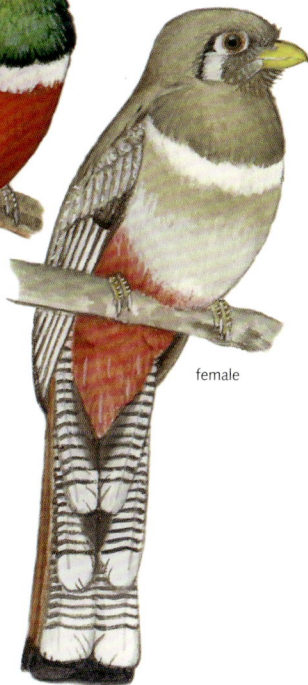

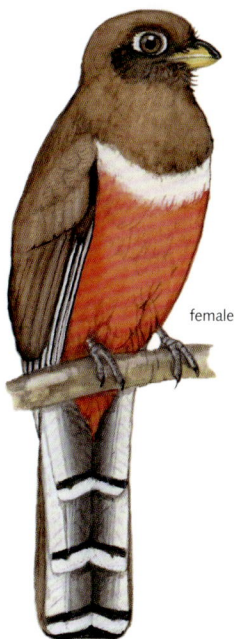

male

Collared
Trogon

female

Black-headed Trogon (Trogón Cabecinegro)   *Trogon melanocephalus*
10.5 in (27 cm). Only Nicaraguan trogon that has an **unbarred undertail with very broad white tips**; in poor light, sometimes confused with Gartered Trogon, but note large white tips on underside of tail feathers and **pale blue orbital ring** (yellow orbital ring on male Gartered and white on female Gartered). Female is further distinguished from female Gartered Trogon by thin orbital ring and unbarred wing coverts. Common in Pacific, rare to uncommon on eastern slopes of Northern Highlands; to 4,400 ft (1,350 m). Uncommon in Caribbean; to 3,300 ft (1,000 m). Although it tolerates open areas more so than other trogons, prefers dry forest and edge, second growth, woodland, and scrub. Forages from mid-canopy to canopy. Feeds in pairs or in small groups that congregate during the breeding season. Delivers an accelerating series of rapid-fire *chúp* notes (typically lasting 2–3 seconds) that ends abruptly.

Gartered Trogon (Trogón Violáceo)   *Trogon caligatus*
9.5 in (24 cm). On male, black on head becomes **metallic blue** (difficult to see in poor light) on surrounding areas; **yellow orbital ring** and **pale wing coverts** distinguish it from Black-headed Trogon. Female distinguished from female Black-headed Trogon by **broad, elliptical white orbital ring** and **heavily barred undertail**. Uncommon to locally common in lowlands and foothills, rare in Northern Highlands; to 4,400 ft (1,350 m). Inhabits most forested habitats, occasionally visiting cloud forest (especially during dry season); generally found alone or in pairs, sometimes in small groups. Delivers a series of emphatic, somewhat grating, monotone *weh* whistles, rapidly repeated (4 notes per second) at length, with uniform speed and pitch.

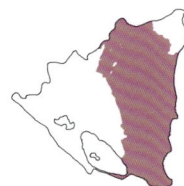

Black-throated Trogon (Trogón Gorginegro)   *Trogon rufus*
10 in (25 cm). Combination of **yellow underparts** and **yellow bill** is unique among the trogons; also note dark culmen on female. **Broad, elliptical white orbital ring** on female is similar to that of female Gartered Trogon, but brown head and upperparts set it apart. Locally common in Caribbean lowlands and foothills; to 2,600 ft (800 m). Inhabits humid lowland forest and secondary forest. Forages alone or in pairs, from understory to mid-canopy. Usually whistles 3 downslurred *feur* notes; sounds very similar to Collared Trogon but each note is slightly longer. Also delivers a short, bubbly trill.

Black-headed Trogon

male

female

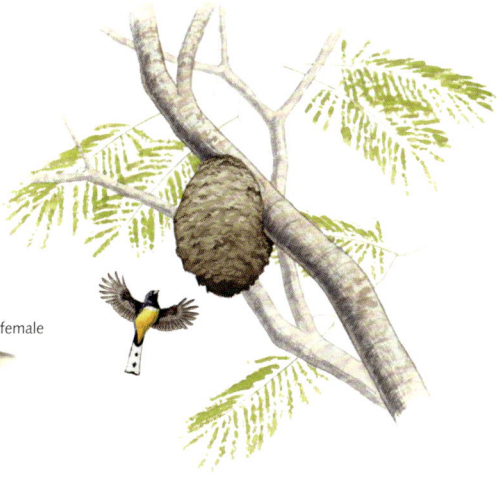

A Black-headed Trogon about to enter its termitarium cavity nest.

male

female

Gartered Trogon

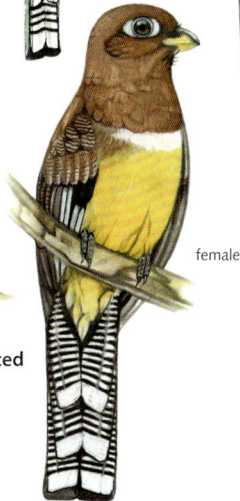

male

female

Black-throated Trogon

Resplendent Quetzal (Quetzal)　　　　　　　　*Pharomachrus mocinno*
15.5 in (40 cm); uppertail coverts on male add 24 in (61 cm). Largest of the trogons. Has a bushy crest (less prominent on female). Iridescent male is unmistakable; extremely **long uppertail coverts** and **white undertail** are unique. On female, **red is reduced to vent and undertail coverts**. Local and common in Saslaya NP; once common throughout Northern Highlands, but now only locally rare because of deforestation (Tepesomoto-Pataste NR, Sierra Dipilto-Jalapa, Miraflor NR, Yali Volcano NR, Arenal NR, Dantanlí-El Diablo NR, Kilambé NR, and Peñas Blancas NR); above 3,300 ft (1,000 m) most of the year, but descends to 1,600 ft (500 m) on the eastern slopes after breeding. Dependent on cloud forest (breeding season) and humid lowland forest (non-breeding season). Prefers unfragmented forest with steep elevation gradient; makes short, seasonal migrations, in which it changes elevation in search of fruiting *aguacatillo* (Family Lauraceae). Alone or in pairs, perches in mid-canopy and canopy. Males vocalize during breeding season, making it the easiest time to find them. Often calls with a forlorn, hollow, descending *peóh* or a nasal, 2-syllable *wa-há*. Flight call consists of a spirited and rolling *wack-a-wack-a*. **NT**

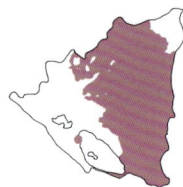

Slaty-tailed Trogon (Trogón Colinegro)　　　　　　　*Trogon massena*
12.5 in (32 cm). **Slaty, unpatterned undertail** and **orange bill** (upper mandible is dark on female) are both diagnostic. Similar to Lattice-tailed Trogon (very rare), but **red-orange orbital ring** and dark iris further distinguish it. Locally abundant in Saslaya NP and Bosawas Biosphere Reserve; common in rest of Caribbean lowlands and foothills; locally uncommon from eastern portion of Rivas Isthmus to Cardenas (south of Lake Nicaragua); otherwise accidental in Pacific lowlands and foothills (1 historical record from Mombacho Volcano, March 1962); very rare on eastern slopes of Northern Highlands; to 3,300 ft (1,000 m). Found in forest, forest edge, forest patches, second growth, and, occasionally, open areas with scattered trees. Forages in mid-canopy and canopy. Usually solitary; sometimes in pairs during breeding season. Vocalizes with a very fast and continuous chuckle (*háháhá…*) that sometimes goes on at great length. Also gives a more defined, crowlike monotone *cah cah cah…* (4 notes per second); sound often carries a considerable distance.

Lattice-tailed Trogon (Trogón Ojiblanco)　　　　　　*Trogon clathratus*
12 in (30 cm). The only Nicaraguan trogon with a **pale white iris**. Similar to sympatric Slaty-tailed Trogon, but note iris color, yellow bill (upper mandible is dark on female), **finely white-barred undertail**, and lack of orbital ring. Rare in extreme southern Caribbean lowlands and foothills; above 500 ft (150 m). Inhabits humid lowland forest, forest edge, and forest patches. Solitary, but can be seen in pairs during breeding season. Feeds in mid-canopy, sometimes descending to understory. Rapidly repeats 10–11 *wah* notes that crescendo before final note abruptly drops in pitch (entire series lasts 1.5 seconds). Reaches northern most distribution in Nicaragua. Endemic to Central American Caribbean Slope EBA.

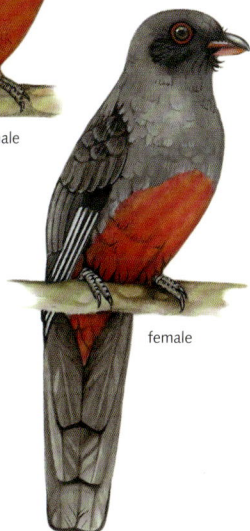

Slaty-tailed Trogon

male

female

male

Resplendent Quetzal

female

male

female

Lattice-tailed Trogon

These beautiful birds occur only in the Neotropics, with the greatest diversity in Central America. All 6 species in Nicaragua have black masks, relatively large heads, and a long, stout bill (black with decurved culmens) specialized for capturing large insects and small vertebrates. They are most famous for their long, racket-tipped tails, which they move side-to-side, like a pendulum. Despite their colorful plumage, they are inconspicuous as they perch within the forest, quietly and upright, waiting for prey to pass by. One exception is the Turquoise-browed Motmot, which is often found in semi-open habitat. Hoarse vocalizations, most often heard at dawn and late afternoon, betray their presence; also look for burrows dug in earth banks that they use for nests.

Tody Motmot (Guardabarranco Enano)   *Hylomanes momotula*

6.5 in (17 cm). **Small** and has a **short tail** (no rackets). Distinct head pattern includes **short, light blue superciliary** and **prominent white malar stripe**. Rare in northern Caribbean foothills and highlands (locally uncommon in Saslaya NP); from 1,000 to 3,300 ft (300 to 1,000 m). Three historical records from Peñas Blancas NR (1909) in Northern Highlands. Found in primary forest with dense understory; perches quietly in understory at eye level, moving tail in pendulum fashion; easily overlooked. Feeds on insects, spiders, and snails. Typically whistles a lively *wháh wháh wháh…*; delivered for extended periods, with 1 note per second. When it is delivered with abbreviated notes and at a faster pace, the call may resemble pygmy-owl toots.

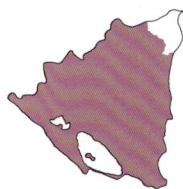

Lesson's Motmot (Guardabarranco Coroniazul)   *Momotus lessonii*

16 in (41 cm). Only motmot with **red iris**; **intense turquoise-blue ring adorns the black crown**. Black mask is more extensive than on smaller Keel-billed Motmot (p. 208), and lack of rufous distinguishes it from smaller, sympatric Turquoise-browed Motmot. Common in Northern Highlands; to 6,000 ft (1,800 m). Common in Caribbean lowlands and foothills. Uncommon in Pacific foothills and highlands (especially on Sierras Managua, Mombacho Volcano, San Cristóbal Volcano, and Casita Volcano), rare in lowlands. Found in a variety of forest habitats. Perches from understory to mid-canopy. Sallies to ground to catch large insects, spiders, and lizards; may accompany army ant swarms. Most often heard delivering a soft, double-noted *hoop-hoop*. Birds sometimes respond with a faster, higher-pitched *húp-húp*. Both calls are reminiscent of vocalizations made by owls or ground-doves.

Turquoise-browed Motmot   *Eumomota superciliosa*
(Guardabarranco Común)

13.5 in (34 cm). **Rufous patch** behind eye and on back is distinctive, and distinguishes it from larger Lesson's Motmot. Bare shafts on rackets are longer than on other motmots. Abundant in Pacific; uncommon to common in Northern Highlands; uncommon from east of Lake Nicaragua to eastern slopes of Sierra Chontaleña; to 5,000 ft (1,500 m). Prefers dry forest and edge, gallery forest, woodland, and urban areas. Perches at eye level. Moves tail in pendulum fashion; sallies to capture insects. Commonly heard calling with a deep, raspy *quawh*; spacing between notes varies and call is sometimes delivered as *quawh quawh*. National bird of Nicaragua.

Tody
Motmot

Lesson's
Motmot

Turquoise-browed
Motmot

A Turquoise-browed Motmot
leaving its bank-side nest burrow.

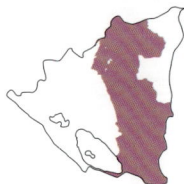

Rufous Motmot (Guardabarranco Canelo Mayor) *Baryphthengus martii*
20.5 in (52 cm). Similar to sympatric Broad-billed Motmot but much larger and **rufous extends to lower belly**. Note lack of greenish chin. Uncommon in Caribbean, but locally common in Saslaya NP and Indio Maíz BR; to 3,300 ft (1,000 m). Inhabits humid lowland forest; perches in canopy but descends to understory to follow army ant swarms. Calls with resonant, low-pitched *hududu* and a high-pitched *whoop-whoop*, often caroled rapidly together; normally heard at dawn.

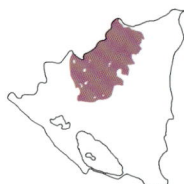

Keel-billed Motmot (Guardabarranco Picoancho) *Electron carinatum*
12.5 in (32 cm). **Mostly green** (but underparts vary to rufous), with **rufous forehead and short blue superciliary**. Similar in size, form, and voice to Broad-billed Motmot, but coloration is notably distinct; distinguished from larger Lesson's Motmot by black iris and less extensive black mask. Locally uncommon on eastern slopes of Northern Highlands and in Caribbean; from 700 to 4,000 ft (200 to 1,200 m). Inhabits shady interior of humid lowland forests. Forages on insects at mid-canopy and canopy; easily overlooked. Croaks a raspy *awh, awh, awh…*; approximately 3 seconds between each call. Call is slightly lower-pitched and hoarser than very similar call of Broad-billed Motmot. **VU**

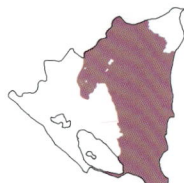

Broad-billed Motmot *Electron platyrhynchum*
(Guardabarranco Canelo Menor)
12.5 in (32 cm). Much smaller than similar Rufous Motmot, but **rufous extends only to mid-breast**. Note **greenish chin**. Keel-billed Motmot is similar in size, form, and voice, but differs notably in coloration. Uncommon to locally common in Caribbean; generally to 3,000 ft (900 m) but occasionally to 3,600 ft (1,100 m). Locally very rare on eastern slopes of Sierra Isabelia and Sierra Dariense of Northern Highlands, to 3,900 ft (1,200 m). Prefers humid lowland forest and tall second growth. Perches and calls from mid-canopy and canopy, descending to understory to feed. Croaks a raspy *awh, awh, awh…* (time between notes is approximately 3 seconds); only slightly higher-pitched than very similar call of Keel-billed Motmot. Also gives a dry, soft contact call: *ah*.

Rufous
Motmot

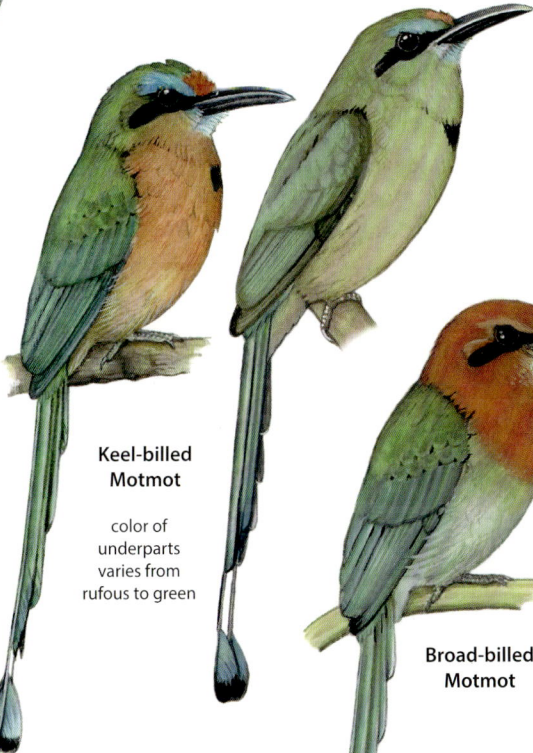

Keel-billed
Motmot

color of
underparts
varies from
rufous to green

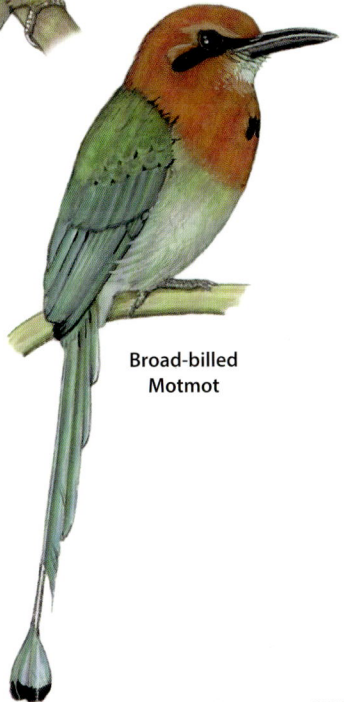

Broad-billed
Motmot

A cosmopolitan family with about 90 species worldwide. Nicaragua is home to all six of the species that occur in the Americas. Kingfishers have large heads, short necks and feet, and stout, long, pointed bills for catching fish. Upperparts are blue-gray or metallic green; underparts show varying amounts of white and rufous. If water is present, there is possibly a kingfisher nearby, perched quietly and searching for prey. Look for burrow-nests in cliff sides, near water.

Ringed Kingfisher (Martín Pescador Collarejo)   *Megaceryle torquata*
16 in (41 cm). Largest Nicaraguan kingfisher. **Extensive rufous underparts** distinguish it from smaller Belted Kingfisher (also with Blue-gray upperparts). Female has **rufous underwing coverts** visible in flight. Common in lowlands, uncommon in foothills, and rare on western slopes of Northern Highlands, including Lake Apanás; to 3,300 ft (1,000 m). Conspicuously perches on prominent lookouts near rivers, streams, lakeshores, ponds, estuaries, mangroves, and coastlines, while looking for prey. Delivers a rapid, rattling set of staccato notes (*ki-ki-ki-ki-ki-ki*), often in flight. Call is louder than that of Belted Kingfisher.

Belted Kingfisher (Martín Pescador Norteño)   *Megaceryle alcyon*
12.5 in (32 cm). **Mostly white underparts** distinguish it from larger Ringed Kingfisher (also with blue-gray upperparts). Winter resident (mid-Aug to April); common in lowlands and foothills; to 1,300 ft (400 m). Locally uncommon on western slopes of Northern Highlands and Lake Apanás and its tributaries; to 4,000 ft (1,200 m). Perches high in trees and other lookouts near rivers, lakeshores, estuaries, and coastlines. Hovers before catching fish, crustaceans, small mammals, and reptiles. Delivers a rapid, rattling set of staccato notes (*chi-chi-chi-chi-chi-chi*), often in flight. Call is quieter than that of Ringed Kingfisher and more electronic sounding.

Green Kingfisher (Martín Pescador Verde)   *Chloroceryle americana*
8 in (20 cm). **White spots and bars on wings** and white on base of outer rectrices (seen in flight) distinguish it from Amazon Kingfisher (also with green upperparts). Common countrywide; to 5,000 ft (1,500 m). Prefers streams with forest edge, lakeshores, and ponds. Perches on low branches over water or on emergent rocks, from which it plunges into water to capture fish. Delivers a soft and unassuming staccato *tek-tek-tek…* or *tek-tek tek-tek tek-tek…*, with much variation in speed of delivery.

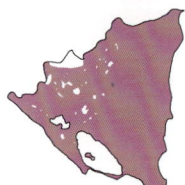

Amazon Kingfisher (Martín Pescador Pechicanelo)   *Chloroceryle amazona*
11 in (28 cm). Larger size and **lack of white spots on wings** distinguish it from Green Kingfisher (also with green upperparts). Abundant in Caribbean lowlands (uncommon in foothills); 1,600 ft (500 m). Common in Pacific, rare to uncommon at Lake Apanás and other aquatic habitats in Northern Highlands; to 3,300 ft (1,000 m). Favors rivers, lakeshores, mangroves, and estuaries. Perches on branches and wires over water, from which it plunges into water to capture fish. Delivers a dry, scratchy *keh, keh, keh*, with noticeable separation between notes. Sings a long series of rapid, sweet notes: *chi-chi-chi…* (higher-pitched) and *chur-chur-chur…* (lower-pitched).

Ringed
Kingfisher

female

male

Belted
Kingfisher

female

male

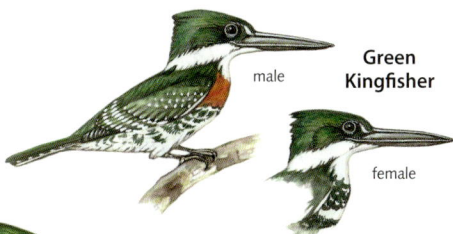

Green
Kingfisher

male

female

Amazon
Kingfisher

male

female

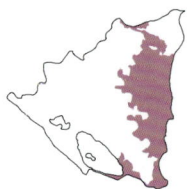

Green-and-rufous Kingfisher (Martín Pescador Bicolor)   *Chloroceryle inda*
9.5 in (24 cm). Combination of green upperparts and **rich rufous underparts** is diagnostic. Notably larger than American Pygmy Kingfisher. Rare in Caribbean lowlands; locally uncommon in Guatuzo Plains, shady tributaries of San Juan River, Indio Maíz BR, and swamps surrounding Bluefields Bay. Perches in densely vegetated swamps, along slow-moving streams, and at ponds; plunges for small fish and aquatic insects. Calls with staccato notes; makes a *chit-chit* or a *chit-chit-chu* (third note is lower-pitched). Also emits a buzzy downslurred *cheeu*. Song consists of a jumble of piercing, buzzy notes.

American Pygmy Kingfisher (Martín Pescador Pigmeo)   *Chloroceryle aenea*
5.5 in (14 cm). Green upperparts. **Very small size** makes it unmistakable, **white vent and belly** further distinguish it from Green-and-rufous Kingfisher. Uncommon to locally common in lowlands. Prefers streams with forest edge, ponds, and mangroves, where it inconspicuously perches on low twigs over water before plunging for small fish. Easily overlooked because of its shy behavior. Calls with a very short and crisp *twik*. Sings a fast *ch-ch-ch-cheer* that is downslurred at the end.

JACAMARS Galbulidae

A Neotropical family of metallic green and rufous birds with long bills and long tails. Strictly arboreal, jacamars perch motionless with their bills at an upward angle. Adept sit-and-wait insectivores, they capture butterflies and dragonflies by piercing their wings with long pointed bills.

Rufous-tailed Jacamar (Jacamar Colirrufo)   *Galbula ruficauda*
9 in (23 cm). **Green breast** and **rufous undertail** distinguish it from larger Great Jacamar; also note very long, **slender, straight bill**. Common in Caribbean lowlands and foothills; to 2,600 ft (800 m) in Saslaya NP and 3,300 ft (1,000 m) in Musún NR. Rare to uncommon on eastern slopes of the Northern Highlands; to 4,300 ft (1,300 m). Prefers humid lowland forest and edge, gallery forest, and tall second growth. Alone or in pairs, perches in mid-canopy with bill pointed upward, moving head side-to-side to survey for insects. Yelps a sprite, rising, quick *wheeú*. In song, quickly repeats this note, with increasing speed and ever higher pitch; ends with a descending trill or with notes simply tapering off.

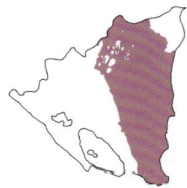

Great Jacamar (Jacamar Mayor)   *Jacamerops aureus*
12 in (30 cm). Rufous breast and **blue-black undertail** distinguish it from smaller Rufous-tailed Jacamar; also note very long, **stout bill** (with decurved upper mandible). Rare in Caribbean lowlands and foothills (locally common in Saslaya NP, Bosawas Biosphere Reserve, and Indio Maíz BR); to 2,000 ft (600 m). Limited to humid lowland forest, where it is easily overlooked as it perches motionlessly in canopy; often in pairs. Song is distinctive; whistles a downslurred, drawn-out note with piercing quality (up to 3 seconds in length), sometimes with a noticeable hitch in the tone.

female

male

**Green-and-rufous
Kingfisher**

female

male

**American Pygmy
Kingfisher**

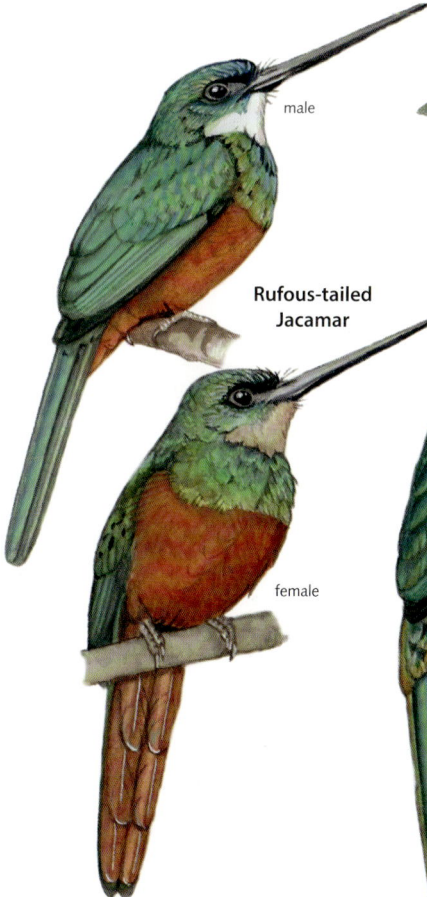

male

**Rufous-tailed
Jacamar**

female

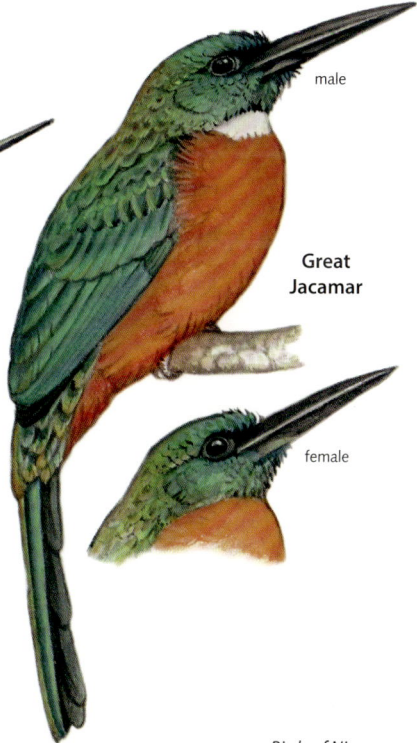

male

**Great
Jacamar**

female

A Neotropical family associated with lowland and foothill forests. The name *puffbird* derives from a puffy look caused by dense feathering. Their sturdy, compact appearance is enhanced by large heads, stout bills (slightly decurved), and short, graduated tails. These birds perch silently and motionless for long periods of time, waiting for passing large insects (sometimes lizards and small frogs).

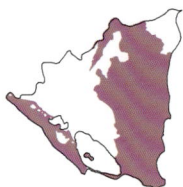

White-necked Puffbird (Buco Collarejo) — *Notharchus hyperrhynchus*

10 in (25 cm). Notably larger than Pied Puffbird; **white forehead and collar**, **red iris**, and **dark undertail** further distinguish it. Locally common throughout lowlands and foothills; to 1,500 ft (450 m). Inhabits dry forest and humid lowland forest, forest edge, and tall second growth. Perches from mid-canopy to canopy, where it remains motionless (only moving head) before sallying to catch insects; prepares prey for consumption by beating it against branches. Produces a constant and shrill *wibibibibi…* (lasting 3–6 seconds); reminiscent of high-pitched electronic beeping. Calls with a thin, wheezy *wheeú*.

Pied Puffbird (Buco Enano) — *Notharchus tectus*

6 in (15 cm). Notably smaller than White-necked Puffbird; **thin white superciliary**, dark iris, **white scapular patch**, and **white tips on undertail** further distinguish it. Locally uncommon from Guatuzos WR to San Juan River and its tributaries. Reaches northernmost distribution in Nicaragua. Prefers humid lowland forest, forest edge, tall second growth, and semi-open areas with scattered trees. Perches motionless in mid-canopy and canopy, where it can be easily overlooked. Actively moves in pairs or groups during breeding season. Pipes a squeaky *witdididit-deee di-dit di-dit di-dit…*, and calls with a thin, downslurred *peeút*.

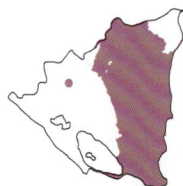

White-whiskered Puffbird (Buco Bigotudo) — *Malacoptila panamensis*

7.5 in (19 cm). Has a unique **puffy appearance** and a **large, chunky head**. Both male (cinnamon) and female (brown) have dark streaks on breast. Common in Caribbean lowlands and foothills; to 2,600 ft (800 m). Historical record from San Rafael del Norte, Jinotega (1905), suggests it once was found (at least accidentally) to 3,300 ft (1,000 m) on eastern slopes of Northern Highlands. Perches motionlessly in humid lowland forest, adjacent forest patches, and secondary forest. Accompanies mixed-species flocks and follows army ants swarms. Gives a reedy, descending whistle: *weeeo* (1 second in length).

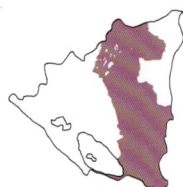

White-fronted Nunbird (Buco Cariblanco) — *Monasa morphoeus*

11 in (28 cm). Unmistakable. Note **white forecrown and chin**, **red-orange bill**, and **slate-gray body**. Uncommon to locally common in Caribbean lowlands and foothills; to 2,300 ft (700 m). Prefers humid lowland forest, adjacent forest remnants, and tall second growth. Alone or in small flocks, perches at mid-canopy and canopy. Delivers a fast, almost querulous *chuá-chuá-chu chuachuachu*, with last phrase given at rapid-fire pace; breaks up into an agitated chatter when groups chorus together.

White-necked
Puffbird

Pied
Puffbird

female

male

White-whiskered
Puffbird

White-fronted
Nunbird

A group of White-fronted
Nunbirds singing in unison.

These quintessential Neotropical birds are known for the long, hollow bills that they use for picking hard-to-reach fruit. The color and pattern of the vibrant bills are important keys for distinguishing among members of the family. Toucans primarily feed on fruit, but complement their diets with insects, bird eggs, lizards, and other small vertebrates (sometimes predating nest hatchlings of other species). They move within the canopy in pairs or flocks. Toucans often incessantly vocalize from canopy perches, descending to mid-canopy from time to time.

Northern Emerald-Toucanet   *Aulacorhynchus prasinus*
(Tucancito Verde Norteño)

14 in (36 cm). Mostly **emerald-green** body, **white throat**, and **rufous undertail** make this toucan unmistakable. Common in Northern Highlands, Quirragua NR, Musún NR, and Saslaya NP; above 3,300 ft (1,000 m). Reaches southernmost distribution in Nicaragua. Occurs in cloud forest and edge, where it forages at mid-canopy; frequently moves in small flocks in search of fruiting trees or nests to predate. Barks at length a hoarse *ekh! ekh! ekh!…* (3 calls per second); also communicates with soft, gravelly grunts.

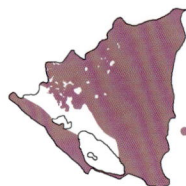

Collared Aracari (Tucancito Collarejo)   *Pteroglossus torquatus*

16 in (41 cm). Only Nicaraguan toucan with **extensive yellow underparts divided by a black band**; note **black spot on breast**. Locally uncommon in Pacific lowlands and foothills and common in Northern Highlands; to 4,400 ft (1,350 m). Common in Caribbean lowlands and foothills; to 2,600 ft (800 m). Moves in pairs or small flocks within a variety of forest habitats, forest edge, second growth, and coffee plantations. Emits a sharp, very high-pitched *tsu-WEET*, repeated at variable intervals.

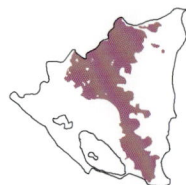

Yellow-eared Toucanet (Tucancito Pechinegro)   *Selenidera spectabilis*

15 in (38 cm). Only Nicaraguan toucan that has **black from throat to belly**. Male has **flaring yellow ear patch**; female has **chestnut crown and nape**. Bicolored bill is similar to that of Yellow-throated Toucan, but the two are distinguished by size and body color. Locally common in Saslaya NP and throughout Caribbean; from 1,000 to 3,900 ft (300 to 1,200 m), occasionally descends lower. Very rare on eastern slopes of Northern Highlands. Forages in humid lowland forest canopy, but will drop to mid-canopy. Repeatedly croaks a gruff, rhythmic *ba-dzát* as it moves its head up and down with each call; delivery is lower-pitched and slower (1 call per second instead of 2) than that of Keel-billed Toucan.

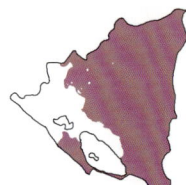

Keel-billed Toucan (Tucán Pico Iris)   *Ramphastos sulfuratus*

20 in (50 cm). **Multi-colored bill** (green, orange, and turquoise with wine-colored tip) earns this bird its Spanish name, translated as Rainbow-billed Toucan, and sets it apart from the larger Yellow-throated Toucan. Common in Caribbean and on eastern slopes of Northern Highlands; uncommon in northern region of Sierra Chontaleña; rare in Pacific from Sierras Managua and south; to 4,900 ft (1,500 m). Prefers humid lowland forest, cloud forest, secondary forest, and forest edge. Found in pairs or small flocks in the canopy as it feeds on fruiting trees. Continuously croaks a grating *dreezt!*; the repeated notes vary in pitch or sometimes progressively rise in pitch; higher-pitched and faster delivery (2 calls per second instead of 1) than that of Yellow-eared Toucanet.

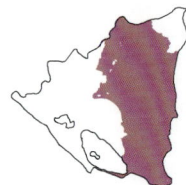

Yellow-throated Toucan   *Ramphastos ambiguus*
(Tucán Pechiamarillo Norteño)

22 in (56 cm). Largest toucan in Nicaragua. The **bicolored bill** (yellow and chestnut) distinguishes it from smaller Keel-billed Toucan. Bill is similar to that of smaller Yellow-eared Toucanet, but the two are distinguished by size and body color. Common in Caribbean; uncommon in Musún NR (and possibly in nearby Quirragua NR); accidental on eastern slopes of Northern Highlands; to 3,300 ft (1,000 m). Favors primary forest, but also found in secondary forest and forest edge. Forages at mid-canopy in pairs or small flocks. Whistles a far-carrying and iconic *wee-óh oh oh* phrase, often repeated. **NT**

Northern
Emerald-Toucanet

female

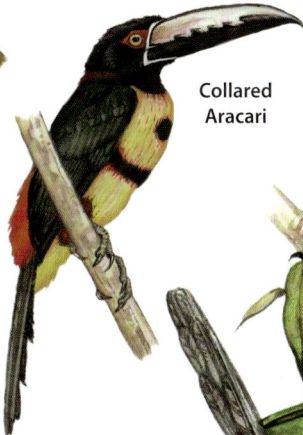

Collared
Aracari

Yellow-eared
Toucanet

male

A family group of
Collared Aracaris moving
through cecropia trees.

Keel-billed
Toucan

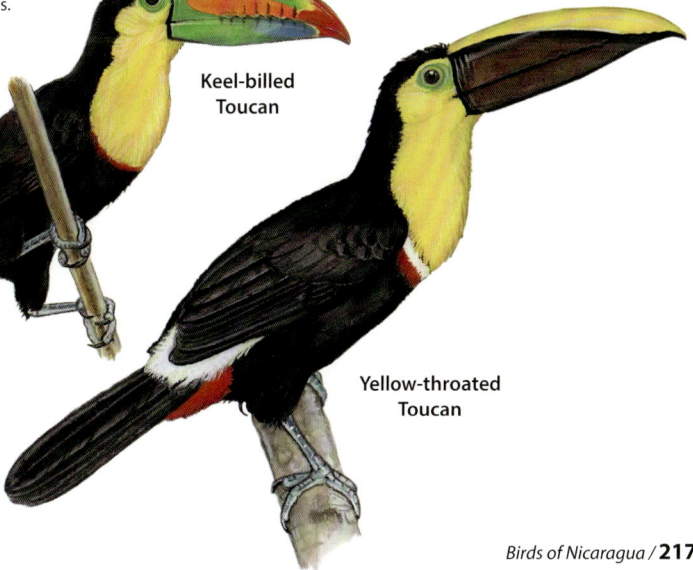

Yellow-throated
Toucan

A family that is found almost worldwide; woodpeckers are notably absent from Australasia. Of the 16 species in Nicaragua, only 1 is migratory; for 4 of the species, Nicaragua lies at one terminus of their distribution. Strong, pointed bills and thick, spongy skulls (to absorb the shock of pecking wood) allow them to drill into tree trunks, and with their long, barbed tongues they are able to artfully extract exposed larvae. Strong legs and feet and sharp claws allow them to climb vertically up tree trunks, a habit that is also aided by stiff central rectrices. Primarily insectivorous, woodpeckers complement their diet with fruit and even sap. A clue to their presence is the sound of drumming on wood, and experienced birders can even identify a given species by the cadence and strength of its drumming.

Acorn Woodpecker (Carpintero Careto)   *Melanerpes formicivorus*
8.5 in (22 cm). Only Nicaraguan woodpecker with **solid black upperparts** and **streaked underparts** (not barred). Facial pattern suggests that of a clown. In flight, note **3 white patches** (2 on the wings and 1 on the rump). Locally common in Northern Highlands; above 2,000 ft (600 m). Uncommon in the Mosquitia and as far south as El Rama (3 historical records, 1953 and 1954). Generally associated with oaks (*Quercus*), but also seen in highland pine and pine-oak forest and adjacent cloud forest, and lowland pine savanna. Very noisy and social; often seen in groups. Feeds on—and collects—nuts, which are stored in cavities drilled into tree snags at colony locations; also eats fruit and preys on insects caught in flight. Often heard vocalizing with a nasal, churring cackle (*á-ká-ká-ka*); also throws in short churrs. Colonies produce a cacophony of noises. Drum pattern primarily consists of many, rapid-fire taps, evenly spaced and 1–3 seconds in total length.

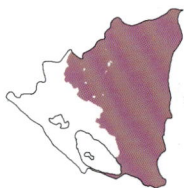

Black-cheeked Woodpecker (Carpintero Carinegro)   *Melanerpes pucherani*
7.5 in (19 cm). Combination of **elongated, postocular white spot** and **black mask extending to nape** distinguishes it from other *Melanerpes* woodpeckers in Nicaragua. Common in Caribbean; to 3,300 ft (1,000 m). Rare on eastern slopes of Northern Highlands; to 4,300 ft (1,300 m). Favors humid lowland forest and edge, secondary forest, second growth, and open areas with scattered trees. In canopy, forages alone or in pairs for insects, fruit, and nectar. Repeats a tremulous and screeching *cheeá*. Drum pattern primarily consists of many even rapid-fire taps (1–3 seconds in total length), slightly faster than Hoffmann's Woodpecker.

Hoffmann's Woodpecker (Carpintero Nuquigualdo)   *Melanerpes hoffmannii*
7.5 in (19 cm). Easy to mistake for larger Golden-fronted Woodpecker, but distinguished by **golden-yellow nape**; also, **upperparts are more coarsely barred**. Abundant in Pacific; uncommon in dry intermontane valleys of Northern Highlands; due to deforestation (it prefers open areas), becoming common in some places in the western Caribbean; to 4,300 ft (1,300 m). Prefers dry forest, scrub, open areas with scattered trees, and urban parks. Usually present in pairs; forages from mid-canopy to canopy for insects, larvae, and fruit. Repeats a harsh, scratchy *cheeeA*, and delivers a long, rolling churr, also with a harsh, scratchy quality. Drum pattern mainly consists of many rapid-fire taps, evenly spaced and lasting 1–3 seconds in total length; pace of drumming is slightly slower than that of Black-cheeked Woodpecker.

Golden-fronted Woodpecker (Carpintero Frentigualdo) *Melanerpes aurifrons*
9.5 in (24 cm). Easy to mistake for smaller Hoffmann's Woodpecker, but note **red nape** and more distinctly barred **upperparts**. Uncommon in Northern Highlands (rare on eastern slopes). Recently documented in Saslaya NP (March, 2016); above 1,800 ft (550 m). Reaches southernmost distribution in Nicaragua. Favors secondary forest, second growth, open areas with scattered trees, forest edge, and scrub. Forages from mid-canopy to canopy for nuts, fruit, and insects. Repeats a lively *che-a-á*, as well as a gruff, scratchy, low-pitched *ché-a*. Drum pattern mainly consists of many rapid-fire taps, evenly spaced and lasting 1–3 seconds in total length.

female

male

**Acorn
Woodpecker**

male

**Black-cheeked
Woodpecker**

female

male

**Hoffmann's
Woodpecker**

male

**Golden-fronted
Woodpecker**

female

female

Yellow-bellied Sapsucker (Carpintero Chupador) *Sphyrapicus varius*
8.5 in (22 cm). Conspicuous **large, white patch** along folded wing is diagnostic in all plumages. Winter resident (Oct to April). Uncommon in Northern Highlands and Caribbean; rare in Pacific foothills. Visits a variety of forest habitat and edge, but prefers highland pine and pine-oak forest and lowland pine savanna. Drills horizontal holes into tree trunks to harvest sap and the insects attracted to it; also catches insects in flight. Calls with a whiny *meyáh*. Drums with a rapid, continuous, unique double-rap pattern.

Ladder-backed Woodpecker (Carpintero Cremoso) *Picoides scalaris*
7 in (18 cm). Distinctive facial pattern is created by **black eye line and malar stripe that join on sides of neck**, leaving a white ear patch; also note barred back and black spots on underparts. Two disjunct populations occur in Nicaragua, both with strict habitat associations. Uncommon and local in mangroves of northwest Pacific coast (including Estero Real NR and Isla Juan Venado WR); the other population is uncommon in lowland pine savanna in the Mosquitia; to 330 ft (100 m). Reaches southernmost distribution in Nicaragua. Drills into tree trunks to find and extract insects and larvae; also eats fruit. Call is a short *cheat!*, usually given once but sometimes delivered in a short, rapid series. Drumming is intermittent and sporadic, without a definitive pattern.

Smoky-brown Woodpecker (Carpintero Atabacado) *Picoides fumigatus*
6.5 in (16 cm). Combination of **uniform brown plumage** (except for red crown on male) and **pale face** is distinctive. Locally common on eastern slopes of Northern Highlands; to 5,000 ft (1,500 m). Locally common in Caribbean; to 3,300 ft (1,000 m). Accidental on Sierras Managua, with only 1 record (Chocoyero-El Brujo WR, Dec 2012). Found in cloud forest, humid lowland forest, secondary forest, forest edge, and in open areas with scattered trees. Forages on small branches and vines in canopy within forest, but descends to lower levels at forest edge; accompanies mixed-species flocks. Calls include a sharp, squeaky *weep!* and a short *tsenk*; also delivers a nasal, pulsating *rírírírírírírírí*. Exhibits rapid-fire drumming, evenly spaced and lasting 1 second.

Hairy Woodpecker (Carpintero Serranero) *Picoides villosus*
7 in (18 cm). **Broad white central patch on unbarred black back** is diagnostic; also note **black mask, white superciliary and malar stripe**, and tawny underparts. Uncommon in Northern Highlands; above 4,000 ft (1,200 m). Favors highland pine and pine-oak forest, cloud forest, and forest edge. Forages in canopy for insects, fruit, and nuts; hikes up tree trunks while maintaining an erect posture. Calls with a loud and sharp *wit!*; also delivers a rapid rattle. Exhibits rapid-fire drumming, evenly spaced and lasting 1 second.

Yellow-bellied Sapsucker

male

female

male

Ladder-backed Woodpecker

female

male

Smoky-brown Woodpecker

female

male

Hairy Woodpecker

female

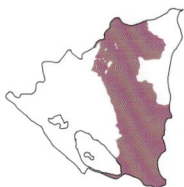

Rufous-winged Woodpecker (Carpintero Alirrufo) *Piculus simplex*
7 in (18 cm). Plumage color and pattern is similar to that of Golden-olive Woodpecker, but note **pale whitish-blue iris, olive-green face,** and **rufous in primaries** (inconspicuous). Uncommon in Caribbean lowlands and foothills; to 2,600 ft (800 m); also wanders onto eastern slopes of Northern Highlands, where it is accidental. Found in humid lowland forests. Forages in canopy for insects and larvae; accompanies mixed-species flocks. Incessantly repeats an emphatic, downslurred *eeáw* (delivered once or twice per second); jaylike in quality and intensity. Exhibits rapid-fire drumming, evenly spaced and lasting 1 second.

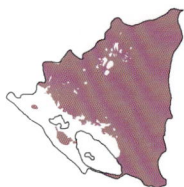

Golden-olive Woodpecker (Carpintero Alidorado) *Colaptes rubiginosus*
8.5 in (22 cm). Plumage color and pattern is similar to that of Rufous-winged Woodpecker, but note **dark iris** contrasting with **buffy-white face** and lack of rufous in primaries. Common in Northern Highlands; to 5,000 ft (1,500 m). Common in Caribbean lowlands and foothills (more likely in foothills); to 2,000 ft (600 m) but higher in Musún NR. Uncommon in Pacific foothills (San Cristóbal-Casita NR, Sierras Managua, and Mombacho Volcano); above 1,300 ft (400 m). Found in cloud forest, highland pine and pine-oak forest, humid lowland forest, secondary forest, and second growth. Usually alone; forages in canopy for insects, larvae, and fruit. Call is a loud, sharp *fíuw*; also makes a shrill and rapid trill. Drums with rapid-fire pace, evenly spaced and lasting 1–2 seconds.

Northern Flicker (Carpintero Norteño) *Colaptes auratus*
11 in (28 cm). Unmistakable. **Cinnamon crown and nape** and **black semicircular breast patch** are distinctive. Locally common in Northern Highlands; above 3,300 ft (1,000 m). Reaches southernmost distribution in Nicaragua. A highland pine and pine-oak forest specialist, although it occasionally visits cloud forest. Forages on ground, extracting ants, beetles, and larvae from the soil; also eats fruit and nuts. When alarmed, perches erect on horizontal branch. Sings a high-pitched, rapid *wí-wí-wí…* (about 3 seconds in length); and also gives a raptorlike *peeur* call. Drumming pattern consists of rapid-fire, evenly spaced raps that sound soft in comparison with the drumming of other woodpeckers.

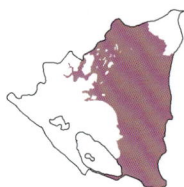

Olivaceous Piculet (Carpinterito Oliváceo) *Picumnus olivaceus*
3.5 in (9 cm). **Tiny** compared to other woodpeckers; unmistakable. Note **lemon-yellow central rectrices** on black tail and **white speckling on dark head**. Very rare on eastern slopes of Northern Highlands and in area southwest of Lake Nicaragua (as far west as Cardenas, Rivas Dept.), locally uncommon in Guatuzo Plains and at the San Juan River headwaters; otherwise rare in Caribbean lowlands and foothills; to 2,300 ft (700 m). Prefers woodland, humid lowland forest edge, old plantations, and tall second growth. Alone or in pairs, forages actively in mid-canopy and canopy; often seen hanging upside down from small branches, twigs, and vines (without using tail for support). Accompanies mixed-species flocks. Sings with a very high-pitched, thin, descending, trilled whistle (1.5 seconds in length); call is a brief *tsit*.

Rufous-winged Woodpecker

male

female

Golden-olive Woodpecker

female

male

Northern Flicker

male

female

Olivaceous Piculet

female

male

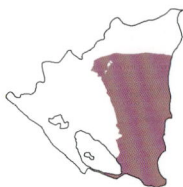

Cinnamon Woodpecker (Carpintero Canelo) *Celeus loricatus*
8 in (21 cm). Similar to larger Chestnut-colored Woodpecker, but black scaling on **whitish underparts** and **only 1 color on head and upperparts** distinguish it; **crest is shorter**. Locally common in Saslaya NP; uncommon in Guatuzos WR and Indio Maiz BR, otherwise rare in Caribbean; to 3,300 ft (1,000 m). Reaches northernmost distribution in Nicaragua. Favors humid lowland forest. Forages alone in the canopy for termites and ants. The easiest way to detect this bird is to listen for its distinctive call. Whistles a descending, high-pitched song (*PEE-PEE-PEE-pit-pu*), with the first 3 notes loud and evenly spaced and the final 2 abrupt and less emphatic. Exhibits a rapid-fire drumming pattern, evenly spaced and lasting less than 1 second.

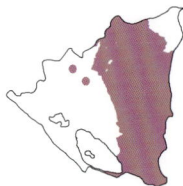

Chestnut-colored Woodpecker (Carpintero Castaño) *Celeus castaneus*
9 in (23 cm). Similar to smaller Cinnamon Woodpecker, but distinguished by black scaling on **dark chestnut-brown underparts** and the contrast between **pale buffy head** and **chestnut-brown upperparts**; also note that **crest is shaggy and longer**. Uncommon in Caribbean lowlands and foothills (rare in Guatuzo Plains); to 2,600 ft (800 m). Scattered records suggest it is accidental on eastern slopes of Northern Highlands (Quilalí, Jan 1909; Peñas Blancas NR, June and May 1909 and March 2013); to 4,000 ft (1,200 m). Prefers humid lowland forest, secondary forest, and forest edge. Forages alone or in pairs from mid-canopy to canopy for termites and ants, and also fruit; accompanies mixed-species flocks. Song is a distinctive *peó-wi-wi*: the first note slides down in pitch and is followed by 2 nasal, staccato barks; the same notes may also be given individually. Drumming is given in a rapid-fire, evenly spaced, pattern.

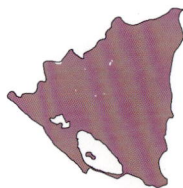

Lineated Woodpecker (Carpintero Crestirrojo) *Dryocopus lineatus*
13.5 in (34 cm). At first glance, appears similar to Pale-billed Woodpecker, but **white stripe across black face** readily distinguishes it; **white stripes on scapulars do not connect on back**. Common in Northern Highlands and Caribbean; uncommon in Pacific; to 5,600 ft (1,700 m). Prefers forest and second growth; frequently found feeding on ants in *Cecropia* trees. Forages in pairs from mid-canopy to canopy. Flight pattern is very strong and undulating. Repeats an emphatic *WIC-WIC-WIC…* (2–4 seconds in length). Deep sounding, rapid-fire drumming is evenly spaced; it trails off in intensity toward end.

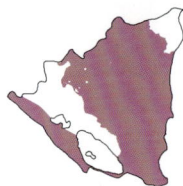

Pale-billed Woodpecker (Carpintero Picoplata) *Campephilus guatemalensis*
14 in (36 cm). Initially appears similar to Lineated Woodpecker, but **entirely red head** readily distinguishes it; **white stripes on scapulars almost connect on back** (forming a V). Common in Caribbean; to 4,400 ft (1,350 m). Common on eastern slopes of Northern Highlands and uncommon in Pacific; to 5,000 ft (1,500 m). Found in most forest habitats and associated edge, secondary forest, and second growth, and, to a lesser extent, in highland pine and pine-oak forest. Pairs forage in canopy, searching for beetles, ants, larvae, and fruit. Gives a somewhat dry and nasal *whét whét wheatá* that sometimes dissolves into a jumbled chatter. Distinctive, deep, double-rap drumming pattern carries far distances.

female

male

Cinnamon Woodpecker

male

Chestnut-colored Woodpecker

female

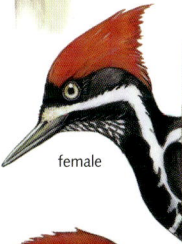

female

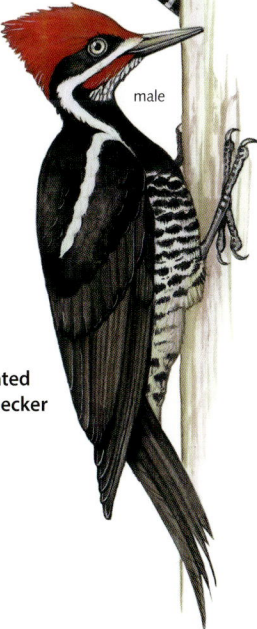

female

male

Lineated Woodpecker

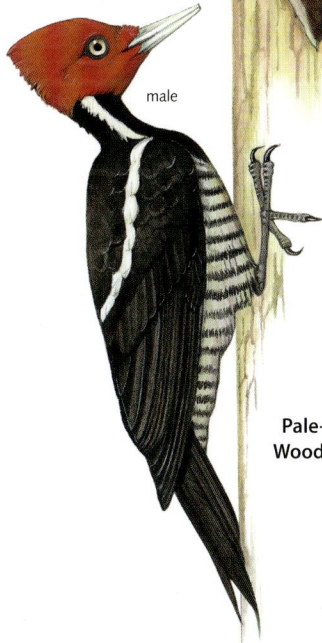

male

female

Pale-billed Woodpecker

A diverse family of diurnal raptors. Nicaragua is home to 3 groups found only in the Neotropics—forest falcons, caracaras, and the monotypic Laughing Falcon—along with typical falcons of the cosmopolitan genus *Falco*. In general, forest-falcons are ambush specialists of the forest understory; caracaras scavenge for carrion and easy prey in open areas; the Laughing Falcon is a snake specialist; and members of the genus *Falco* dominate the skies, where they maneuver with deadly speed. Forest-falcons and members of the genus *Falco* exhibit reverse sexual dimorphism, with females having larger bodies than males. Take note that some forest-falcons might be confused with *Accipiter* hawks because of similar morphology and shared habitats; additionally, in silhouette members of the genus *Falco* can resemble a jaeger, gull, pigeon, or even—at a distance—a swallow.

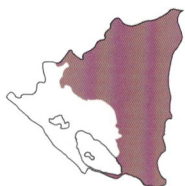

Barred Forest-Falcon (Halcón Barreteado)   *Micrastur ruficollis*

13 in (32 cm). On adult, note **fine dark barring on underparts** and gray upperparts (female has brown mantle). **Yellow orbital skin** and **long, graduated tail** help distinguish it from smaller Tiny Hawk (p. 176). Immature often has pale nuchal collar (on brown upperparts) and sporadic brown barring on underparts. Uncommon on eastern slopes of Northern Highlands and in Caribbean foothills and highlands, rare in Caribbean lowlands. Prefers humid lowland forest and cloud forest. Ambushes prey within understory and mid-canopy. Active at dawn and dusk—and furtive—the Barred Forest-Falcon is heard more often than it is seen. Continually barks a nasal *ehr!* that is repeated at intervals of 1–2 seconds.

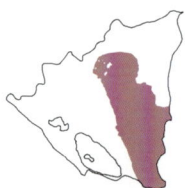

Slaty-backed Forest-Falcon (Halcón Dorsigrís)   *Micrastur mirandollei*

16 in (41 cm). **Slaty-gray upperparts** and **white underparts** could suggest a Semiplumbeous Hawk (p. 180), but note **yellow cere and orbital skin** and narrow white bands on long, black tail. Rare to locally uncommon in Caribbean lowlands and foothills, with documented populations in Saslaya NP, Río San Juan WR, Indio Maíz BR; to 2,600 ft (800 m). Reaches northernmost distribution in Nicaragua. Secretive within humid lowland forest and adjacent second growth; active at dawn and dusk, when it is heard more often than seen. Sings a series of hollow, nasal notes, starting with quickly delivered (2–3 calls per second) *oaw! oaw! oaw!…*, which transitions into a softer and slower *uráw uráw uráw…* .

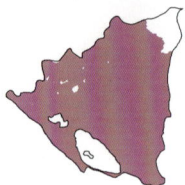

Collared Forest-Falcon (Halcón Collarejo)   *Micrastur semitorquatus*

21 in (53 cm). Note striking contrast between **pale (white or buffy) cheeks and nuchal collar** and **black upperparts**. Immature has brown upperparts and heavy mottling and barring on pale underparts. Rare dark morph is black overall, with white barring on belly, flanks, and thighs and narrow white bands on tail. Common in Pacific and Northern Highlands (perhaps absent in Sierra Dipilto-Jalapa), and rare in Caribbean; to 4,800 ft (1,450 m). Ubiquitous in a variety of forest and scrub habitats, forest edge, and secondary growth. Hunts in understory and on ground, by ambushing prey or actively pursuing it. Proclaims a nasal, mournful, *OAWH*, often repeated, but with several seconds between repetitions; each note is drawn-out and subtly rises and falls in pitch.

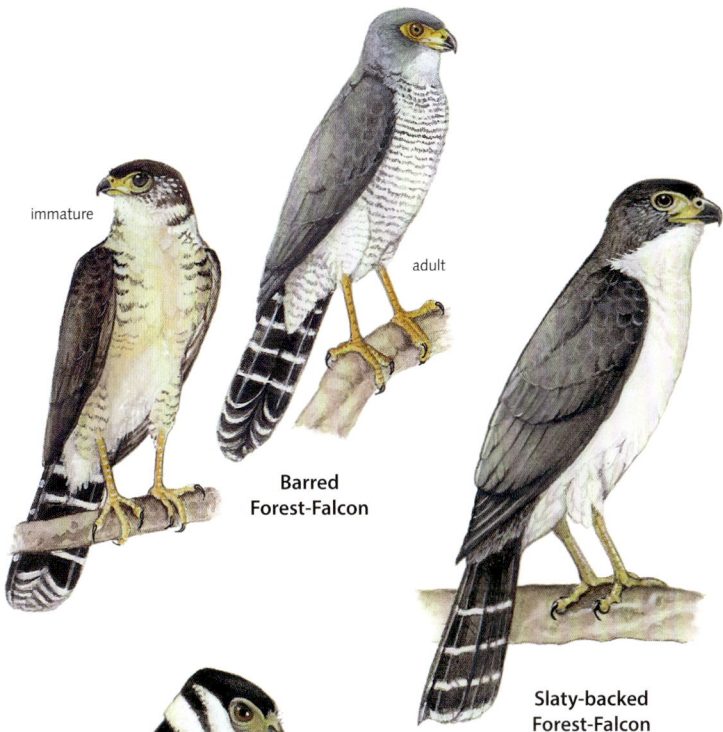

immature

adult

**Barred
Forest-Falcon**

**Slaty-backed
Forest-Falcon**

adult
pale morph

**Collared
Forest-Falcon**

immature

adult
dark morph

Laughing Falcon (Guaco)   *Herpetotheres cachinnans*
20.5 in (52 cm). **Broad black mask** wraps around white head; body has a stocky appearance. In flight, note short, broad wings; also note long tail with black and white bands. Common countrywide; to 4,600 ft (1,400 m). Occurs in areas with fragmented forest, often at forest edge perched on conspicuous snags or dead branches from which it scans the ground for snakes. The Laughing Falcon is heard more than it is seen. Vocalization begins with a subtle, rising series of nasal *oaw* notes, then escalates with an incessant repetition of *OAW! OAW! OAW!…*, toward the end of which the notes become intensely loud. Sometimes birds sing duets, causing even more raucous noise.

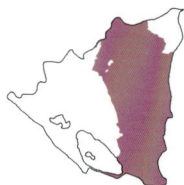

Red-throated Caracara (Caracara Avispero)   *Ibycter americanus*
22 in (56 cm). Although superficially cracid-like, this species is unmistakable; note **bare, red skin on face and throat** and mostly black body. White belly, thighs, and vent are prominent in flight. Rare in Caribbean lowlands and foothills. Found in humid lowland forest, lowland pine savanna, and forest edge. Tears open wasp and bee nests to feed on larvae; often seen in small groups. Makes a deep, guttural *raw raw RAAWH* that shares the grating quality of a macaw but is not as robust; call is delivered in a variety of changing, chaotic forms.

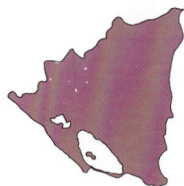

Crested Caracara (Caracara Crestado)   *Caracara cheriway*
22.5 in (57 cm). Large bulky body and distinct plumage make it unmistakable. **Black crown** terminates with a **short bushy crest** and sharply contrasts with **white collar** (nape, face, and throat) and reddish-orange facial skin and base of bill. White outer primaries on long, straight black wings and predominantly white tail and coverts are conspicuous in flight. Immature has muted coloration, and is browner overall. Common in Pacific and Caribbean lowlands and uncommon in Northern Highlands; to 5,300 ft (1,600 m). Patrols open and semi-open habitats such as grasslands, aquaculture and salt ponds, beaches, marshes, scrub and thorn forest, highland pine forest, lowland pine savanna, and humid lowland forest edge. Opportunistic foraging behavior often takes it to the ground for carrion; also captures live prey. Generally silent.

Yellow-headed Caracara (Caracara Cabecigualdo)   *Milvago chimachima*
16 in (41 cm). **Dark, narrow postocular stripe** is the only marking on **buffy head and underparts**. In flight, buffy wing linings and pale base of primaries contrast with black secondaries. Buffy parts on immature are heavily streaked with brown. Vagrant, but its range continues to expand north from Costa Rica. Most records are from the lowlands that lie close to the Costa Rican border, although there are recent records from the Northern Highlands and northern Caribbean. Prefers open and semi-open habitats such as grasslands, marshes, scrub, and forest edge, where it feeds opportunistically on carrion and other prey. Screams a wheezing *weee-áh*.

**Laughing
Falcon**

**Red-throated
Caracara**

adult

**Crested
Caracara**

immature

adult

adult

**Yellow-headed
Caracara**

adult

immature

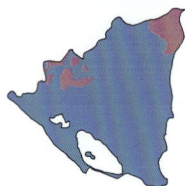

American Kestrel (Cernícalo Americano) *Falco sparverius*

10 in (25 cm). **Colorful, bold head pattern** and **rufous upperparts and tail** are diagnostic. Male upperwing coverts are blue-gray; also note broad, black subterminal tail band. On female, upperwing coverts are rufous; tail shows several narrow black bands. Three subspecies occur. *Tropicalis* and *nicaraguensis* subspecies: Paler overall and mostly lack rufous crown patch. *Tropicalis* subspecies is uncommon in highland pine and pine-oak forest and adjacent cloud forest in the Northern Highlands (excluding Sierra Dariense); *nicaraguensis* subspecies is uncommon in lowland pine savanna of the Mosquitia. *Sparverius* subspecies: Winter resident (Oct to early April); common in Pacific and Northern Highlands and rare in Caribbean. The American Kestrel perches on poles, fence posts, wires, and short trees in open and semi-open habitats. While perched it pumps its tail; often seen gracefully hovering before pouncing on prey. Gives a fast, shrill, series of notes: *klí-klí-klí…* (similar to that of Bat Falcon but slightly faster-paced).

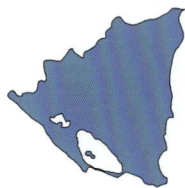

Merlin (Esmerejón) *Falco columbarius*

11 in (28 cm). Three subspecies occur in Nicaragua, but the *columbarius* subspecies (Taiga Merlin) is the only one expected with any frequency. Male has bluish-gray upperparts; female and immature have brown upperparts; all plumages have **thin pale superciliary**, pale-white throat, indistinct moustachial stripe, and **heavy streaking on pale underparts** (with rufous tinge). In flight, note uniform dark underwings, pale cinnamon vent and undertail coverts, and white bars on black tail. *Suckleyi* subspecies (Black Merlin) is much darker and *richardsonii* subspecies (Prairie Merlin) is much paler. Rare winter resident (mid-Sept to April), countrywide. Hunts with brute force and speed in open areas such as grasslands, agricultural fields, marshes, and coastlines. Generally silent.

Peregrine Falcon (Halcón Peregrino) *Falco peregrinus*

17 in (43 cm). Helmeted appearance is created by **dark crown, nape, and broad malar stripe**. Except for white throat and breast, underparts are black barred. Long, rounded tail is also barred black-and-white. Immature has underparts with heavy brown streaking. Uncommon winter resident (Sept to May), countrywide; to 4,300 ft (1,300 m). There is one record (Cosigüina Peninsula, Aug 2016) of *cassini* subspecies, a SA vagrant that has cinnamon tinged underparts. Powerful aerial hunter over agricultural fields, marshes, lakes, rivers, and coastlines, where it engages in high speed chases of medium-sized birds. Generally silent.

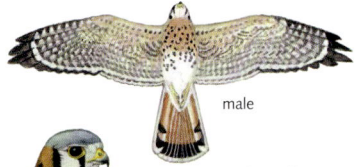

male

American Kestrel

male
tropicalis/nicaraguensis ssp.

female
sparverius ssp.

male
sparverius ssp.

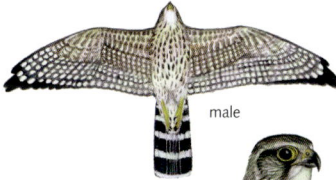

male
columbarius ssp.

male

male
suckleyi ssp.

female/immature
columbarius ssp.

Merlin

immature

adult

Peregrine Falcon

adult

Bat Falcon (Halcón Murcielaguero) *Falco rufigularis*
10 in (25 cm). Elegant appearance derives in part from **black half-hood** and **white throat, upperbreast, and sides of neck**. In flight, could be confused at a distance with a White-Collared Swift (p. 72). Immature shows diffuse cinnamon coloration on the white throat, upper breast, and sides of neck and has black barring on cinnamon undertail coverts. Despite the size difference, easily confused with larger Orange-breasted Falcon, but note: smaller head; wings with narrower base; longer and narrower tail that is squared and slightly notched; and bars on black breast band that are paler, straighter, and less distinct. Some individuals have variable cinnamon coloration on what is normally white plumage. Uncommon in Northern Highlands and Caribbean and rare in Pacific; to 4,600 ft (1,400 m). Frequents forest edge, lowland pine savanna, agricultural fields, freshwater marshes, and rivers. Aerially forages in the open or above adjacent canopy. Often perches exposed on snags or trees before dashing to capture prey in flight. Gives a fast and shrill *kee-kee-kee…* (similar to that of American Kestrel but slightly slower-paced).

Aplomado Falcon (Halcón Bigotudo) *Falco femoralis*
16 in (40 cm). Note bold, handsome head pattern: **black malar stripe** and **white superciliary** connect on back of head. Dark breast band separates white throat and breast from cinnamon underparts. In flight, note pale leading edge on black underwings and long, black tail with narrow white bars. Immature has dark streaking on breast. Uncommon in the Mosquitia and accidental through rest of lowlands. A lowland pine savanna specialist. Ambushes prey from perch or from hovering position; also forages aerially at grassland fires waiting for escaping prey. Utters squealing screams that often rise in pitch.

Orange-breasted Falcon (Halcón Pechicanelo) *Falco deiroleucus*
15 in (38 cm). **Cinnamon lower breast and sides of neck** contrast with **black half-hood** and **white throat and upper breast**. Immature has paler coloration, light black streaking on breast, and heavy spotting on thighs. Despite the size difference, easily confused with Bat Falcon, but note: larger head; wings with broader base; shorter and broader tail with a graduated tip; and bars on black breast band that are heavier, wavier, and more distinct. Very rare (possibly extirpated from the country); only 2 historical records exist (Matagalpa Dept. at 4,000 ft [1,200 m], June 1891; Mosquitia, 1962). Would occur in forested areas and possibly at vertical cliffs for nesting. Delivers a rapid, high-pitched, squeaky *eh!-eh!-eh!…*; also gives a single, short, squeaky *ehr!* **NT**

Bat Falcon

Aplomado Falcon

Orange-breasted Falcon

While identification of parrots is fairly straightforward with perched birds, you most often see parrots in flight between roosts and feeding sites. On birds that are in flight or back-lit, note body size and tail length and shape. Habitat loss and the pet trade have depleted populations throughout the Neotropics.

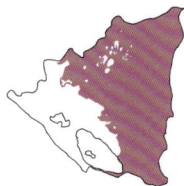

Olive-throated Parakeet (Perico Pechiolivo)   *Eupsittula nana*

9 in (23 cm). Green on most of body but note olive hue on throat and breast and yellow-green on lower underparts; has a conspicuous **white orbital ring** with a teardrop shape. In flight, note long, pointed tail and contrast between **green wing linings** and **dark blue flight feathers**. Common throughout Caribbean lowlands and foothills; to 2,000 ft (600 m). Occurs on the eastern slopes of Northern Highlands, where it is uncommon in foothills and rare in highlands; uncommon in Quirragua NR and Musún NR; to 4,400 ft (1,350 m). Prefers forest edge, secondary forest, and other disturbed areas with scattered trees, but also found in humid lowland forest and in cloud forests. Calls with a variety of *kree* notes delivered with a gravelly, grating quality, often in groups of 2 or 3 notes. Lower-pitched and duller than that of Orange-fronted Parakeet; call is not as loud as that of *Psittacara* species, and with a raspier tone.

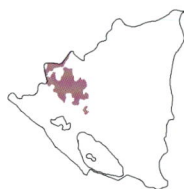

Green Parakeet (Perico Gorgirrojo)   *Psittacara holochlorus*

11 in (28 cm). Shows varying amounts of **reddish-orange on throat and upper breast**; some birds only have reddish-orange mottling, immature birds sometimes entirely lack it. **Brown orbital ring** blends in with eye color. In flight, note long, pointed tail and contrast between **green wing linings** and **yellow flight feathers**. Abundant on Sierra Dipilto-Jalapa and rare to uncommon throughout rest of Northern Highlands; generally above 3,000 ft (900 m), but accidental birds descend to lower elevations. Reaches southernmost distribution in Nicaragua. Move in flocks, sometimes very large, between cloud forest and highland pine and pine-oak forest. Gives a combination of screeching *ka ka* calls and metallic *KREE KREE* shrieks (louder than calls of *Eupsittula*).

Orange-fronted Parakeet (Perico Frentinaranja)   *Eupsittula canicularis*

9 in (23 cm). Combination of **orange forehead and blue midcrown** is unique, as is conspicuous **yellow orbital ring**. In flight, note long, pointed tail and contrast between **yellow-green linings** and **dark blue flight feathers**. Abundant throughout Pacific and on western slopes of Northern Highlands; mainly found below 2,600 ft (800 m) but does wander up to 4,600 ft (1,400 m). Rare in western regions of the Caribbean (following deforestation). Prefers thorn forest, scrub, and a variety of open and disturbed areas with scattered trees; tends to move into deforested areas. Also found in dry forest. Call is a metallic and grating *kree*, often delivered in groups of 2 and 3 notes. Higher-pitched and sharper than that of Olive-throated Parakeet; call is not as loud as *Psittacara* calls, and with a raspier tone.

Pacific Parakeet (Perico Verde)   *Psittacara strenuus*

12.5 in (31 cm). Uniform green upperparts contrast only slightly with yellow-green underparts; sometimes shows some red, scattered flecks on throat or breast. **Pale-gray orbital ring** is distinct. In flight, note long, pointed tail and contrast between **green wing linings** and **yellow flight feathers**. Locally abundant on Cosigüina Peninsula and at Chocoyero-El Brujo WR, Masaya Volcano NP, Granada, and Ometepe Island; generally common in the Pacific. Forages in a variety of forest types (including secondary forest) and scrub, edges, open areas, and cities. Populations at Chocoyero-El Brujo WR and Volcán Masaya NP use cliff cavities for colonial roosting and breeding. Gives a combination of screeching *ka ka* calls and metallic *KREE KREE* shrieks (louder than calls of *Eupsittula*). Reaches southernmost distribution in Nicaragua. Endemic to North Central America Pacific slope EBA.

Crimson-fronted Parakeet (Perico Frentirrojo)   *Psittacara finschi*

11 in (28 cm). **Red leading edge of wing and wing linings** and **red forecrown** distinguish it from the two other *Psittacara*. Abundant in Caribbean lowlands and foothills. Common on eastern slopes of Northern Highlands; rare to uncommon in Pacific, as far north as León; to 4,900 ft (1,500 m). Prefers forest edge, secondary forest, and agricultural and urban areas. When foraging and roosting, it is often seen in very large flocks. Over the past 2 to 3 decades, this species has continued to expand northward, following in the path of deforestation. Repeats a scratchy, nasal, metallic *ka ka ka*…, usually in sets of 5–6 notes; call is slightly raspier than those of other two *Psittacara* and louder than those of *Eupsittula*.

Olive-throated Parakeet

A pair of Orange-fronted Parakeets at their termitarium cavity nest.

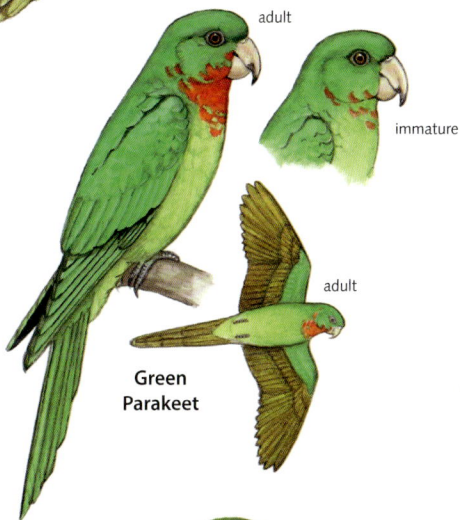

adult

immature

adult

Green Parakeet

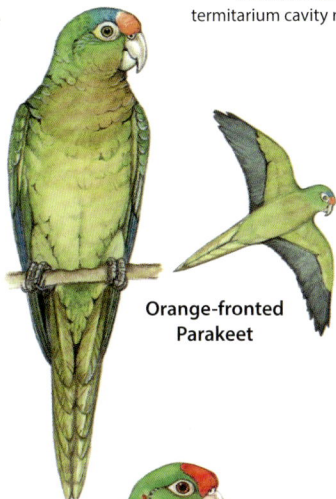

Orange-fronted Parakeet

Pacific Parakeet

Crimson-fronted Parakeet

Barred Parakeet (Chocoyo Listado) *Bolborhynchus lineola*
6.5 in (16 cm). Smallest of the Nicaraguan parrots. On perched birds, note **fine black barring** over much of its green body. In flight, note short, pointed tail. Rare throughout Northern Highlands; above 4,300 ft (1,300 m). Has a patchy distribution due to its preference for cloud forests, which themselves have a patchy distribution; moves within cloud forest in search of seeding *Chusquea* bamboo or fruits on trees in the family Lauraceae. Mainly seen flying in pairs or small groups high above the canopy. Gives a soft and rising *twi* call; when passing flocks are vocalizing, the result is a rich chatter.

Orange-chinned Parakeet (Chocoyo Barbinaranja) *Brotogeris jugularis*
7 in (18 cm). Small **orange chin** is typically not seen at a distance or in flight; **brown scapular** is easily seen on perched birds. In flight, note short, pointed tail and contrast between **yellow-green wing linings** and **blue-green flight feathers**. Abundant in Pacific and common in dry intermontane valleys of Northern Highlands; mostly in lowlands and foothills, but locally as high as 4,600 ft (1,400 m). Common in Caribbean lowlands and foothills. Ubiquitous throughout a variety of Pacific habitats, but only occurs in disturbed areas, open areas, forest edge, and secondary forest in the rest of the country. Often seen feeding or flying in sizable flocks; note very fast wingbeats and an undulating rise-and-fall flight pattern. Utters a shrill, piercing *twee* note or a double-noted *twee-twa* that has a distinctive up-down pitch pattern. Also mixes in a scratchy, scolding, chattering *chi-chi-chi*… .

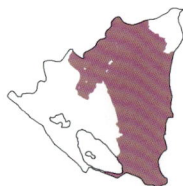

Brown-hooded Parrot (Loro Cabecipardo) *Pyrilia haematotis*
8.5 in (22 cm). Note **white orbital ring and lore**, **brown head**, and (less noticeable) **red ear patch**. In flight, red axillaries on blue underwings are diagnostic. Locally common in Caribbean lowlands and foothills; to 2,600 ft (800 m). Locally rare on eastern slopes of the Northern Highlands; to 4,900 ft (1,500 m). Forages in canopy of cloud forest and humid lowland forest. In small groups, flies between forest edges with a fast, teetering quality. Call is a squeaky, metallic, upslurred *kreel*; also makes a *kreel kree-yáh*; and a variety of other short calls, all with a shrill quality.

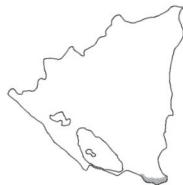

Blue-headed Parrot (Loro Cabeciazul) *Pionus menstruus*
9.5 in (24 cm). **Blue head** makes perched birds unmistakable. In flight, distinguished from *Amazona* parrots by deep, floppy wingbeats. White-crowned Parrot also has **red vent and undertail coverts**, and in poor light conditions it can be difficult to note the difference in head colors; green flight feathers (blue-green on White-crowned) help distinguish it. Very rare on southern edges of Indio Maíz BR and Río San Juan WR. This species continues to move northward from Costa Rica; breeding birds have yet to be documented in Nicaragua. Utters a sharp, slurred *cheeut!* that is very similar to call of White-crowned Parrot.

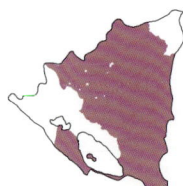

White-crowned Parrot (Loro Gorgiblanco) *Pionus senilis*
9.5 in (24 cm). Combination of **white crown**, red orbital ring, and blue head is unmistakable. In flight, distinguished from *Amazona* parrots by deep, floppy wingbeats. Blue-headed also has **red vent and undertail coverts**, and in poor light conditions it can be difficult to note the difference in head colors; blue-green flight feathers (green on Blue-headed) help distinguish it. Common in Caribbean lowlands and foothills; to 2,600 ft (800 m). Locally uncommon on Sierras Managua and rare in adjacent lowlands; very rare in Rivas Isthmus. Uncommon on eastern slopes of Northern Highlands, in northern region of Sierra Chontaleña, and in Quirragua NR and Musún NR; to 4,900 ft (1,500 m). Forages in canopy of cloud forest and humid lowland forest and forest edge, but also visits nearby agricultural fields and plantations. Utters harsh, and often shrill, electric shrieks: *see-úp!* or *see-íp!*; very similar to call of Blue-headed Parrot.

Barred
Parakeet

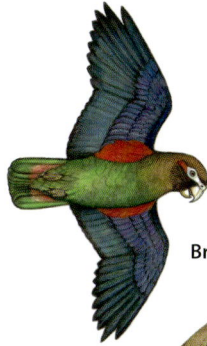

Brown-hooded
Parrot

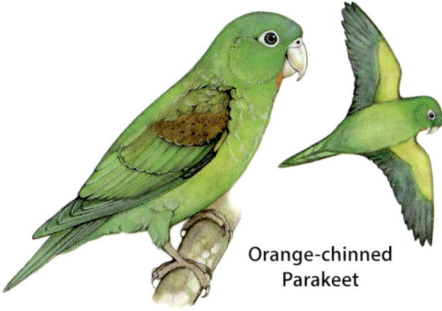

Orange-chinned
Parakeet

Blue-headed
Parrot

White-crowned
Parrot

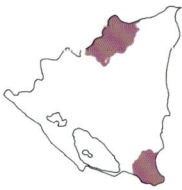

Great Green Macaw (Lapa Verde Mayor)   *Ara ambiguus*

30.5 in (77 cm). Strikingly beautiful and unmistakable. Body is primarily green but note broad **red forehead** and **blue flight feathers**. It is uncommon in the lowlands and foothills of Bosawas NR; it is also uncommon in the lowlands and foothills of Punta Gorda NR, Indio Maíz BR, and Rio San Juan WR, and the surrounding areas of the southeastern region of the Caribbean. Found in humid lowland forest; requires the majestic *almendro* tree (*Dipteryx panamensis*) for both food and nesting. Raucous screams are often heard before the bird flies into view; it belts out a forced and hoarse *RAAK!* that is very similar to the screams of the Scarlet Macaw but slightly harsher and more erratic. **EN**

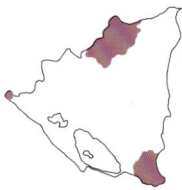

Scarlet Macaw (Lapa Roja)   *Ara macao*

34 in (86 cm). Extravagantly colored and unmistakable. Note predominantly **red body that shows yellow, green, and blue**. Locally uncommon on Cosigüina Volcano NR. It is uncommon in the Bosawas Biosphere Reserve; it is also uncommon in Punta Gorda NR, Indio Maíz BR, and Rio San Juan WR, and surrounding areas of the southeastern region of the Caribbean. Found in cloud forest and humid lowland forest; historically it occurred in lowland pine savanna in the Mosquitia. Raucous screams are often heard before it flies into view; it belts out a forced and hoarse *RAAK! RAAK!* that is very similar to that of the Great Green Macaw, but slightly less harsh and more stable.

White-fronted Parrot (Loro Frentiblanco)   *Amazona albifrons*

10.5 in (26 cm). Combination of **narrow blue midcrown, broad white forecrown, small red mask, and yellow bill** is unique. In flight, note short, squared tail and dark blue flight feathers; on male, red patch on upperwing primary coverts is diagnostic. Abundant throughout Pacific, common in Northern Highlands, and uncommon in western region of Caribbean; moving farther east, it becomes accidental; to 6,200 ft (1,900 m). Ubiquitous in a variety of habitats but prefers open areas with scattered trees; has begun to colonize deforested lands in the Caribbean. In flight, screams a harsh *AK-AK-AK…*, sometimes delivered in distinct douplets. When perched, gives a high-pitched, squeaky *arek*.

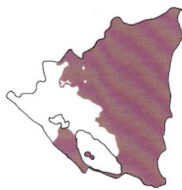

Red-lored Parrot (Loro Frentirrojo)   *Amazona autumnalis*

13.5 in (35 cm). Only *Amazona* parrot with **red forehead and lores**; yellow on face is more prominent in northern populations. In flight, note short, square tail with yellow-tipped outer rectrices; red patch on outer secondaries of upperwing distinguishes it from White-fronted Parrot. Common in Caribbean lowlands and foothills; locally common on Ometepe Island; locally rare from Sierras Managua and south; very rare on eastern slopes of Northern Highlands; to 4,300 ft (1,300 m). Utters a variety of shrill, quavering high-pitched shrieks that rise and fall in pitch. In flight, screams a *CHEP! CHEP!*; the cacophony of passing flocks can be deafening.

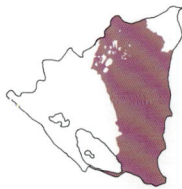

Mealy Parrot (Loro Verde)   *Amazona farinosa*

15.5 in (39 cm). Largest of the *Amazona* parrots. At close range, note **bold, white orbital ring** and **light blue on midcrown and nape**. In flight, also note short, square tail with broad yellow-tipped rectrices and red patch on outer secondaries of upperwing. Common in Caribbean lowlands and foothills; to 2,000 ft (600 m). Accidental on eastern slopes of Northern Highlands. Requires humid lowland forest, where it forages at canopy level and often frequents forest edge. Flies with shallow, stiff wingbeats. When perched, utters squeaky, complex squabbling. In flight, screams a harsh, electric *CHUP! CHUP! CHUP!…* (similar to that of Red-lored Parrot but deeper). **NT**

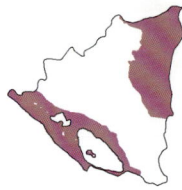

Yellow-naped Parrot (Loro Nuquiamarillo)   *Amazona auropalliata*

14 in (36 cm). Nicaragua has two distinct and disjunct populations; in both populations, **large yellow nape** is diagnostic. Some individuals (of both populations) have yellow on the forehead. *Auropalliata* subspecies: Has less yellow on nape and forehead and lacks red on the leading edge of the wing. Common throughout Pacific lowlands and foothills, uncommon on southeastern borders of Lake Nicaragua and San Juan River headwaters, and accidental on Sierra Amerrisque; to 3,000 ft (900 m). *Parvipes* subspecies: Often has more extensive yellow on nape and forehead and has red on the leading edge of wing. Common in the Mosquitia, stretching west to borders of Cola Blanca NR and Bana-Cruz NR and south to Pearl Lagoon. Found in lowland pine savanna, humid lowland forest edge, and semi-open areas. Flies with shallow, stiff wingbeats. Makes a nasal, downward-hitched squeal: *heee-húp*. Also emits a rhythmic *wall-úp*. **VU**

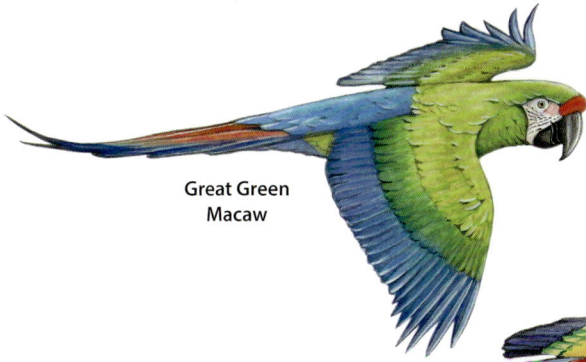

Great Green Macaw

Scarlet Macaw

adult male

adult male

immature

White-fronted Parrot

northern populations with more yellow on face

Red-lored Parrot

Mealy Parrot

auropalliata ssp.

parvipes ssp.

auropalliata ssp.

Yellow-naped Parrot

The members of this strictly Neotropical family dwell in the understory and mid-canopy, where many participate in one of the great dramas of the Neotropical rainforest. As massive army ant swarms raid the ground and understory in search of prey, the antbirds follow along to feed on the arthropods that the army ants stir up. Not all species rely on army ants as beaters, but those that do are so dependent on the ants that if the ants were to disappear they would most likely follow. Antbirds are generally clad in somber colors to blend in with the vegetation in their low light environment. Variation in body size and form is hinted at by the various suffixes attributed to members of the family, including -shrike, -vireo, and -wren.

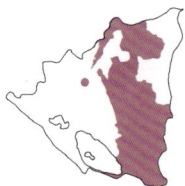

Fasciated Antshrike (Batará Lineado)   *Cymbilaimus lineatus*
7 in (18 cm). On male, **black crown** and **red iris** distinguish it from smaller male Barred Antshrike; also note **narrower black-and-white barring**. Female is tawny and has a rufous crown. Uncommon in Caribbean lowlands and foothills; to 1300 ft (400 m). Two historical records suggest it can occur at high elevations on eastern slopes of the Northern Highlands (Peñas Blancas NR, June 1909). Found in humid lowland forest, forest edge, second growth, and along the edges of forest streams. Searches for large insects in dense vegetation of mid-canopy and canopy. Shy and inconspicuous, this species is easily overlooked and perhaps more abundant than the few records indicate. Whistles 4–8 distressed *wheeá* notes (2–3 notes per second); each note rises slightly in pitch, with the entire phrase building intensity throughout.

Barred Antshrike (Batará Búlico)   *Thamnophilus doliatus*
6.5 in (16 cm). On male, **erect crest** and **pale yellow iris** distinguish it from larger male Fasciated Antshrike; also note **broader black-and-white barring**. On female, combination of **cinnamon rufous upperparts and black streaking on face** is distinctive. Common in Pacific and uncommon in Northern Highlands (excluding Sierra Dipilto-Jalapa); to 4,600 ft (1,400 m). Uncommon in Caribbean lowlands and foothills; to 2,000 ft (600 m), but to 3,300 ft (1,000 m) in Musún NR (possibly present in nearby Quirragua NR). Favors dry forest, secondary forest, dense thickets, and humid lowland forest edge. Usually in pairs, forages in dense understory for insects, fruit, and seeds; occasionally follows army ant swarms. Delivers a rapid-fire string of accelerating, chuckling *há* notes, with the last note an abrupt, downslurred *WEH!* (3–5 seconds in total length); flicks tail in unison with each note; more harshly inflected than that of Black-crowned Antshrike. Calls with a strained and nasal *errh*.

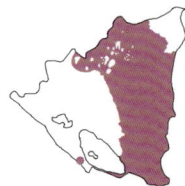

Great Antshrike (Batará Mayor)   *Taraba major*
8 in (20 cm). On both sexes, combination of **large size** and bulky appearance, **bicolored pattern**, and **red iris** is distinctive. Rare on eastern slopes of Northern Highlands (Segoviana Plateau) and in Caribbean lowlands and foothills; accidental in Rivas Isthmus (5 historical records, 1905); to 2,000 ft (600 m). Frequents humid lowland forest edge, secondary forest, gallery forest, scrub, and dense, riparian marshlike vegetation. Usually found in pairs, foraging in dense understory, tangles, and vines for insects and lizards; very elusive and difficult to locate. Delivers an accelerating chuckle consisting of repeated *chúp* notes (typically 2–6 seconds in total length); call similar to that of Black-headed Trogon but notes are delivered more rapidly towards the end (sounds like a bouncing ball nearing the end of its bounce) and often followed by a raspy growl: *whyaa*. Also gives a scolding churr.

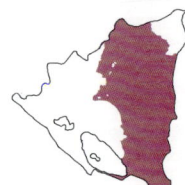

Russet Antshrike (Batará Café)   *Thamnistes anabatinus*
5.5 in (14 cm). Note **stout, hook-tipped bill** and a **dark eye line** that contrasts with **pale buffy superciliary**. Color and foraging behavior can suggest a foliage-gleaner, but it is smaller and lacks spine-tipped tail feathers. Locally uncommon in Caribbean foothills and highlands (Saslaya NP and Musún NR) and rare in lowlands; to 3,300 ft (1,000 m). Actively forages for insects in canopy of humid lowland forest, forest edge, and secondary forest, but will descend to mid-canopy; accompanies mixed-species flocks. Sings a series of plaintive, downslurred notes: *seeuu-su-su-su…* (series usually consists of 4–7 notes that trail off in intensity at end). Calls with a variety of squeaky, high-pitched, chattering noises, including a piercing *swEE!* and dry *whit* notes.

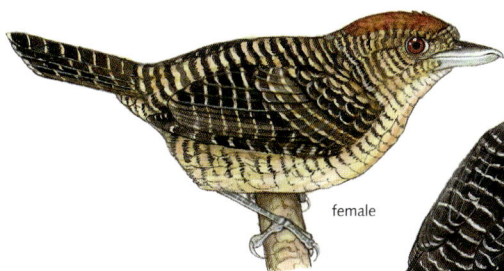

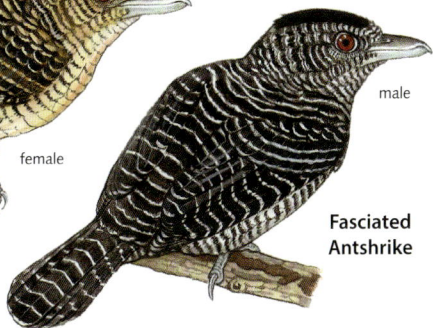

female

male

Fasciated Antshrike

female

male

Barred Antshrike

female

Great Antshrike

male

Russet Antshrike

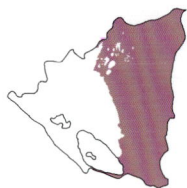

Black-crowned Antshrike (Batará Plomizo)   *Thamnophilus atrinucha*

6 in (15 cm). Male is **stouter and has a more obvious hook on its bill** than slightly smaller Dusky Antbird; also note more extensive white on wings and tips of tail. Female has buffy wingbars, spots on upper wing coverts, **and pale buffy-tipped rectrices**. Common in Caribbean lowlands and rare in foothills; to 2,000 ft (600 m). Rare in extreme southeastern spur of Rivas Isthmus, south of Lake Nicaragua. Prefers humid lowland forest, secondary forest, forest edge, and second growth. Usually in pairs, forages from understory to mid-canopy in dense vine tangles for insects and small lizards. Very aggressive and territorial; joins mixed-species flocks and army ant swarms passing through its territory. Extremely vocal; delivers a fast chuckling series of evenly spaced *há* notes, with the last note an abrupt, upslurred *WEH!* (2.5 seconds in total length); more softly inflected than that of Barred Antshrike. Calls with a nasal *arrh!* that is sometimes followed immediately by a short nasal trill.

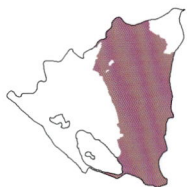

White-flanked Antwren   *Myrmotherula axillaris*
(Hormiguerito Flanquiblanco)

4 in (10 cm). Male's **white flanks** are diagnostic. **Buffy spots on wing coverts** of female distinguish her from female Slaty Antwren; **white throat** and **whitish flanks** distinguish her from Checker-throated Antwren (p. 244). Uncommon in Caribbean lowlands and foothills; to 3,000 ft (900 m). Found in humid lowland forest and adjacent tall second growth. Usually in pairs, actively forages in understory and mid-canopy, gleaning insects and spiders from foliage. Whistles a slow, descending (both in pitch and intensity) series of *fwee* notes, with even spacing. Calls with a squeaky *tí-ta*, the first note noticeably higher in pitch than the second.

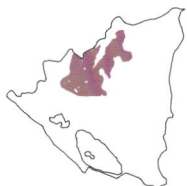

Slaty Antwren (Hormiguerito Pechinegro)   *Myrmotherula schisticolor*

4 in (10 cm). Smaller than Dusky Antbird, and with a shorter tail. On male, throat and breast are blacker than the rest of the body; note lack of white flanks (present on male White-flanked) and white-tipped tail (present on male Dot-winged). **Lack of spots on wing coverts** distinguishes female from female White-flanked Antwren and female Checker-throated Antwren (p. 244). Common on eastern slopes of Northern Highlands; to 4,900 ft (1,500 m). Common in Caribbean foothills and highlands; above 1,300 ft (400 m). Prefers cloud forest, humid lowland forest, and second growth adjacent to forest. Often in pairs, very actively forages for insects in understory and mid-canopy; accompanies mixed-species flocks. Whistles a sharp *swEET swEET*, sometimes followed by a mousy, nasal *nerh-erh*.

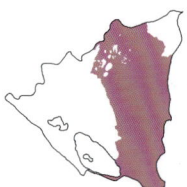

Dot-winged Antwren (Hormiguerito Alipinto)   *Microrhopias quixensis*

4.5 in (11 cm). Female is unmistakable. On male, combination of **broad white wing bar and strongly graduated, white-tipped tail** distinguishes it from male Slaty Antwren and male White-flanked Antwren. Common in Caribbean lowlands and foothills; to 2,000 ft (600 m). Accidental on eastern slopes of Northern Highlands; to 3,900 ft (1,200 m). In pairs or small family groups, actively forages for insects and spiders on vine tangles and in thickets in understory of humid lowland forest, forest edge, and low second growth; joins mixed-species flocks. Sings a series of sharp, rapid notes, slightly rising in pitch before accelerating in speed at end (2 seconds in total length). Also emits a variety of sharply downslurred shrill or plaintive *féur* calls.

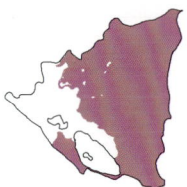

Dusky Antbird (Hormiguero Pizarroso)   *Cercomacroides tyrannina*

5.5 in (14 cm). Larger than Slaty Antwren, and with a longer tail. On male, **bill is more slender and hook on bill is less evident** than on slightly larger male Black-crowned Antshrike; also note less extensive white on wings and tips of tail. Female is bicolored, with brown upperparts and rich tawny underparts. Locally common on Sierras Managua and uncommon south through Rivas Isthmus; to 2,800 ft (850 m). Common in Caribbean lowlands and foothills; generally to 2,000 ft (600 m) but up to 3,300 ft (1,000 m) in Musún NR (possibly present in nearby Quirragua NR). Uncommon on eastern slopes of Northern Highlands and south to northern region of Sierra Chontaleña; to 4,600 ft (1,400 m). In pairs, forages for insects and other arthropods in dense liana tangles and thickets of second growth, humid lowland forest edge, cloud forest edge, river edge, and dry forest. Sings 8–20 sharp, piping *whíp* notes, accelerating in speed and rising in pitch at midpoint before descending in pitch as it reaches abrupt final note.

male

female

Black-crowned Antshrike

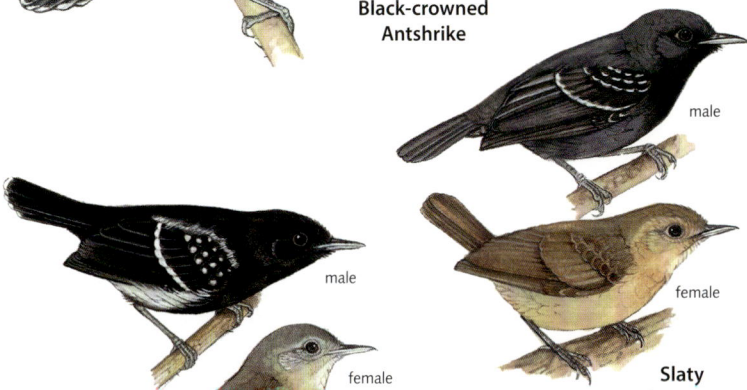

male

female

Slaty Antwren

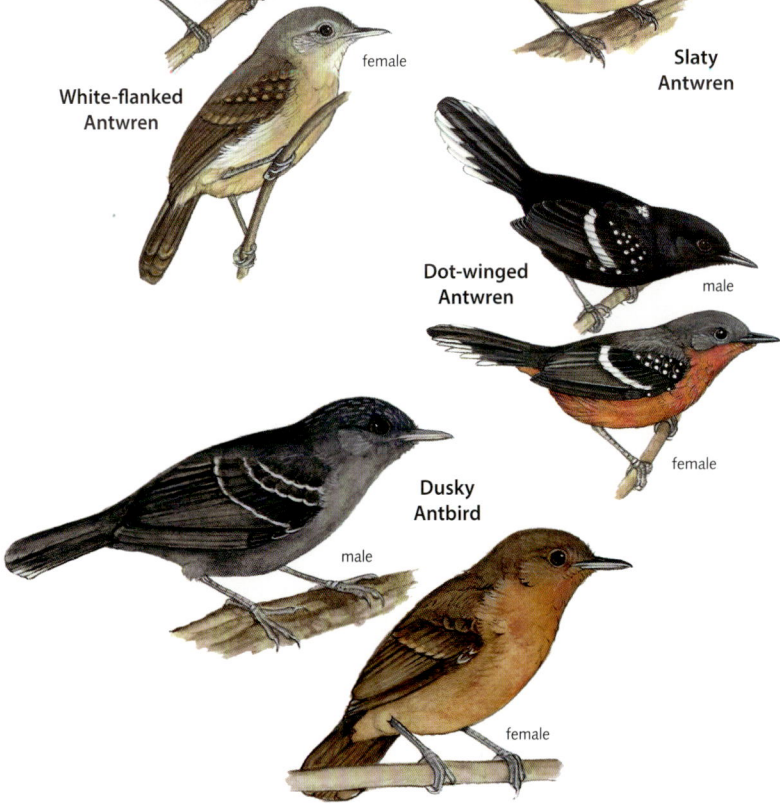

male

female

White-flanked Antwren

Dot-winged Antwren

male

female

Dusky Antbird

male

female

Plain Antvireo (Batarito Cabecigrís) *Dysithamnus mentalis*
5 in (12 cm). Male is olive-gray, with pale yellow underparts; note dusky ear patch. Female can be confused with Tawny-crowned Greenlet (p. 298), but note **dark eye with narrow white eye ring** and **shorter tail**. Plainer than Streak-crowned Antvireo. Rare in northern Caribbean foothills and highlands, but locally common in Musún NR and Saslaya NP; rare on eastern slopes of Northern Highlands; from 1,300 to 4,300 ft (400 to 1,300 m). Usually in pairs, forages for insects in understory and mid-canopy of humid lowland forest, cloud forest, and secondary forest; joins mixed-species flocks. Whistles a clear ascending phrase that accelerates to a trill at the end. Calls with a soft and nasal *weyáh* and makes a brief, downslurred, whiny *erh*.

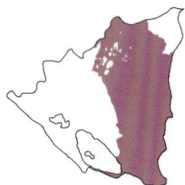

Streak-crowned Antvireo (Batarito Pechirrayado) *Dysithamnus striaticeps*
4.5 in (11 cm). **Pale grayish iris** and **streaked crown, throat, and breast** distinguish it from Plain Antvireo. **Wing bars** and streaked breast distinguish female from Tawny-crowned Greenlet (p. 298). Locally common in Saslaya NP and Indio Maíz BR, otherwise rare to uncommon in Caribbean lowlands and foothills; to 2,000 ft (600 m). Dwells in humid lowland forest and adjacent tall second growth with dense understory. Pairs join mixed-species flocks in understory. Whistles a clear series of notes that rise in pitch and accelerate in speed before descending in pitch, terminating with a trill-like note (3.5 seconds in total length). Calls with a plaintive *féu* or a double noted *féu-fi*, with the first note higher in pitch than the second. Endemic to Central American Caribbean slope EBA.

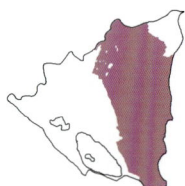

Checker-throated Antwren (Hormiguerito Café) *Epinecrophylla fulviventris*
4.5 in (11 cm). On male, **black-and-white checkered throat** is distinctive; note **pale iris** on both sexes. On female, lack of white on throat and flanks further distinguishes it from female White-flanked Antwren (p. 242); **buffy spots on blackish wing coverts** further distinguishes it from female Slaty Antwren (p. 242). Uncommon in Caribbean lowlands and foothills, but locally common in foothills of Bosawas Biosphere Reserve; to 2,600 ft (800 m). Inhabits understory of humid lowland forest and adjacent dense second growth. Usually in pairs, actively forages by gleaning insects and spiders from dead leaf clusters; joins mixed-species flocks and opportunistically attends army ant swarms. Delivers a series of spirited *swit* notes, repeated several times and varying in speed.

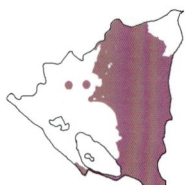

Spotted Antbird (Hormiguero Moteado) *Hylophylax naevioides*
5 in (12 cm). Dapper male is unmistakable. On female, **spotted breast** and **bold wing bars** distinguish it from female White-flanked Antwren (p. 242) and female Checker-throated Antwren. On both sexes, note **buffy terminal band** on tail. Common in Caribbean; uncommon in extreme southeastern spur of Rivas Isthmus, of Pacific; to 3,300 m (1,000 m). Historical records suggest the species once occurred on the eastern slopes of the Northern Highlands (San Rafael del Norte, April 1907; Peñas Blancas, May and June 1909). Found in pairs; joins mixed-species flocks. Forages in understory of humid lowland forest, but also follows army ant swarms to adjacent second growth. Repeats a distinct whistle (*feee-ba*), with the first note higher in pitch than the second, 9–11 times in quick succession; each series lasts 3.5 seconds and crescendos at midpoint before losing intensity and speed; entire song has an insect-like quality. Call is a short, nasal *twát*.

Plain Antvireo

female

female

Streak-crowned Antvireo

male

male

female

Checker-throated Antwren

male

female

male

male

Spotted Antbird

Antibirds at an army ant swarm feeding on arthropods as they attempt to flee the ants. Left to right: male Spotted, Bicolored (p. 247), female Spotted, and Ocellated (p. 247).

Zeledon's Antbird (Hormiguero Inmaculado Occidental)   *Myrmeciza zeledoni*
7.5 in (19 cm). **Uniform color** (black on male, dark brown on female) and **larger size** distinguish it from all other antbirds with blue orbital skin. Accidental; 2 country records (Indio Maíz BR, 2007). Found in humid lowland forest, second growth, and forest edge. Usually in pairs; joins army ant swarms. Sings 8–9 piping, evenly spaced whistles (*FEE*), each inflected slightly downward.

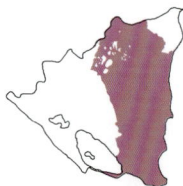

Chestnut-backed Antbird (Hormiguero Dorsicastaño)   *Myrmeciza exsul*
5.5 in (14 cm). Note **slaty-black head** (duller on female) contrasting with **chestnut-brown upperparts.** Common in southern Caribbean lowlands and foothills; rare to uncommon in northern Caribbean lowlands and foothills; to 2,000 ft (600 m). Forages for insects, spiders, lizards, and frogs in understory—sometimes on ground—of humid lowland forest and edge, second growth, dense thickets, and streamside tangles. Follows army ant swarms and mixed-species flocks when they pass through its territory. Secretive and best detected by voice; song is an explosive *TWEE-TU or TWEE-TWEE-TU*, made frequently and with an up-down pitch pattern.

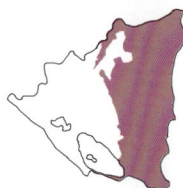

Bare-crowned Antbird (Hormiguero Frentiazul)   *Gymnocichla nudiceps*
6.5 in (16 cm). On male, combination of **extensive bare blue skin on face and crown** and **narrow white wing bars** is diagnostic. On female, **rufous underparts** and **rufous wing bars** distinguish it from other antbirds with blue orbital skin. Uncommon in Caribbean lowlands and foothills; to 1,300 ft (400 m). Favors dense understory of humid lowland forest edge, secondary forest, and swamp thickets. Usually in pairs, joins mixed-species flocks and often follows army ant swarms passing through its territory. Shy, inconspicuous, and easily overlooked; more often heard than seen. Sings 8–14 clear, piping, downslurred whistles (*feó*); slightly accelerates toward end of series.

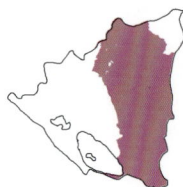

Bicolored Antbird (Hormiguero Bicolor)   *Gymnopithys bicolor*
5.5 in (14 cm). Only antbird with blue orbital skin that also has mostly **white underparts** (which contrast with **chestnut upperparts**). Common in Caribbean; to 3,300 ft (1,000 m). Occurs in humid lowland forest. Forages almost exclusively at army ant swarms; several birds may follow the swarm into adjacent second growth. Sings a distinctive series of high-pitched *swee* notes that sharply arc in speed, volume, and pitch at song's midpoint before quickly falling. Also gives a short, descending churr.

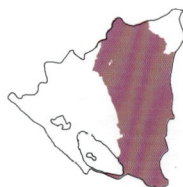

Ocellated Antbird (Hormiguero Ocelado)   *Phaenostictus mcleannani*
8.5 in (21 cm). Beautifully **scaled body** is unmistakable. Extravagant appearance is enhanced by **chestnut collar** and **blue orbital skin**. Uncommon in the Caribbean; locally common in foothills of Bosawas Biosphere Reserve; very rare southeast of Lake Nicaragua; to 3,600 ft (1,100 m). Found in humid lowland forest and secondary forest. Pairs or small groups regularly follow army ant swarms; this is the dominant species in the mixed-species flocks that follow army ant swarms. Sings a distinctive series of piercing *swee* notes that slowly ascend in pitch before quickly dropping, transforming into a trill, and finishing with scratchy, metallic notes. Also gives a soft, insect-like trill.

female

**Zeledon's
Antbird**
male

female

**Chestnut-backed
Antbird**

male

female

**Bare-crowned
Antbird**

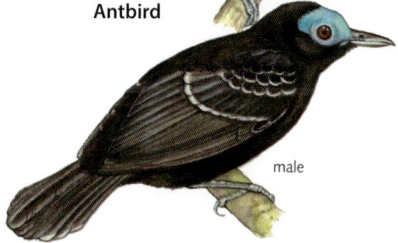
male

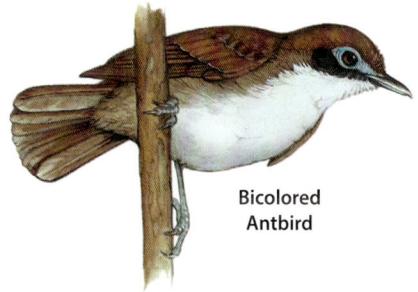
**Bicolored
Antbird**

**Ocellated
Antbird**

Wing-banded Antbird (Hormiguero Alifranjeado)   *Myrmornis torquata*
6.5 in (16 cm). Unmistakable. **Long bill** and **short tail** give it the appearance of an antthrush or antpitta, but **3 conspicuous buffy wing bars** prevent confusion. Rare in the northern Caribbean, to 4,000 ft (1,200 m), but locally uncommon to common in foothills of Musún NR, Saslaya NP, and Bosawas NR. Inhabits humid lowland forest; most easily found on hill slopes. Exposes insects below leaf litter by flipping over leaves with its bill. Sings a repeated series of evenly spaced, upslurred *fee* notes that rise in pitch and intensity throughout. Also gives a short, descending churr.

ANTPITTAS   Grallariidae

Antpittas are more often heard than seen. They patrol the forest floor for invertebrates and, if flushed, will often fly up to a low perch before quickly retreating beyond view.

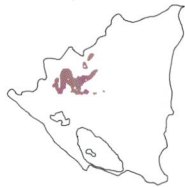

Scaled Antpitta (Tororoí Escamado)   *Grallaria guatimalensis*
7 in (18 cm). Note **scaled head and mantle** and **buffy malar stripe**; it is a comparatively large antpitta. Rare on Sierra Isabelia and Sierra Dariense of Northern Highlands and in Quirragua NR and Musún NR; from 3,300 to 5,200 ft (1,000 to 1,600 m). Hops on ground of cloud forest and highland pine-oak forest. It moves leaves in search of invertebrate prey, lunging after it when found. Typically solitary and furtive. Makes a short, rapid series of deep, hollow, quavering toots that increase in intensity.

Streak-chested Antpitta (Tororoí Pechirrayado)   *Hylopezus perspicillatus*
5 in (13 cm). Combination of **heavy black streaks on breast, belly, and flanks; conspicuous buffy eye ring**; and tawny spots on wing coverts distinguish it from Thicket Antpitta. Rare throughout Caribbean lowlands and foothills; to 1,600 ft (500 m). Prefers humid lowland forest, where it hops on the ground and runs in short spurts in search of invertebrate prey hiding in leaf litter. While foraging, habitually puffs breast feathers and intermittently flicks wings. Delivers a series of 7–8 whistles that descend in pitch; begins with high-pitched slurred *fi* notes and ends with lower-pitched, more sluggish *fuw* notes.

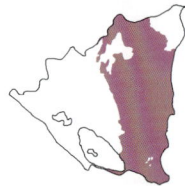

Thicket Antpitta (Tororoí Pechicanelo)   *Hylopezus dives*
5 in (13 cm). Combination of **paler streaks on breast and belly**; indistinct buffy eye ring; and lack of spots on wing coverts distinguishes it from Streak-chested Antpitta. The cinnamon wash on the breast becomes bold on the flanks. Uncommon throughout Caribbean lowlands and foothills; accidental on eastern slopes of Northern Highlands; generally to 1,600 ft (500 m) but accidental to 3,300 ft (1,000 m). Prefers dense and thicket-like vegetation (ie. forest streams, tree falls, forest edge, and young second-growth forests) within or adjacent to humid lowland forest. Gives 11–12 piping whistles (2 seconds in length) that build in intensity and pace and end abruptly.

ANTTHRUSHES   Formicariidae

Rarely leaving the ground—and with cryptic coloration and secretive behavior—antthrushes are seldom seen. As is suggested by their name, they follow army ant swarms to feed on the invertebrates that flee the ants.

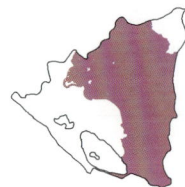

Black-faced Antthrush (Formicario Carinegro)   *Formicarius analis*
7.5 in (19 cm). As this is the only antthrush in the country, its long legs and **cocked tail** suffice to distinguish it from other secretive ground-dwelling forest birds. White loral spot stands out from **black face and throat**; note contrasting rufous patch on sides of neck. Common on eastern slopes of Northern Highlands and throughout Caribbean; to 4,600 ft (1,400 m). Walks slowly and in methodical fashion on ground of cloud forest and humid lowland forest, flipping leaf litter over in search of invertebrate prey. Occasionally found on the edges of army ant swarms. Sings a *PEEP pip-pip-pip-pip* phrase of rich, strained whistles; first note is slightly emphasized and higher pitched than the following sequence of montone notes, which can vary in number.

female

male

Wing-banded Antbird

Scaled Antpitta

Streak-chested Antpitta

Thicket Antpitta

A Scaled Antitta on its nest in the fork of a low tree in the forest.

With tail cocked, a Black-faced Antthrush forages on the forest floor.

Black-faced Antthrush

The members of this Neotropical family are insectivorous birds with cryptic coloration and patterns. There are 3 subfamilies, each with a distinct foraging niche. Woodcreepers (subfamily Dendrocolaptinae) forage by climbing vertically on the bark of trees, a behavior supported by sharp toe nails and spine-tipped rectrices. Foliage-gleaners (subfamily Furnariinae) specialize in searching for prey in dead leaf clusters, epiphytes, and vine tangles, in the understory and mid-canopy. Leaftossers (family Sclerurinae) forage on the ground and are known for creating a noisy commotion as they flip over leaf litter in search of prey.

Tawny-throated Leaftosser (Tirahojas Gorgirrojo)   *Sclerurus mexicanus*

6 in (15 cm). **Tawny-rufous throat and breast** are diagnostic; **rufous rump** adds additional color. Similar to slightly larger Scaly-throated Leaftosser, but it lacks tawny-rufous throat and breast and rufous rump (and there is little range overlap). Similar to Ruddy Foliage-gleaner (p. 258), but note longer and slenderer bill and shorter tail. Uncommon in Caribbean foothills and highlands; locally common in Saslaya NP; above 2,600 ft (800 m). Rare on eastern slopes of Northern Highlands; above 3,900 ft (1,200 m). Found in cloud forest and humid lowland forest. Forages alone or in pairs on the ground; hops about while tossing leaves aside with bill. Although very secretive, it is frequently seen perched on exposed branches for several minutes after it is flushed. Sings a series of thin whistles (*seá*) that rise and fall in pitch; decreases in intensity as song progresses, all with a whiny quality. Call is a sharp *twík!*

Scaly-throated Leaftosser   *Sclerurus guatemalensis*
(Tirahojas Barbiescamado)

6.5 in (17 cm). **Whitish mottling on throat** and **dull brown plumage** distinguish it from slightly smaller Tawny-throated Leaftosser (little range overlap). Coloration is similar to that of much smaller Nightingale Wren (p. 322), but note longer bill and tail. Rare in Caribbean lowlands and uncommon in foothills and highlands, but locally common in Saslaya NP; to 3,300 ft (1,000 m). According to historical records, accidental on Rivas Isthmus south of Lake Nicaragua (San Emilio, April 1905) and on eastern slopes of Northern Highlands (Peñas Blancas NR, 1926). Found in humid lowland forest. Alone or in pairs, forages noisily by tossing leaves aside with bill to expose insects. Generally inconspicuous, but can be located by its rustling of leaves. Sings a series of 13–15 clear, emphatic whistles (*twít*) that descend in pitch while increasing in speed and volume (total length of series is 3 seconds); sometimes seemingly endlessly repeats the series with practically no pause between repetitions. Calls with a sharp and rising *TWIT!*

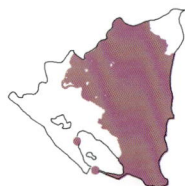

Plain Xenops (Piquivuelto Común)   *Xenops minutus*

4.5 in (12 cm). Note **buffy superciliary, white malar stripe**, and **upturned lower mandible**. **Plain (unstreaked) underparts** rule out the possibility of the hypothetical Streaked Xenops (p. 418). Uncommon in Caribbean; to 3,300 ft (1,000 m). Uncommon on eastern slopes of Northern Highlands; to 4,600 ft (1,400 m). Accidental in southern Pacific lowlands and foothills; documented by historical records (San Emilio, Rivas, April 1905; Mombacho Volcano NR, March 1962 and April 1966). Favors humid lowland forest, cloud forest, tall second growth, and open areas with scattered trees. Forages with acrobatic flair on mid-canopy branches and vine tangles. Delivers a high-pitched song (1 second in length) that consists of short, trilled *pít* notes that sharply rise to a piercing quality, then decrease in pitch, intensity, and speed. Calls with an abrupt and sharp *PIP!*

Tawny-throated Leaftosser

On the forest floor, a Tawny-throated Leaftosser tosses leaves in search of prey.

Scaly-throated Leaftosser

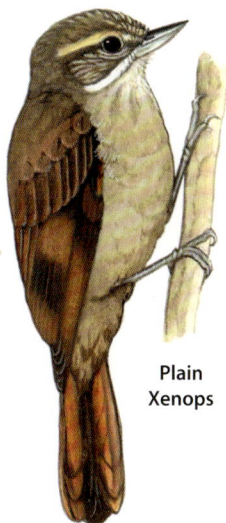

Plain Xenops

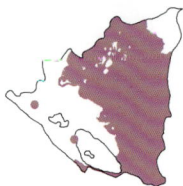

Olivaceous Woodcreeper (Trepatronco Oliváceo)   *Sittasomus griseicapillus*
6.5 in (16 cm). **Unpatterned plumage** and **gray head, breast and belly** distinguish it from slightly smaller Wedge-billed Woodcreeper; also note rufous undertail coverts. Common on eastern slopes of Northern Highlands, otherwise accidental; to 4,600 ft (1,400 m). Uncommon in Caribbean lowlands and foothills; to 2,000 ft (600 m). Accidental in the Pacific (historical records from Mombacho Volcano, Jan 1953; Casita Volcano, Nov 1961); locally uncommon in Rivas Isthmus. Forages for insects by climbing up tree trunks and branches in a linear or spiral fashion, then dropping down to base of next tree to restart the process. Sings a reedy, electric trill (1 second in length) with a smooth rise and fall in pitch. Calls with a sharp and agitated *svit!*

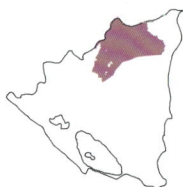

Long-tailed Woodcreeper (Trepatronco Colilargo)   *Deconychura longicauda*
8 in (20 cm) with spines included; 7 in (18 cm) without spines. Only other woodcreeper with **spotted breast** is larger Spotted Woodcreeper (p. 256), but note narrow superciliary, **unstreaked back**, and buffy throat; **bill is relatively short and thin**. Also note long spines on tail. Can look plain under poor light conditions. Locally uncommon in Saslaya NP and other foothill regions of Bosawas Biosphere Reserve; rare in rest of northern Caribbean; from 1,100 to 3,300 ft (350 to 1,000 m). Found in humid lowland forest. Usually solitary; sometimes joins mixed-species flocks. Sings a long series of clear and forceful whistles; series starts slowly, speeds up at midpoint, and then ends slowly (4–15 seconds in length).

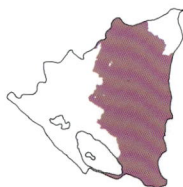

Wedge-billed Woodcreeper   *Glyphorynchus spirurus*
(Trepatronco Pico-Cuña)
6 in (15 cm). Combination of **buffy-white streaking on breast** and **wedge-shaped bill** (note upturned lower mandible) distinguishes it from slightly larger Olivaceous Woodcreeper. Longer tail, spine-tipped rectrices, and lack of white malar stripe distinguish it from Plain Xenops (p. 250). Common in most of Caribbean but uncommon in Guatuzos WR; accidental on eastern slopes of Northern Highlands; to 3,600 ft (1,100 m). Inhabits humid lowland forest and cloud forest. In understory and mid-canopy, climbs up tree trunks in linear or spiral fashion, gleaning and pecking for insects. Flies from one tree to the next with undulating flight. Sings a series of squeaky, undulating notes that ascend in pitch and lengthen throughout song (1–2 seconds in length). Calls with a harsh *chip!*

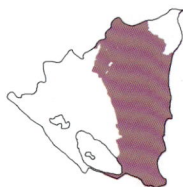

Plain-brown Woodcreeper (Trepatronco Pardo)   *Dendrocincla fuliginosa*
8.5 in (22 cm). Combination of **drab brown plumage, dark malar stripe, and pale, gray-buffy face** distinguishes it from other unpatterned *Dendrocincla* woodcreepers. Uncommon in Caribbean lowlands, but locally common in foothills of Bosawas Biosphere Reserve; to 2,600 ft (800 m) but occasionally wanders up to 4,250 ft (1,300 m) in cloud forests of Sierra Isabelia and Sierra Dariense. Dwells in humid lowland forest and adjacent old second growth; forages in understory and mid-canopy. Delivers a lengthy, rapid-fire rattle, with sporadic hitches throughout the song; notes become slower paced as the song finishes. Calls with a sharp *WEET!*

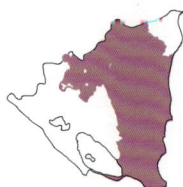

Tawny-winged Woodcreeper (Trepatronco Alirrufo)   *Dendrocincla anabatina*
8 in (20 cm). **Faint buffy superciliary and throat** and **tawny flight feathers** (with dusky tips) distinguish it from other unpatterned *Dendrocincla* woodcreepers. Common on eastern slopes of Northern Highlands and Caribbean (but only 1 record in Sierra Dipilto-Jalapa, Jalapa, Jan 1955); rare south of Lake Nicaragua (Guatuzos WR); to 4,400 ft (1,350 m). Favors humid lowland forest, cloud forest, and adjacent tall second growth. Occurs from understory to mid-canopy. Solitary; usually follows army ant swarms. Delivers an evenly spaced and monotone twá-twá-*twá*…, at great length, sometimes seemingly without end. Calls with a sweet and squeaky *tu-tí.*

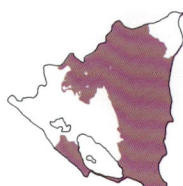

Ruddy Woodcreeper (Trepatronco Rojizo)   *Dendrocincla homochroa*
8.25 in (21 cm). **Uniform rufous plumage** distinguishes it from other *Dendrocincla* woodcreepers that are also unpatterned. **Shaggy, feathered crown and throat** and **broad grayish lore** are unique. Rare to locally uncommon in southern Pacific. Uncommon in Northern Highlands; to 4,600 ft (1,450 m). In Caribbean, rare in lowlands and uncommon in foothills and highlands; to 3,900 ft (1,200 m). Occurs in humid lowland forest, cloud forest, dry forest, and secondary forest. Sings a machine-gun-like rattle that sometimes slows to a sputter at the end (length varies from 2 to 20 seconds or more). Calls with a plaintive and downslurred *seeuw.*

Olivaceous Woodcreeper

Long-tailed Woodcreeper

Wedge-billed Woodcreeper

Plain-brown Woodcreeper

Tawny-winged Woodcreeper

Ruddy Woodcreeper

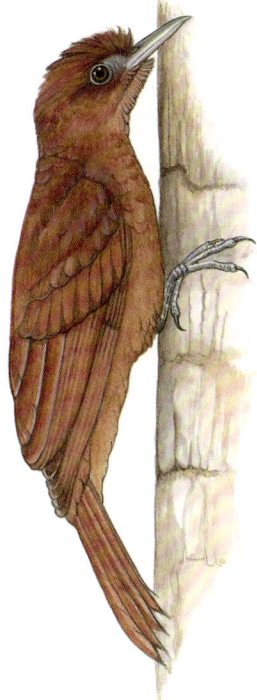

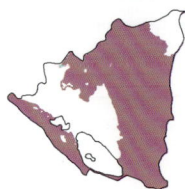

Northern Barred-Woodcreeper
Dendrocolaptes sanctithomae
(Trepatronco Barreteado)
11 In (28 cm). A unique woodcreeper with **fine barring on entire body** and no streaking (very rare Black-banded Woodcreeper has streaking); also note **long, stout dark bill**. Common on eastern slopes of Northern Highlands and in Caribbean; to 4,300 ft (1,300 m). Uncommon in Pacific lowlands and foothills; to 2,300 ft (700 m). Forages in understory and mid-canopy of cloud forest, humid lowland forest, and dry forest. Accompanies army ant swarms with other Furnariidae; less often joins mixed-species flocks composed of species from other families. Whistles with a sense of urgency, making a clear and sometimes whiny *dweeAT!-dweeAT!-dweeAT!-dweeAT!*, with each note upslurred (2 seconds in length); tone and spacing between notes is consistent but loudness increases throughout series.

Black-banded Woodcreeper
Dendrocolaptes picumnus
(Trepatronco Vientribarreteado)
10.5 in (27 cm). **Fine black barring** on lower underparts is similar to that on more common Northern Barred-Woodcreeper, but **buffy streaking on head and breast** distinguish it. Very rare in Northern Highlands with only 2 records (Arenal NR, March 1996; Peñas Blancas NR, March 2013); above 4,000 ft (1,200 m). In Nicaragua, seems to be limited to cloud forest, but occurs in highland pine and pine-oak forest in other countries. Forages at mid-canopy and canopy, where it is mostly silent and solitary; easily overlooked and could be more numerous than the few records suggest. Delivers a slow-paced, bubbly, descending chatter.

Brown-billed Scythebill (Picoguadaña Andino)   *Campylorhamphus pusillus*
9.5 in (24 cm). **Decurved, long, slender bill** is diagnostic. There is a single report from the junction of the Bartola and San Juan rivers. Forages at all canopy levels, and even descends to ground. Joins mixed-species flocks in understory. Sings with a combination of trills and whistles; whistles are typically delivered in sets of 4–5 jerky, spirited notes.

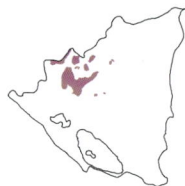

Strong-billed Woodcreeper
Xiphocolaptes promeropirhynchus
(Trepatronco Picofuerte)
12.5 in (32 cm). Largest Nicaraguan woodcreeper. **Long massive bill has slightly decurved culmen**. Also note **buffy-white throat** bordered by indistinct **malar stripe**. Uncommon on eastern slopes of Northern Highlands (including Sierra Dipilto-Jalapa), Musún NR (probably in nearby Quirragua NR), and Saslaya NP; above 3,300 ft (1,000 m). Found alone, less often in pairs, in cloud forest and highland pine and pine-oak forest. Forages from understory to canopy; climbs tree trunks and large branches, inconspicuously inspecting for insects. Joins mixed-species flocks of other Furnariidae, forest flycatchers, ant-tanagers, and Slaty Antwrens. Has an interesting repertoire of vocalizations. Whistles a repeated, clear *per whit* that rhythmically changes from low to high pitch with each repetition. Also whistles a repeated, slurred *pío-WHIP* that starts with a nasal quality before ending on a piercing note. Also gives a distinctive *meeee-CHEK!*, the first note of which sounds like the meow of a cat, and the second forceful and scratchy.

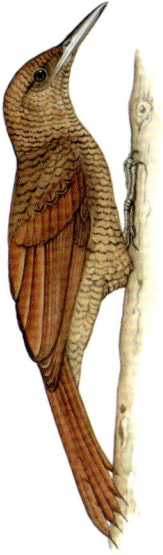

Northern Barred-Woodcreeper

Three common species of woodcreeper at a lowland forest ant swarm. Left to right: Two Ruddy (p. 253), Tawny-winged (p. 253), and Northern-barred.

Black-banded Woodcreeper

Strong-billed Woodcreeper

Brown-billed Scythebill

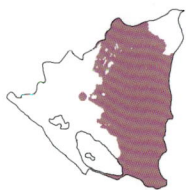

Cocoa Woodcreeper (Trepatronco Gorgicrema)   *Xiphorhynchus susurrans*
9 in (23 cm). **Long, stout, straight bicolored bill** (dark upper mandible and pale lower mandible) best key to distinguishing it from other streak-backed woodcreepers; has **buffy throat** and dark, narrow indistinct malar stripe. **Buffy superciliary** further distinguishes it from smaller Streak-headed Woodcreeper. Common in Caribbean lowlands and foothills; to 2,300 ft (700 m). Occasionally wanders onto eastern slopes of Northern Highlands (particularly at Esperanza Verde RSP); to 3,900 ft (1,200 m). Prefers humid lowland forest edge, light gaps in forest, old second growth, riparian forest, and mangroves. Sings a continuous series of evenly spaced and loud *WHIT* whistles that reach maximum pitch and speed at midpoint of song (song is 3–4 seconds in length). Calls with a loud and clear *CHEU* and a softer, muffled *cheeú* (repeated in duplets or triplets).

Ivory-billed Woodcreeper (Trepatronco Piquiclaro)   *Xiphorhynchus flavigaster*
9.5 in (24 cm). **Long, stout, straight pale bill** best distinguishes it from other streak-backed woodcreepers. Has bolder **black-margined, buffy streaks on back and breast** than does Cocoa Woodcreeper (sympatric south of Lake Nicaragua). Further distinguished from smaller Streak-headed Woodcreeper by **narrow dark malar stripe**. Little range overlap with smaller Spot-crowned Woodcreeper. Common in Pacific, rare to uncommon on western slopes and in dry intermontane valleys of Northern Highlands; to 4,600 ft (1,400 m). Forages on branches of canopy in dry forest, second growth, and forest patches. Alone or in pairs; occasionally with mixed-species flocks. Sings a lengthy series of emphatic *WIT* whistles (variable in length) with a slight rise in pitch at onset before slowly descending in pitch and speed. Calls with a spirited *piáo*.

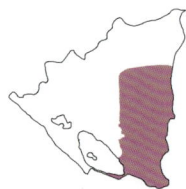

Black-striped Woodcreeper (Trepatronco Pinto) *Xiphorhynchus lachrymosus*
9.5 in (24 cm). Extensive **dark-margined, white-buffy streaks** on head, back, and breast give it an unmistakable pattern. Generally rare in southern Caribbean lowlands and foothills, as far north as Alamikamba NR; but common in Punta Gorda NR and Indio Maíz BR; to 1,600 ft (500 m). Reaches northernmost distribution in Nicaragua. Alone or in pairs, forages on tree trunks and branches from mid-canopy to canopy; joins mixed-species flocks. Whistles a continuous, descending whinny composed of short *whí* notes; series descends in pitch with final notes (typically 3–3.5 seconds in length). Calls with an abruptly upslurred *feur-UR!*

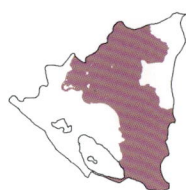

Spotted Woodcreeper (Trepatronco Manchado)   *Xiphorhynchus erythropygius*
9 in (23 cm). Only other woodcreeper with **spotted breast** is smaller and more slender-bodied Long-tailed Woodcreeper, but note **buffy eye ring**, **streaked back**, and spotted throat; **straight bill is long and stout** (2-toned color). Spots on breast rule out Cocoa Woodcreeper (streaking on breast). Uncommon on eastern slopes of the Northern Highlands and in the Caribbean, but locally common in Saslaya NP; to 4,600 ft (1,400 m). Forages on large branches and tree trunks in understory and mid-canopy of cloud forest, humid lowland forest, and adjacent second growth. Alone or in pairs, joins mixed-species flocks and follows army ant swarms. Sings 2–3 mournful *feeeur* whistles, each cascading down from the previous. Call is a soft and squeaky *peea*.

Streak-headed Woodcreeper   *Lepidocolaptes souleyetii*
(Trepatronco Cabecirrayado)
8 in (20 cm). **Streaked crown** (not spotted) and **distinctly streaked back** distinguish it from Spot-crowned Woodcreeper. **Slender, pinkish, decurved bill** and lack of buffy superciliary distinguish it from larger Cocoa Woodcreeper. Common in Pacific and Caribbean; uncommon in Northern Highlands; to 4,600 ft (1,400 m). Found in a variety of forest habitats, open areas with scattered trees, and shaded plantations (including coffee). Gives a soft, short, descending trill (1–2 seconds in length).

Spot-crowned Woodcreeper (Trepatronco Coronipunteado)   *Lepidocolaptes affinis*
8.5 in (21 cm). **Spotted crown** (not streaked) and **indistinct streaking on back** distinguish it from Streak-headed Woodcreeper; sometimes shows a dusky malar stripe. Rare to uncommon in Northern Highlands; above 4,000 ft (1,200 m). Also likely in Caribbean foothills in Saslaya NP, Cola Blanca NR, and throughout Bosawas Biosphere Reserve. Frequents cloud forest, secondary forest, and tall second growth; usually alone but occasionally found in pairs. Forages at all forest levels, spiraling up tree trunks in search of insects before dropping to base of nearby tree. Sings a very short, high-pitched *feeíp-fee*.

Cocoa
Woodcreeper

Ivory-billed
Woodcreeper

Black-striped
Woodcreeper

Spotted
Woodcreeper

Streak-headed
Woodcreeper

Spot-crowned
Woodcreeper

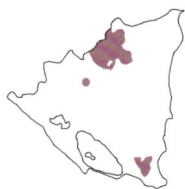

Striped Woodhaunter (Trepamusgo Rayado)   *Automolus subulatus*
7.5 in (19 cm). **Buffy streaking on head, upper back, and breast** and **buffy throat** resemble that of a woodcreeper, but this species does not climb on tree trunks. Locally uncommon in Caribbean foothills (Saslaya NP); from 1,000 to 3,000 ft (300 to 900 m). Accidental in Caribbean lowlands and on eastern slopes of Sierra Isabelia and Sierra Dariense of Northern Highlands; documented by historical records (Peñas Blancas, 1926). Alone or in pairs, forages in mid-canopy to canopy of humid lowland forest; actively gleans insects, spiders, frogs, and small lizards from dead leaf clumps, epiphytes, and bromeliads. Delivers a high-pitched and strident triplet: *speck! speck! speck!* (sometimes with 4 notes or continuous delivery). Also gives a churry rattle.

Scaly-throated Foliage-gleaner   *Anabacerthia variegaticeps*
(Trepamusgo de Anteojos)
6.5 in (16 cm). Distinguished from larger Buff-throated Foliage-gleaner by **tawny spectacles** (not pale buffy) and **dark crown and cheek**; lacks pale buffy throat seen on Buff-throated. Very rare on eastern slopes of Northern Highlands (more likely on Sierra Dipilto-Jalapa); above 3,300 ft (1,000 m). From mid-canopy to canopy of cloud forest, inspects bromeliads, dense foliage, vine tangles, and ferns for prey; joins mixed-species flocks. Typical song pattern is a series of dry, forceful, evenly delivered *chep* notes (entire series fluctuates erratically in pitch); sometimes sings with no variation in pitch. Call is a squeaky, buzzy *zEEp*.

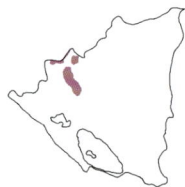

Ruddy Foliage-gleaner (Hojarasquero Rojizo)   *Clibanornis rubiginosus*
8 in (20 cm). Note **long, stout bill** and **narrow pale orbital ring**; also note **chestnut-brown upperparts** and **tawny-rufous underparts**. Has similar coloration to that of smaller Tawny-throated Leaftosser (p. 250), but note shorter bill and longer tail. Rare in Northern Highlands; from 3,600 to 6,200 ft (1,100 to 1,900 m). Alone or in pairs, forages in understory thickets of cloud forest, adjacent tall second growth, and highland pine and pine-oak forest; searches for small arthropods, frogs, and lizards; occasionally joins mixed-species flocks. Extremely secretive. Delivers a throaty, ascending *kruáh*, the first syllable of which has a wavering quality; sometimes given in duplets or triplets.

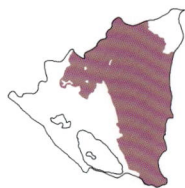

Buff-throated Foliage-gleaner   *Automolus ochrolaemus*
(Hojarasquero Gorgicrema)
7.5 in (19 cm). **Pale buffy spectacles and throat** distinguish it from smaller Scaly-throated Foliage-gleaner; also lacks dark crown and cheek seen on Scaly-throated. Locally common on eastern slopes of Northern Highlands; locally common in Saslaya NP (foothills and highlands); and rare in Caribbean lowlands and uncommon in foothills and highlands; to 4,300 ft (1,300 m). Alone or in pairs, forages on ground and in understory of humid lowland forest, cloud forest, forest edge, and tall second growth; joins mixed-species flocks. Sings a descending chatter-like trill that slows in pace toward the end (1 second in length). Calls with a sharp *spik!* or *spet!*

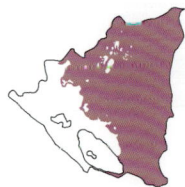

Slaty Spinetail (Colaespina Apizarrado)   *Synallaxis brachyura*
6 in (15 cm). Bold **rufous crown and wing coverts** distinctively contrast with **slate-gray plumage**. Common in Caribbean lowlands and foothills; to 2,000 ft (600 m). Rare on eastern slopes of Northern Highlands; to 4,300 ft (1,300 m). Very rare in northern region of Sierra Chontaleña; above 3,000 ft (900 m). Favors open areas with low thickets, overgrown pastures, second growth, and river edge. Difficult to see in tangled thickets, but can be located by its frequently given, dry, chuckling, descending rattle (each rattle 1 second in length).

**Scaly-throated
Foliage-gleaner**

**Striped
Woodhaunter**

**Ruddy
Foliage-gleaner**

**Buff-throated
Foliage-gleaner**

immature

adult

**Slaty
Spinetail**

With 66 species in the country, this is the largest family of birds in Nicaragua; most are residents but some are migrants from North and South America. Flycatchers occur only in the Americas. The diverse members of this family are known for their upright posture and for acrobatic sallies to catch aerial insects. Size varies from the miniscule pygmy-tyrants to the Fork-tailed Flycatcher, whose tail exceeds the length of its body. Some species have colorful crown patches or ornate crests, but these are often concealed except when the birds are alarmed. Distinguishing among some members of this family is a famously difficult challenge—even for experienced birders. Identifying call notes is often one of the best ways—or only way—to identify some species.

Yellow-bellied Tyrannulet (Mosquerito Cejiblanco)   *Ornithion semiflavum*
3.5 in (9 cm). Tiny and has a very short tail. Broad white superciliary extends well beyond the eye and contrasts with **gray crown** (note brown crown on Brown-capped Tyrannulet). Uncommon in Caribbean lowlands and foothills; to 1,800 ft (550 m). Found in humid lowland forest and forest edge; also ventures out to adjacent second growth. Actively picks invertebrate prey and berries from mid-canopy and canopy foliage. Occasionally joins mixed-species flocks. Whistles 6–8 high-pitched notes in continuous succession; the series slightly descends in pitch as the notes become shorter, and the final notes are sometimes inflected sharply upward. Also sings a sharply whistled, 4-noted *feeu-fi-fi-fee*. Makes a shrill, arcing *feeo* call.

Brown-capped Tyrannulet (Mosquerito Gorricafé)   *Ornithion brunneicapillus*
3.5 in (9 cm). Tiny and has a very short tail. Broad white superciliary extends only a little beyond the eye and contrasts with **brown crown** (note gray crown on Yellow-bellied Tyrannulet). Very rare in Guatuzo Plains and along San Juan River. Reaches northernmost distribution in Nicaragua. Actively picks invertebrate prey and berries from mid-canopy and canopy foliage in humid lowland forest and forest edge; also ventures out to adjacent second growth. Occasionally joins mixed-species flocks. Sings a high-pitched *fwEET fi-fi-fi-fidih*; the first note is noticeably upslurred and separated by a pause from the series that follows, which descends in pitch. Sometimes simply delivers a single *fwEET* note.

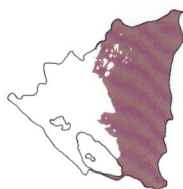

Yellow Tyrannulet (Mosquerito Amarillo)   *Capsiempis flaveola*
4 in (10 cm). Note two-toned appearance (olive upperparts and yellow underparts); **yellow superciliary, faint black eye line**; and two yellow wing bars and yellow wing edgings. Locally common in Guatuzo Plains and along San Juan River; locally uncommon in Caribbean lowlands and foothills; to 2,000 ft (600 m). Prefers thickets in humid lowland forest edge and second growth. Active foraging behavior resembles that of vireos and warblers but note upright posture when perched. Gives several squeaky *wee-it* notes that devolve into a jumbled twitter.

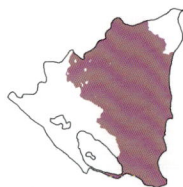

Paltry Tyrannulet (Mosquerito Cejigrís)   *Zimmerius vilissimus*
4 in (10 cm). White superciliary varies in intensity. Distinguished from Tennessee Warbler (p. 366) by **yellow edging on wing coverts and secondaries**, pale iris, and shorter, thicker bill. From below, note yellow tinged flanks, vent, and undertail coverts on mostly gray underparts. Common in Caribbean; rare on eastern slopes of Northern Highlands; to 3,900 ft (1,200 m). Found in a variety of forest habitats, and forest edge. Forages actively from mid-canopy to canopy, often with tail cocked; favors mistletoe berries. At dawn, whistles a sweet *feeo few-few*. Call is a repeated, unvarying whistle: *feeip!, feeip!, feeip!...* .

Yellow-bellied
Tyrannulet

Brown-capped
Tyrannulet

Yellow
Tyrannulet

Paltry
Tyrannulet

Northern Beardless-Tyrannulet
(Mosquerito Chillón)

Camptostoma imberbe

4 in (10 cm). A small, drab bird. Shorter bill than on *Contopus* flycatchers; pale wing bars are less conspicuous than on *Empidonax* flycatchers. Often raises bushy crest. Common throughout Pacific, rare on western slopes of Northern Highlands and in extreme southwestern corner of Caribbean; to 3,900 ft (1,200 m). Prefers arid habitats such as dry forest, thorn forest, and scrub; also visits gallery forest, and forest edge and second growth of other forest types. Sallies for insects and berries, from understory to canopy. Rapidly pumps tail while singing. Whistles a song of 5–7 sharp, falling *fee* notes. Gives a plaintive *feeou* call, which sometimes dissolves into a scratchy tone.

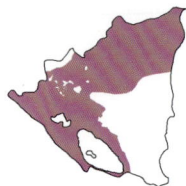

Greenish Elaenia (Elenia Coronigualda)
Myiopagis viridicata

5.5 in (14 cm). Very easily confused with Yellow-olive Flycatcher (p. 266) and Yellow-margined Flycatcher (p. 266), but note **slender bill with a flesh-colored base on the lower mandible**; longer tail; and (generally) more upright posture. Iris is dark. Broken eye ring and slender bill distinguish it from Eye-ringed Flatbill. Uncommon in Caribbean lowlands and foothills. Generally rare in the Pacific, but locally uncommon on Cosigüina Peninsula, Sierras Managua, Los Pueblos Plateau, and Rivas Isthmus; rare in the Northern Highlands (absent from Sierra Dipilto-Jalapa); to 4,300 ft (1,300 m). Frequents forest edges and adjacent second growth, where it sallies to pick insects off foliage. Song consists of alternating *fee-oyu* (downslurred) and *fee-ayee* (upslurred) squeaky whistles, with noticeable pauses between whistles. Call is a buzzy, rising trill.

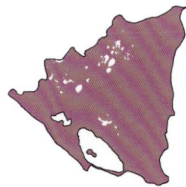

Yellow-bellied Elaenia (Elenia Copetona)
Elaenia flavogaster

6 in (16 cm). When raised, the **prominent crest** reveals a white crown patch; when crest is not raised, unique unkempt look of crown is caused by longer feathers in the front of the crest. Common in the Pacific and Northern Highlands; to 4,600 ft (1,400 m). Generally common in the Caribbean lowlands and foothills, but abundant in the Mosquitia; to 2,000 ft (600 m). Forages for insects and fruit on forest edge and in a variety of open areas; often in pairs. Rapidly repeats a harsh, scratchy *tidi beeu*. Calls with an agitated, slightly arcing *weeeur*.

Mountain Elaenia (Elenia Montañera)
Elaenia frantzii

6 in (16 cm). Note **rounded head**; pale, narrow eye ring; two wing bars and extensive wing edgings (noticeably broader on the tertial feathers). Birds tend to be green in its northern distribution and brownish-gray in its southern distribution. Common in the Northern Highlands and locally uncommon in Quirragua NR, Musún NR, and Saslaya NP; above 3,300 ft (1,000 m). Locally common on Mombacho Volcano and Maderas Volcano and surrounding foothills. Primarily found in cloud forest edge and adjacent second growth, as well as in shaded coffee plantations. Sallies to pick insects from foliage or grab berries. Whistles a short *pí-urr* (sometimes buzzy), with a noticeable up-down pitch pattern.

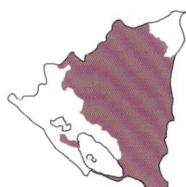

Ochre-bellied Flycatcher (Mosquero Oliváceo)
Mionectes oleagineus

5 in (13 cm). Note **ochraceous underparts**, from belly to undertail coverts. Also note uniform olive head, **very slender bill, and large dark eye**. Common on eastern slopes of Northern Highlands, northern region of Sierra Chontaleña, and throughout Caribbean. Very rare and local on Sierras Managua and Mombacho Volcano. Found in understory of forest and forest edge. Usually solitary but sometimes joins mixed-species flocks. Flicks one wing above its back, and then the other. Repeats an abrasive *wheet!, wheet!, wheet!…*; this sometimes transitions into a scratchy, nasal *pah pah pah…*, and then reverts to the opening phrase.

Sepia-capped Flycatcher
(Mosquero Cabecipardo)

Leptopogon amaurocephalus

5 in (13 cm). Note **brown crown** and **dark ear patch**. Also note **cinnamon wing bars and wing edgings**. Rare on eastern slopes of Northern Highlands and in the Caribbean; to 4,600 ft (1,400 m). Frequents understory of humid lowland forest, cloud forest, and secondary forest. Flicks one wing above its back, and then the other. At dawn, sings a rich *swi-da swi-da swi-da*. Calls include a harsh *pwik!* and a squeaky, descending trill.

Northern
Beardless-Tyrannulet

Greenish
Elania

crest
raised

Yellow-bellied
Elaenia

Mountain
Elaenia

Ochre-bellied
Flycatcher

Sepia-capped
Flycatcher

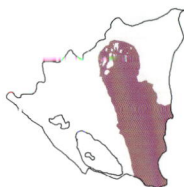

Black-capped Pygmy-Tyrant (Mosquerito Colicorto)   *Myiornis atricapillus*
2.5 in (6 cm). **White spectacles and miniscule size** make it unmistakable. Has odd proportions; bill is relatively long and the tail so short that it is barely visible from a distance. Uncommon in Caribbean lowlands and foothills; to 2,100 ft (650 m). Reaches northernmost distribution in Nicaragua. Inhabits mid canopy and canopy of humid lowland forest and secondary forest. Often inactive and extremely small, this bird is notoriously difficult to see. Repeats a high-pitched, insect-like *krít*, occasionally darting from branch to branch between calls.

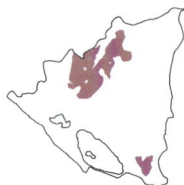

Scale-crested Pygmy-Tyrant (Mosquerito Crestipinto)   *Lophotriccus pileatus*
4 in (10 cm). Tiny, with a **hunched and plump appearance**. Ornate **black-and-rufous crest** is noticeable even when not raised. Uncommon in northern Caribbean and rare on eastern slopes of Northern Highlands; from 1,500 to 4,600 ft (450 to 1,400 m). Uncommon in southern Caribbean foothills. Forages in understory of cloud forest, humid lowland forest, forest edge, and second growth. Excitedly repeats a *prrit! prrit! prrit!…*, which sometimes devolves into a gravelly rattle.

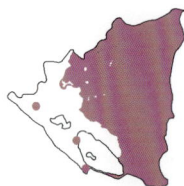

Northern Bentbill (Picotorcido Norteño)   *Oncostoma cinereigulare*
4 in (10 cm). **Bill is decurved towards the tip**. Pale iris and eye ring on gray head create a large-eyed appearance. Common on eastern slopes of Northern Highlands, northern region of Sierra Chontaleña, and in the Caribbean; accidental in Pacific (records from San Cristóbal-Casitas NR, Mombacho Volcano, and southern Rivas Isthmus) and on western slopes of Northern Highlands; to 4,300 ft (1,300 m). Found in forest edge and second growth thickets, where it perches in the understory. Makes a muffled, buzzy *bzzrrrt*, mainly montone and longer than buzz of Slate-headed Tody-Flycatcher.

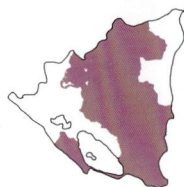

Stub-tailed Spadebill (Piquichato Norteño)   *Platyrinchus cancrominus*
4 in (10 cm). Tiny and plump, with a short tail and **very wide, flat bill**. Note **complex pale facial markings**. From below, white throat noticeably contrasts with light brown breast, which helps distinguish it from Golden-crowned Spadebill (with yellowish underparts). Rare and local in the southern Pacific; uncommon on eastern slopes of Northern Highlands, and northern region of Sierra Chontaleña; generally rare in the Caribbean, but locally common in Quirragua NR, Musún NR, and Saslaya NP; to 4,600 ft (1,400 m). Inhabits forest understory, where it is heard more often than it is seen. Sputters a bubbly *wi-dí-dí-da-da*. Calls with an excited *tí-di!*

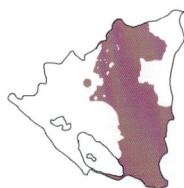

Golden-crowned Spadebill (Piquichato Coronado)   *Platyrinchus coronatus*
3.5 in (9 cm). If seen, **red-orange crown** is diagnostic. Tiny and plump, with a short tail and a **very wide, flat bill**. Pale yellow face shows **complex black facial markings**. From below, yellowish underparts distinguish it from Stub-tailed Spadebill (with light brown breast). Generally uncommon in Caribbean lowlands and foothills, but locally common in Saslaya NP; to 2,600 ft (800 m). Accidental on eastern slopes of Northern Highlands; one record from Peñas Blancas NR (May 1909). Perches inconspicuously in understory of humid lowland forest and adjacent secondary forest, where it is heard more often than it is seen. Occasionally joins mixed-species flocks. Song is a shrill, trilling whistle that wavers in pitch before its upslurred ending.

Black-capped
Pygmy-Tyrant

Scale-crested
Pygmy-Tyrant

crest
raised

Northern
Bentbill

Stub-tailed
Spadebill

Golden-crowned
Spadebill

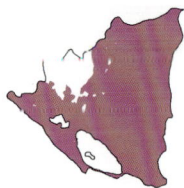

Slate-headed Tody-Flycatcher (Espatulilla Gris) *Poecilotriccus sylvia*

3.5 in (9 cm). Long, flat bill is characteristic of tody-flycatchers. **White supraloral** contrasts with large, dark gray head. Common in the Caribbean, uncommon and local in the Pacific, and very rare on Sierra Dariense; to 3,400 ft (1,050 m). Forages energetically in dense vegetation of understory, forest edge, second growth, and thickets. Makes a soft *tih* note immediately followed by a muffled, buzzy *prrrr prrr*, but may give either alone; shorter than buzz of Northern Bentbill.

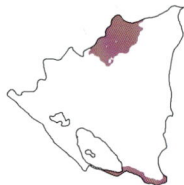

Black-headed Tody-Flycatcher *Todirostrum nigriceps*
(Espatulilla Cabecinegra)

3 in (8 cm). Long, flat bill is characteristic of tody-flycatchers. **White throat** readily distinguishes it from Common Tody-Flycatcher; also note shorter tail. Uncommon in Bosawas Biosphere Reserve and locally common in Rio San Juan WR; to 3,800 ft (1,150 m). Actively forages in canopy of humid lowland forest edge and secondary forest. Sings repeated short, high-pitched whistles: *tih tih tih…* .

Common Tody-Flycatcher (Espatulilla Común) *Todirostrum cinereum*

3.5 in (9 cm). Long, flat bill is characteristic of tody-flycatchers. **Yellow iris** stands out on dark head; entirely yellow underparts and white tips on black tail help distinguish it from Black-headed Tody-Flycatcher. Often holds long, graduated tail in cocked position. Common in Pacific and Northern Highlands; to 4,800 ft (1,450 m). Common throughout Caribbean lowlands and foothills. Active and conspicuous in secondary forest, forest edge, second growth, as well as in gardens in urban environments. Sings a high-pitched, rapid *ti-di-di*, repeated at erratic intervals. Calls with a sharp *chip!*

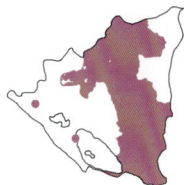

Eye-ringed Flatbill (Piquiplano de Anteojos) *Rhynchocyclus brevirostris*

6 in (16 cm). **Broad white eye ring** and **wide, flat bill** distinguish it from smaller Greenish Elaenia (p. 262). Locally common in Saslaya NP and Indio Maíz BR but rare to uncommon in rest of Caribbean; uncommon on Sierra Isabelia and Sierra Dariense in Northern Highlands; accidental in Pacific highlands, as supported by historical records from San Cristóbal Volcano and Mombacho Volcano; to 4,900 ft (1,500 m). Surveys inconspicuously for prey from understory and mid-canopy perches in cloud forest, humid lowland forest, and secondary forest. Often solitary, but will join mixed-species flocks. Occasionally gives a harsh, ascending, slightly grating *zweeEET!*

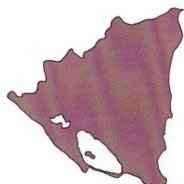

Yellow-olive Flycatcher (Piquiplano Azufrado) *Tolmomyias sulphurescens*

5.5 in (14 cm). **Pale iris** distinguishes it from Yellow-margined Flycatcher; subtle differences include paler gray head, less defined eye ring, and slightly narrower wing edgings. White superciliary is similar to that on Greenish Elaenia (p. 262), but note pale iris (dark on Greenish Elaenia) and wide flat bill. Common countrywide. Found in most forested habitats, where it typically forages from understory to mid-canopy. Makes a high-pitched, ascending *tseep!* that has a slightly hissing quality.

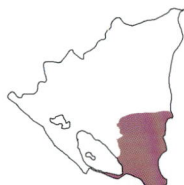

Yellow-margined Flycatcher (Piquiplano Aliamarillo) *Tolmomyias assimilis*

5 in (13 cm). **Dark iris** distinguishes it from Yellow-olive Flycatcher; subtle differences include darker gray head, more defined eye ring, and slightly broader wing edgings. White superciliary is similar to that on Greenish Elaenia (p. 262), but note wide flat bill. Uncommon on southern Caribbean lowlands and foothills; to 2,000 ft (600 m). Reaches northernmost distribution in Nicaragua. Occupies mid-canopy and canopy of humid lowland forest and secondary forest; often joins mixed-species flocks. Gives 4–6 short, shrieking whistles: *sweeo, swee swee swee…*; there is a pause between the first note and the ascending notes that follow.

Slate-headed
Tody-Flycatcher

Black-headed
Tody-Flycatcher

Common
Tody-Flycatcher

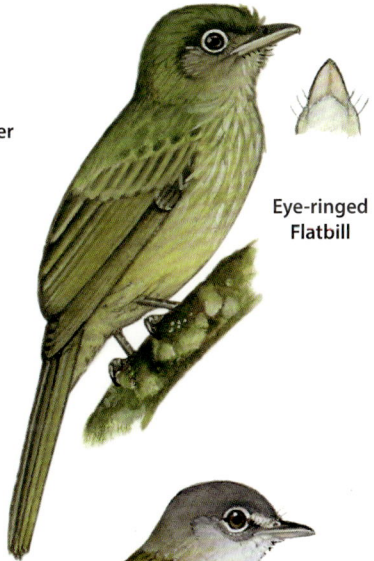

Eye-ringed
Flatbill

Yellow-olive
Flycatcher

Yellow-margined
Flycatcher

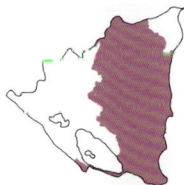

Royal Flycatcher (Cazamoscas Real Norteño)   *Onychorhynchus coronatus*
7 in (18 cm). **Hammerhead profile** is diagnostic. Ornate **fan-shaped crest** is unmistakable when seen, which is not often. When head is not visible, note white-tipped wing coverts and flight feathers (white tips look like spots), as well as tawny rump and uppertail coverts. Uncommon in Caribbean, very rare in southern Rivas Isthmus, and accidental on eastern slopes of Northern Highlands; to 3,900 ft (1,200 m). Perches in understory of secondary forest and forest edge; often found near streams. Call is a shrill *fi-át*; the second note quickly drops in pitch.

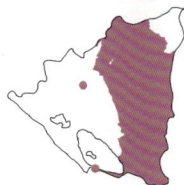

Ruddy-tailed Flycatcher (Mosquerito Colirrufo)   *Terenotriccus erythrurus*
4 in (10 cm). Petite, with a short bill. Note **gray head with slightly bushy crown**; rufous wings, rump, and tail; and cinnamon underparts. Common in Caribbean; to 3,600 ft (1,100 m); accidental in Rivas Isthmus (San Emilio, April 1904) and eastern slopes of Northern Highland (La Sombra RSP, July 2013). Found in humid lowland forest, cloud forest, and forest edge; perches in understory while waiting for flying insects. Whistles a high-pitched *swiu SWEE!*; the first note slides down in pitch and is followed by the explosive second note.

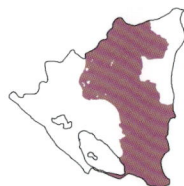

Sulphur-rumped Flycatcher   *Myiobius sulphureipygius*
(Mosquerito Rabiamarillo)
4.5 in (12 cm). Conspicuous **yellow rump** distinguishes it from Tawny-chested Flycatcher; also note uniformly dark wing (Tawny-chested has wing bars). Generally uncommon in Caribbean, but locally common in Saslaya NP; rare and local on eastern slopes on Sierra Isabelia and Sierra Dariense of Northern Highlands; to 3,900 ft (1,200 m). Found in humid lowland forest and cloud forest. Actively forages in understory, often displaying colorful rump; individuals join mixed-species flocks. Call is a dry *pwik!*

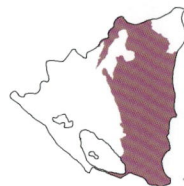

Tawny-chested Flycatcher   *Aphanotriccus capitalis*
(Mosquerito Pechileonado)
5 in (13 cm). Note **white throat and spectacles** and **dark gray head**. Also note tawny wing bars and wing edgings on secondaries. Olive mantle and rump distinguish it from Sulphur-rumped Flycatcher. No distribution overlap with similar Tufted Flycatcher. Uncommon in Caribbean lowlands and foothills; to 1,300 ft (400 m). Found in understory of humid lowland forest, forest edge, and secondary forest; associates with streams and bamboo thickets. Whistles a fast *ji-ji-ji-WEE-ah*, becoming buzzier at the end. Call is an emphatic *jeeu!* Endemic to Central American Caribbean slope EBA. **VU.**

Tufted Flycatcher (Mosquero Moñudo)   *Mitrephanes phaeocercus*
5 in (13 cm). Distinguished by **tufted crest** on **ochraceous head**. No distribution overlap with similar Tawny-chested Flycatcher. Common on Sierra Dipilto-Jalapa; very rare and local on Sierra Isabelia and Sierra Dariense; above 4,300 ft (1,300 m). Prefers light gaps within cloud forest and forest edge, where it returns to same conspicuous open perch after sallying for aerial insects. Sings a soft, nasal series of 2–5 *twee* notes. Calls with an abrupt *pip!*

female

male
(crest raised)

Royal
Flycatcher

Ruddy-tailed
Flycatcher

Sulphur-rumped
Flycatcher

Tawny-chested
Flycatcher

Tufted
Flycatcher

[Genus *Contopus* is characterized by a dark head and upperparts, a slight but noticeable crest on the hindcrown, and barely noticeable eye ring. Members of the genus reliably sally from and return to exposed perches.]

Olive-sided Flycatcher (Pibí Collorto) · Contopus cooperi

7.5 in (19 cm). Morphology is unique among pewees: Note large head and long, thick bill; long wing projection makes tale look exaggeratedly short. **White on throat and below separates dark plumage** (giving the appearance of a vest). **White tufts on sides of rump** are diagnostic if seen. Mostly black bill (yellow at base of lower mandible) distinguishes it from Greater Pewee (which lacks vest). Passage migrant (mid-Aug to Oct and April to May); common on Sierras Managua and in the Northern Highlands; to 5,600 ft (1,700 m). Also a rare winter resident on Sierras Managua and in the Northern Highlands. Within forests, perches on exposed snags and branches, to which it returns after sallying for aerial insects. Gives a quick, peppy *pip pip pip!* **NT**

Greater Pewee (Pibí Mayor) · Contopus pertinax

8 in (20 cm). **Tufted crest** distinguishes it from the other pewees. Dark upperparts contrast with pale gray underparts (which fade to pale ochre on belly and below). Distinguished from Olive-sided Flycatcher by **orange lower mandible** and lack of vest on underparts. Locally common in Northern Highlands; from 3,300 to 5,900 ft (1,000 to 1,800 m). Reaches southernmost distribution in Nicaragua. Primarily found in highland pine and pine-oak forest and forest edge; occasionally found in cloud forest. Sings at dawn with alternating *fee* and *fee-eywít!* whistles. Calls with a repetition of unequally spaced *pip* notes; also gives a single, plaintive *feeo* whistle.

Western Wood-Pewee (Pibí Occidental) · Contopus sordidulus

6.5 in (17 cm). Exceedingly difficult to distinguish from Eastern Wood-Pewee; vocalization is perhaps best way of doing so. It is important to consider several subtle features: Lower wing bar is generally brighter than upper wing bar; the length of the primary projection is equal to the distance that the tail extends beyond the tip of the primary; Western perches with tail in line with back and wings; lower mandible is mostly or entirely dusky; and underparts are generally darker. Primary projection is noticeably longer than that of Tropical Pewee. Passage migrant (Aug to mid-Nov and April to mid-May); uncommon in Pacific and Northern Highlands, and very rare in Caribbean; to 4,900 ft (1,500 m). Found in a wide variety of habitats, but prefers highland pine and pine-oak forest and cloud forest. Call is a buzzy *weeu*, which rises and falls in pitch.

Eastern Wood-Pewee (Pibí Oriental) · Contopus virens

6.5 in (17 cm). Exceedingly difficult to distinguish from Western Wood-Pewee; vocalization is perhaps best way of doing so. It is important to consider several subtle features: Lower and upper wing bars are of equal brightness; the length of the primary projection is generally shorter than the distance that the tail extends beyond the tip of the primary; Eastern perches with tail slightly cocked downward; lower mandible is mostly yellow-orange (has black tip); and underparts are generally lighter. Primary projection is longer than that of Tropical Pewee. Common passage migrant countrywide (Aug to mid-Nov and April to mid-May); very rare winter resident; to 4,900 ft (1,500 m). Found in a wide variety of habitats. Call is a drawn-out, high-pitched *peweee* whistle that gradually rises in pitch; sometimes it is preceded by a staccato *pi* note.

Tropical Pewee (Pibí Tropical) · Contopus cinereus

5 in (13 cm). Greatly resembles the wood-pewees but is distinguished by smaller size and noticeably **dark crown**, **pale lores**, and shorter primary projection. Lower mandible is orange (occasionally black-tipped) and lower belly is yellow-tinged. Common in Northern Highlands (but absent from Sierra Dipilto-Jalapa) and the Caribbean, and uncommon in the Pacific; to 4,900 ft (1,500 m). Inhabits all forest types, forest edge, freshwater marshes, and grasslands with scattered trees and shrubs. Song is a burry trill that quickly rises and falls in pitch. Call is a short, clear *pee!*

Olive-sided
Flycatcher

Greater
Pewee

Western
Wood-Pewee

Eastern
Wood-Pewee

Tropical
Pewee

[Genus *Empidonax* presents many identification challenges, especially with the 6 NA migrants. All birds have drab coloration and pale eye rings and wing bars. A proper indentifcation depends on a combination of subtle characteristics, including head shape, bill length and width, length of primary projection and tail projection, and characteristics of eye ring. But the call note is the surest means of identifying a species.]

Yellow-bellied Flycatcher (Mosquero Ventriamarillo)   *Empidonax flaviventris*
5 in (13 cm). **Underparts yellow**, with olive wash on breast; yellow eye ring is distinct. Only other *Empidonax* with **yellow throat** is Yellowish Flycatcher, but it lacks rounded head and has a yellow breast. Similar to Acadian Flycatcher, but **moderate primary projection** is shorter, bill is wider and shorter, and tail is narrower. Common winter resident countrywide (mid-Aug to mid-May); to 4,900 ft (1,500 m). Prefers understory of primary and secondary forest but can be found in most habitats during migration. Diverse calls include a plaintive, rising *peweep* and a short, emphatic *pwee!*

Acadian Flycatcher (Mosquero Cuellioliváceo)   *Empidonax virescens*
6 in (15 cm). Has **longest primary projection** of the *Empidonax* flycatchers (best field mark to distinguish from Alder Flycatcher and Willow Flycatcher); the shallow sloping forehead often ends with a slight crest. Extensive yellow on underparts, but throat and belly usually are a contrasting white; yellow eye ring is noticeable. Similar to smaller Yellow-bellied Flycatcher, but bill is wider and longer and tail is broader. Winter resident (mid-Sept to early May); rare in Pacific and Northern Highlands and uncommon in Caribbean; to 5,200 ft (1,600 m). Prefers understory light gap thickets in forest and forest edge but can be found in most habitats during migration. Call is a high-pitched, empathic *pweep!*

Alder Flycatcher (Mosquero Norteño)   *Empidonax alnorum*
6 in (15 cm). Virtually indistinguishable from Willow Flycatcher, so best to rely on vocalization. White throat contrasts with brown head and breast; white eye ring is indistinct. Long, wide bill and long, broad tail are similar to those of Acadian Flycatcher, but **moderate primary projection** is shorter. Rare passage migrant (mid-Aug to mid-Oct and mid-April to May) in Pacific, Northern Highlands, and southern extremes of the Caribbean; to 4,300 ft (1,300 m). An accidental winter resident. Can occur virtually in all habitats, but more often found in semi-open areas. Sings a dry, scratchy *ree-beeú*. Calls with a rich *pip!*

Willow Flycatcher (Mosquero Uniforme)   *Empidonax traillii*
6 in (15 cm). Virtually indistinguishable from Alder Flycatcher, so best to rely on vocalization. White throat contrasts with brown head and breast; white eye ring is indistinct. Long, wide bill and long, broad tail are similar to those of Acadian Flycatcher, but **moderate primary projection** is shorter. Winter resident (late Aug to mid-May); uncommon in Pacific and rare in Northern Highlands and western edge of Caribbean; to 4,300 ft (1,300 m). Can occur virtually in all habitats, but more often found in semi-open areas. Sings a buzzy *fítz-breer*. Call is a very short *wip!*

Yellowish Flycatcher (Mosquero Amarillento)   *Empidonax flavescens*
5.5 in (14 cm). Bold **teardrop shaped eye ring** is unique among *Empidonax* flycatchers. Only other *Empidonax* with **yellow throat** is Yellow-bellied Flycatcher, but note crested-head and yellow breast (Yellow-bellied has rounded head and olive wash on yellow breast). Common in Northern Highlands; from 2,600 to 5,900 ft (800 to 1,800 m). Locally uncommon in Saslaya NP; above 2,100 ft (650 m). Found in cloud forest, highland pine and pine-oak forest, forest edges, and adjacent second growth; primarily forages from perch at understory and mid-canopy. At dawn, sings by alternating *swee* (rising in pitch) and *swío* (rising in pitch before abruptly falling) whistles. Throughout rest of the day, only whistles the single *swee* note.

Yellow-bellied
Flycatcher

Acadian
Flycatcher

Alder
Flycatcher

Willow
Flycatcher

Yellowish
Flycatcher

White-throated Flycatcher (Mosquero Gorgiblanco) *Empidonax albigularis*
5 in (13 cm). Prominent **white throat** contrasts with brown head and breast and blends into space below ear coverts. Primary projection is short; pale eye ring is indistinct. Browner upperparts, buffy wing bars, and ochraceous flanks help distinguish it from Alder Flycatcher and Willow Flycatcher (p. 272). Rare on Sierra Isabelia and Sierra Dariense of the Northern Highlands; from 3,300 to 4,900 ft (1,000 to 1,500 m). Found in second growth and grasslands with scattered trees; often found near water, especially freshwater marshes. Sings a scratchy *rhhhRHE!*, which rolls to an explosive ending.

Least Flycatcher (Mosquero Menudo) *Empidonax minimus*
5 in (13 cm). Smallest and grayest of the *Empidonax* flycatchers. Has a rounded head and short primary projection; white eye ring is distinct. Lower mandible of short, slender bill is black-tipped (extent of black varies). Underparts are paler in comparison to Hammond's Flycatcher. Winter resident (Sept to mid-May); uncommon in the Northern Highlands, rare in the Pacific, and very rare in the Caribbean; to 4,900 ft (1,500 m). Frequents forest edges and dry intermontane valleys with scrub. Call is a dry, emphatic *whit!*, sometimes given in erratic succession.

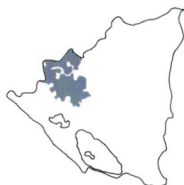

Hammond's Flycatcher (Mosquero de Hammond) *Empidonax hammondii*
5.5 in (14 cm). Only *Empidonax* flycatcher with mostly **dark bill**, which is very short and thin. **White teardrop eye ring** is usually noticeable on gray head, and primary projection is long. Underparts (gray above blending to yellow below) are more colorful than those of Least Flycatcher. Rare winter resident in Northern Highlands (Sept to April); above 2,600 ft (800 m). Reaches southernmost distribution in Nicaragua. Found in highland pine and pine-oak forest, cloud forest, and forest edge, where it forages actively at mid-canopy and canopy. Calls with a rich *pip!*

Buff-breasted Flycatcher (Mosquero Pechicanelo) *Empidonax fulvifrons*
5 in (13 cm). The only *Empidonax* with **cinnamon underparts** (tone varies from bright to dull). Similar to Tufted Flycatcher (p. 268), but distinguished by noticeable eye ring and lack of prominent crest. Uncommon on Sierra Dipilto-Jalapa; above 3,600 ft (1,100 m). Reaches southernmost distribution in Nicaragua. Moves actively in understory and mid-canopy of highland pine forest. Whistles a rich *pío pío pío…*, occasionally interjecting a sharper *PEE-o!* Call is a very short *pik!*

White-throated
Flycatcher

Least
Flycatcher

Hammond's
Flycatcher

Buff-breasted
Flycatcher

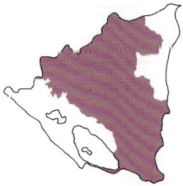

Black Phoebe (Cazamoscas Negro)   *Sayornis nigricans*

6.5 in (17 cm), **Mostly black** but note white on belly and below. Immature has cinnamon wing bars. Common in Northern Highlands; found only above 2,000 ft (600 m) on western slopes, but at all elevations on eastern slopes. Generally uncommon in Caribbean but locally common in Saslaya NP. Requires streams, rivers, and lakeshores, where it conspicuously forages over the water or nearby vegetation; casually pumps tail. Repeats a double-noted *tí-wheeu*, sometimes with a buzzy tone. Call is a sweetly whistled *chip!*

Vermilion Flycatcher (Cazamoscas Rojo)   *Pyrocephalus rubinus*

5.5 in (14 cm). On male, **flamboyant red** is entirely unique. On female, streaked underparts show a red wash on lower belly and below; immature may show a yellow wash. Common in understory and mid-canopy of lowland pine savanna in the Mosquitia; accidental in rest of Caribbean lowlands. Sings a lively *chip-chi-chi-cheeyUP!*, accelerating in speed and rising in pitch at the end. Calls with an energetic *fi!*

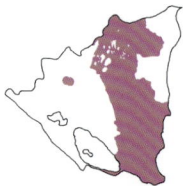

Long-tailed Tyrant (Mosquero Colilargo)   *Colonia colonus*

5 in (13 cm); central rectrices add up to 4 in (10 cm). Long central rectrices (even longer on males) almost double the length of the body. Combination of **white superciliary, gray crown**, and **white, central back stripe** is unique. Uncommon in Caribbean lowlands and foothills; to 2,000 ft (600 m). Accidental on eastern slopes of Northern Highlands (San Rafael del Norte, Dec 1961; El Jaguar RSP, 2006). Found in humid lowland forest and forest edge. Most often seen on high, exposed snags or dead branches, perched over light gaps (from tree falls), streams, and rivers. Call is a sweetly whistled, rising *sweeEE*, with a notable upward inflection at the end. Also delivers a 3-noted *pwi pwEE pí-u*.

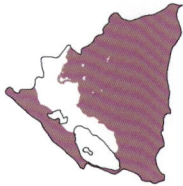

Bright-rumped Attila (Atila Rabiamarilla)   *Attila spadiceus*

8 in (20 cm). Note **large, hook-tipped bill** (lower mandible slightly upturned), **amber iris**, striated, brown head, and **tawny rump**. Locally common on Cosigüina Peninsula, Sierra Maribios, Apacunca GRR, and Rivas Isthmus; locally uncommon on Sierras Managua, and common on eastern slopes of Northern Highlands; to 4,800 ft (1,450 m). Common in the Caribbean; to 3,900 ft (1,200 m). Occupies all levels of forest, forest edge, and second growth and occasionally joins mixed-species flocks or follows army ant swarms; sporadically pumps tail. Sings two distinctly different songs. The first is a series of 8–9 *suey* notes, the entire delivery gradually crescendoing in pitch before falling at the end. The second is an emphatic *wi-di wí-dí-der wí-dí-der WI-DI-DER WI-DI WEEE wip!*

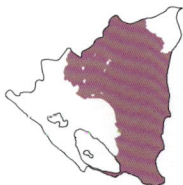

Rufous Mourner (Plañidera Rojiza)   *Rhytipterna holerythra*

8 in (20 cm). Mostly rufous, with cinnamon underparts. Smaller size, more slender body, and no color variation on throat distinguish it from almost identical Rufous Piha (p. 294), which has paler throat in comparison to its body color. Uniform coloration on wings and breast separates it from Speckled Mourner (p. 288), which has darker wing coverts and scaling on the breast. Common in the Caribbean; to 3,300 ft (1,000 m). Very rare on eastern slopes of Northern Highlands; to 4,600 ft (1,400 m). Primarily inhabits humid lowland forest and adjacent secondary growth but occasionally enters cloud forest. Inconspicuously perches on horizontal branches of mid-canopy and canopy while searching for food; occasionally joins mixed-species flocks. Often gives a slow *weeár wéear* double whistle, with the first note rising in pitch and the second note falling. Also whistles a simple, repeated *weear*.

Black
Phoebe

Long-tailed
Tyrant

female

Vermilion
Flycatcher

male

Bright-rumped
Attila

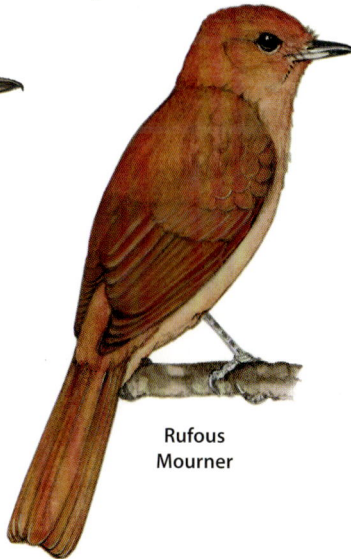

Rufous
Mourner

[The 5 *Myiarchus* flycatchers are remarkably similar. All are slender and have a crested head. Noting body size and subtle coloration differences, relative bill size, and undertail pattern is important for identification. Call notes are helpful, but also present challenges because some species have multiple calls. Some *Myiarchus* cannot be identified beyond genus.]

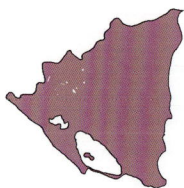

Dusky-capped Flycatcher (Güis Crestioscuro)   *Myiarchus tuberculifer*
6.5 in (17 cm). **Dark crown** is most notable field mark. The **smallest** *Myiarchus*, and with the **slenderest bill**; it is also the only species in the genus that has **rufous edging on both wing coverts and secondaries**. Undertail is mostly brown, with very little rufous. Common countrywide; to 4,900 ft (1,500 m). Found in all kinds of habitats, including in urban areas. Delivers a slightly arcing, plaintive *weeeu* (repeated about every 2 seconds); very similar to that of Nutting's Flycatcher but slightly subtler and more drawn out. Also delivers an excited *pí-EE-ur*, a gruff yelp, and a short, rattling call.

Brown-crested Flycatcher (Güis Crestipardo Mayor)   *Myiarchus tyrannulus*
8 in (21 cm). Virtually identical to Nutting's Flycatcher, only larger, longer-billed, and slightly darker overall, with a gray face. Darker and larger-billed than Ash-throated Flycatcher; browner head and upperparts than on Great Crested Flycatcher. On undertail, rufous extends to the tip. Abundant throughout Pacific lowlands and foothills; to 2,600 ft (800 m). Uncommon on western slopes of Northern Highlands; rare in highlands; to 4,900 ft (1,500 m). Accidental in extreme southwestern Caribbean lowlands. Prefers dry forest, thorn forest, scrub, and mangroves, but is also found in dry intermontane valleys. Song is a series of repeated notes that alternate between a sharply arcing *wheep!* and a double noted *pí-ba* that shifts to a lower pitch on the second note. Also calls with a singular *wheep!* (sometimes a *wheep!-u*).

Ash-throated Flycatcher (Güis Gorgipálido)   *Myiarchus cinerascens*
8 in (20 cm). **Palest** *Myiarchus*; upperparts gray overall but note pale brown crown. Undertail is primarily rufous but has a **black tip**. Distinguished from Nutting's Flycatcher by gray face, sharp contrast between rufous primary edgings and pale yellow secondary edgings, and slightly larger bill. Bill is smaller than that of Brown-crested Flycatcher. Rare winter resident (Sept to April) in Pacific lowlands and western slopes of Northern Highlands; to 3,900 ft (1,200 m). Frequents dry forest, thorn forest, and scrub. Call is a short, gravelly *weeá*.

Nutting's Flycatcher (Güis Crestipardo Menor)   *Myiarchus nuttingi*
7 in (18 cm). Virtually identical to Brown-crested Flycatcher, only smaller, shorter-billed, and slightly paler overall. Distinguished from Ash-throated Flycatcher by brown face, soft contrast between rufous primary edgings and cinnamon secondary edgings, and slightly smaller bill. Rufous on undertail slightly narrows at the tip, but is very similar to that of Brown-crested Flycatcher and Great Crested Flycatcher. Common in Pacific and rare on western slopes of Northern Highlands; to 3,000 ft (900 m). Found in a diversity of habitats. Repeats a slightly arcing, plaintive *weeeu*; very similar to that of Dusky-capped Flycatcher but shorter and slightly more excited. Also incessantly repeats a gravelly *wee-dí*.

Great Crested Flycatcher (Güis Migrador)   *Myiarchus crinitus*
8.5 in (22 cm). **Darkest** of the *Myiarchus*; shows more olive on head and upperparts than does Brown-crested Flycatcher. Also note **pale base of lower mandible**. On undertail, rufous extends to the tip. Winter resident (Sept to May); common in Pacific and uncommon in the Northern Highlands and Caribbean; to 4,600 ft (1,400 m). Found in both primary and secondary forest, as well as at forest edge. Call is a rising *weeep!*

Dusky-capped
Flycatcher

Brown-crested
Flycatcher

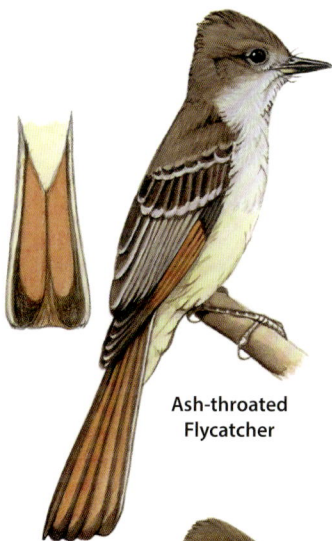

Ash-throated
Flycatcher

Nutting's
Flycatcher

Great Crested
Flycatcher

Great Kiskadee (Güis Común)   *Pitangus sulphuratus*
9.5 in (24 cm). **Rufous edgings on wings and tail** (also seen from below in flight) and brown-hued upperparts distinguish it from Boat-billed Flycatcher; also distinguished by bill that is slenderer and that has a **straight culmen**. Yellow crown patch is not usually visible. Abundant in Pacific and common in Northern Highlands; to 4,900 ft (1,500 m). Common in Caribbean lowlands and foothills; to 2,000 ft (600 m). Ubiquitous in many habitats but prefers open areas; very conspicuous on low perches, from which it drops onto prey. Energetically delivers the onomatopoeic *kis-ka-DEE* or a whining *whit! der.*

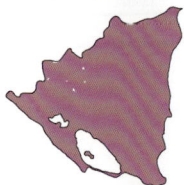

Boat-billed Flycatcher (Güis Picudo)   *Megarynchus pitangua*
9 in (23 cm). **Limited rufous on primary flight feather edgings** (wings and tail appear brown from below in flight) and olive-hued upperparts distinguish it from Great Kiskadee; also distinguished by stout bill with a slightly **decurved culmen**. Red crown patch is not usually visible. Common countrywide; to 5,200 ft (1,600 m). Found in most habitats, including urban areas, but prefers forest and forest edge. Repeats a loud, harsh *tchew!* Also delivers a variety of jumbled shrieks and a squeaky rattle.

Social Flycatcher (Güis Chico)   *Myiozetetes similis*
6.5 in (17 cm). Very similar to White-ringed Flycatcher, but note that **supercillaries do not connect on nape** (though often very close to connecting); also note short, broad bill and less extensive black mask. Red crown patch usually not seen. Common countrywide; to 4,600 ft (1,400 m). Found in most habitats, including urban areas, but prefers open areas and forest edge. Repeats a shrill *chee-át!* whistle. Vocalization is a rolling repetition of *cheer-a-wee-cheer.*

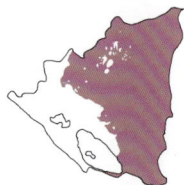

Gray-capped Flycatcher (Güis Cabecigrís)   *Myiozetetes granadensis*
6.5 in (17 cm). Of the flycatchers that have a white throat and a bright yellow belly, this is the only one that shows a **short, white superciliary** on a **pale gray head**. Red crown patch of male usually not seen. Common in Caribbean lowlands and foothills; to 2,000 ft (600 m). Rare on eastern slopes of Northern Highlands; to 4,600 ft (1,400 m). Prefers forest edge, second growth, and agricultural areas; often near water. Sings a 3-noted, burry *tí-twell-áh* phrase, with an up-down-up pitch pattern. Calls with a harsh *CHI!, CHI!, CHI!…*; also makes an excited *weer weer.*

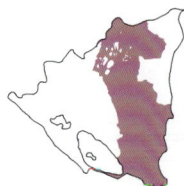

White-ringed Flycatcher (Güis Coroniblanco)   *Conopias albovittatus*
6.5 in (17 cm). **Supercillaries connecting on nape**, longer bill, and more extensive black mask distinguish it from Social Flycatcher. Note conspicuous tertial wing edgings. Yellow crown patch usually not seen. Uncommon in Caribbean lowlands and foothills; to 2,000 ft (600 m). Accidental on eastern slopes of Sierra Isabelia and Sierra Dariense of the Northern Highlands. Moves in canopy of humid lowland forest and forest edge. Gives a variety of squeaky rattles, the most common of which spikes in pitch at the beginning.

Great Kiskadee

Boat-billed
Flycatcher

Social
Flycatcher

Gray-capped
Flycatcher

White-ringed
Flycatcher

Streaked Flycatcher (Cazamoscas Listado)   *Myiodynastes maculatus*
8.5 in (21 cm). Similar to Sulphur-bellied Flycatcher, but a combination of several (somewhat variable) characteristics distinguishes it: larger bill with a conspicuous pale base on lower mandible; malar stripe is narrower and does not connect under the chin; superciliary and moustachial stripe are yellow-tinged; primaries have rufous wing edgings; and underparts are generally whiter. Rare breeding resident, from Rivas Isthmus and south of Lake Nicaragua to Guatuzos WR. Uncommon passage migrant (March to May and Sept to Oct) in the Northern Highlands and rare in the Pacific and Caribbean. Frequents both forest and forest edge, perched from mid-canopy to canopy. Song is an energetic, strained *su-WEET si-ah-wít!* Call is a harsh, squeaky *CHEP!*

Sulphur-bellied Flycatcher   *Myiodynastes luteiventris*
(Cazamoscas Pechiamarrillo)
8 in (20 cm). Similar to Streaked Flycatcher, but a combination of several (somewhat variable) characteristics distinguishes it: smaller bill, often with inconspicuous pale base on lower mandible; malar stripe is broader and often connects under the chin; superciliary and moustachial stripe are white; primaries do not have rufous wing edgings; and underparts are generally yellower. Common breeding migrant (March to Oct), countrywide; to 5,900 ft (1,800 m). Prefers forest and forest edge but possible in a wide variety of habitats. Sings a double-noted, squeaky *zwee-do*; the first note rises sharply and then the second note descends in pitch. Calls with a nasal *pik!*

Piratic Flycatcher (Mosquero Listado)   *Legatus leucophaius*
6 in (16 cm). Combination of **black-and-white striped head** and **heavy, diffuse streaking** on underparts is unique for medium-sized flycatchers. Breeding migrant (mid-Feb to mid-Nov); common in the Caribbean, uncommon in Northern Highlands, and locally rare in the Pacific and on eastern slopes of Sierra Isabelia and Sierra Dariense of the Northern Highlands, and rare on Sierras Managua and surrounding areas; to 4,300 ft (1,300 m). Acquires nest by pestering owners into abandoning it, thus earning its name and explaining why pairs can be found among oropendola and cacique colonies. Song is a strained *swí-u* whistle, which sharply rises and falls in pitch. Also gives a rapid, staccato series: *pi-pi-pi-pi.*

Streaked
Flycatcher

Sulphur-bellied
Flycatcher

Piratic
Flycatcher

A Social Flycatcher (p. 281) protecting its nest from a Piratic Flycatcher.

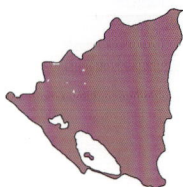

Tropical Kingbird (Tirano Tropical)   *Tyrannus melancholicus*
8.5 in (21 cm). More likely to be seen than very similar Cassin's Kingbird and Western Kingbird. Notable differences include **olive-tinged breast**, longer bill, and **unmarked, dark, notched tail**. Abundant countrywide; to 5,200 ft (1,600 m). Found mostly in open, disturbed habitat. Delivers a high-pitched, ascending rattle that is less than 1 second in length. Also gives a sharp, double-noted *pi pi*.

Cassin's Kingbird (Tirano Gritón)   *Tyrannus vociferans*
9 in (23 cm). Very similar to both Tropical Kingbird and Western Kingbird. Notable differences include **dark gray head and breast**, **white malar stripe and chin**, and **buffy-tipped, black, squared tail**. Very rare winter resident (mid-Oct to April); records from Moropotente NR, Lake Apanás, and Los Pueblos Plateau; from 2,000 to 4,600 ft (600 to 1,400 m). Reaches southernmost distribution in Nicaragua. Prefers grasslands and agricultural fields on exposed plateaus; associates with Tropical Kingbirds. Call is a burry, muffled *breer* that quickly rises and falls in pitch.

Western Kingbird (Tirano Colinegro)   *Tyrannus verticalis*
8.5 in (22 cm). Very similar to Tropical Kingbird and Cassin's Kingbird. Notable differences include **light gray head and breast** and **black, notched tail with outer white rectrices** (may be absent on worn plumage). Winter resident (Oct to early May); common in Pacific and Northern Highlands and uncommon in Caribbean; to 4,900 ft (1,500 m). Perches in semi-open areas such as thorn forest, agricultural fields, second growth, forest edge, as well as freshwater marshes and lagoons; associates with Scissor-tailed Flycatchers. Calls with an agitated *pik!*

Tropical
Kingbird

Cassin's
Kingbird

Western
Kingbird

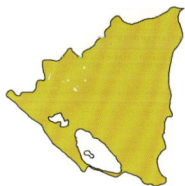

Eastern Kingbird (Tirano Norteño)　　　　　*Tyrannus tyrannus*
8.5 in (22 cm). Note **black upperparts and white underparts** (breast washed with gray); **white terminal band on black tail** is conspicuous. Passage migrant (Sept to mid-Oct and mid-March to mid-May); common in Caribbean, uncommon in Pacific, and rare in the Northern Highlands; to 4,900 ft (1,500 m). Migrates in impressively large flocks (up to hundreds); at stopover sites, feeds on canopy fruits. Call is a shrill *su-WEE*, sometimes followed with high-pitched chatter.

Gray Kingbird (Tirano Costero)　　　　　*Tyrannus dominicensis*
9 in (23 cm). Note large-headed appearance and long, thick bill. **Black wings and mask** contrast with **gray upperparts**; on white underparts, also note slight gray wash across breast. Very rare passage migrant (Sept to Oct and March to April) on Caribbean coastline; records outside the passage months suggest that small numbers stay locally as winter residents or even year-round. Individuals are known to migrate with Eastern Kingbird flocks. The squeaky call begins with *bi-di* and is immediately followed by a slightly descending, slurred trill.

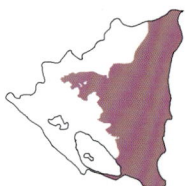

Fork-tailed Flycatcher (Tijereta Sabanera)　　　　　*Tyrannus savana*
7.5 in (19 cm); outer rectrices add up to 6 in (15 cm). **Dramatically long forked-tail** (longer on male) is similar to that on Scissor-tailed Flycatcher, but note **black head**. In flight, also note white wing linings and underparts and all-black tail that flows fluidly. Generally common in the Caribbean lowlands, but locally abundant in the Mosquitia; to 1,600 ft (500 m). Rare on eastern slopes of Siera Isabelia and Sierra Dariense in the Northern Highlands, but locally uncommon around Lake Apanás; accidental in Pacific; to 4,300 ft (1,300 m). Inhabits lowland pine savanna, grasslands, and agricultural fields, where it conspicuously perches on shrubs and fences. Calls with a soft *pik!*

Scissor-tailed Flycatcher (Tijereta Rosada)　　　　　*Tyrannus forficatus*
7.5 (19 cm); outer rectrices add up to 8 in (20 cm). **Very long forked-tail** (longer on male) is similar to that on Fork-tailed Flycatcher, but note **pale gray head**. Further distinguished by **pink wing linings and lower underparts** (seen in flight) and black-and-white pattern on the tail, which is stiff. Abundant winter resident (Oct to early May) in the Pacific and Northern Highlands; uncommon on Amerrisque NR and in extreme southwestern Caribbean lowlands; generally to 3,900 ft (1,200 m) but locally uncommon to 4,600 ft (1,400 m). Inhabits a variety of open areas such as grasslands, agricultural fields, thorn forest, scrub, and urban areas. Perches atop trees and shrubs; several hundred can be seen congregating at night roosts. Calls with an often repeated *pip!*

Eastern Kingbird

Gray Kingbird

Fork-tailed
Flycatcher

Scissor-tailed
Flycatcher

male

Finally granted family status, this group has been placed variously with tyrant flycatchers (Tyrannidae), cotingas (Cotingidae), and manakins (Pipridae). These birds have large heads and short tails and occur only in the Neotropics. They are mainly insectivorous but also eat fruit.

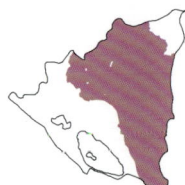

Northern Schiffornis (Saltarín Oliváceo)   *Schiffornis veraepacis*
7 in (17 cm). **Dark olive-brown overall, without markings**. Note pale eye ring. Rare to uncommon on eastern slopes of Northern Highlands; common in Caribbean foothills and highlands (Saslaya NP); uncommon in Caribbean lowlands; to 4,200 ft (1,300 m). Dwells in understory of humid lowland forest. Solitary; feeds on insects, berries, and seeds. With dark coloration and its understory habitat, this bird is more often heard than seen; fortunately, its song is distinctive. Whistles a clear and melodious sliding sound (sometimes slightly piercing) followed by two abrupt notes: *feeuuuu-wít-ú*! (2 seconds in total length).

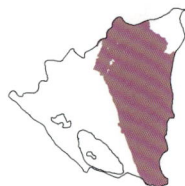

Speckled Mourner (Plañidera moteada)   *Laniocera rufescens*
8 in (20 cm). **Dark wing coverts with tawny wing bars** and inconspicuous **dusky scaling on breast** distinguish it from Rufous Piha (larger), p. 294, and Rufous Mourner (p. 276). Also note **narrow yellowish eye ring** and yellow patch on upper flanks (often concealed). Uncommon in Caribbean; generally to 3,300 ft (1,000 m) but occasionally wanders to 4,400 ft (1,350 m) on eastern slopes of Northern Highlands. Forages for insects and fruit at mid-canopy of humid lowland forest; favors swampy areas and stream edge. Inconspicuous and solitary; can be easily overlooked. Whistles a melodic, but slightly worrisome, arcing *peeo-ít*, with quick and erratic repetition.

Masked Tityra (Titira Carirroja)   *Tityra semifasciata*
8.5 in (21 cm). **Scarlet red orbital skin and base of bill** distinguish both sexes from smaller Black-crowned Tityra. Two subspecies intergrade in Nicaragua: *personata* subspecies (larger), from the north, and *costaricensis* subspecies (smaller), from the south. Common countrywide; to 6,200 ft (1,900 m). Found in the canopy of a wide variety of forest habitats, forest edge, and woodland. Forages in pairs or family groups for large insects and fruit. Utters various ratchet-like grunts that have a croaking quality; harsher than those of Black-crowned Tityra.

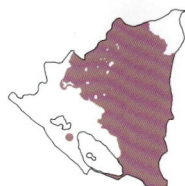

Black-crowned Tityra (Titira Coroninegra)   *Tityra inquisitor*
7.5 in (19 cm). **All-black bill** and **black on crown** rule out larger Masked Tityra; also note **lack of red on face**. Uncommon in Caribbean lowlands and foothills; to 2,600 ft (800 m). Rare on eastern slopes in the Northern Highlands; to 4,400 ft (1,350 m). Accidental in Pacific foothills, with only a single record (Jan 2017). Inhabits humid lowland forest and edge, and tall second growth with emergent trees. Usually in pairs but also in small groups, forages at mid-canopy for fruit and large insects. Utters a variety of scratchy and slightly distressed cackles; softer than those of Masked Tityra.

Northern Schiffornis

Speckled Mourner

female

male

Masked Tityra

A pair of Masked Tityras at their cavity nest.

female

male

Black-crowned Tityra

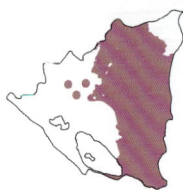

Cinnamon Becard (Cabezón Canelo)   *Pachyramphus cinnamomeus*
5.5 in (14 cm). Has mostly cinnamon and rufous plumage with exception of **dusky lores** and **buffy supraloral stripe**. Uniform rufous crown distinguishes it from other female becards; uniform color of wings further distinguishes it from female Gray-collared Becard; **graduated tail** further distinguishes it from larger female Rose-throated Becard. Uncommon in Caribbean lowlands and foothills; to 2,600 ft (800 m). Historical records suggest it was at least accidental on eastern slopes of Northern Highlands (Arenal NR, 1917; San Rafael del Norte, 1929; Peñas Blancas NR, 1929); to 4,100 ft (1,250 m). Favors humid lowland forest edge, second growth, gallery forest, open areas with scattered trees, and cloud forest edge. Alone or in pairs, forages in canopy, but may descend to nearby second growth in search of insects and berries. Sings a sweet but plaintive, slightly down-slurred *feeur* whistle, typically followed by several rapid-fire notes in the same pitch.

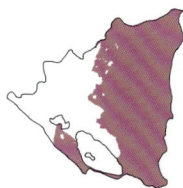

White-winged Becard (Cabezón Aliblanco)   *Pachyramphus polychopterus*
5.5 in (14 cm). On male, **black nape** and **lack of whitish supraloral stripe** distinguish it from male Gray-collared Becard (little range overlap). On female, combination of **pale spectacles, olive upperparts, and cinnamon-rufous edging on dark wing coverts** is distinctive. Common in Caribbean lowlands and foothills; very rare in southern Pacific lowlands and foothills; to 2,600 ft (800 m). Accidental on eastern slopes of Northern Highlands. Prefers second growth adjacent to forest patches; in the Caribbean, can also be found in open areas with scattered trees. Alone or in pairs, forages from mid-canopy to canopy for insects and berries; joins mixed-species flocks. Whistles several *chew* notes, the first followed by a brief pause, which is then followed by a cascading series of notes.

Gray-collared Becard (Cabezón Collarejo)   *Pachyramphus major*
6 in (15 cm). On male, **pale gray nape** and **white supraloral** distinguish it from male White-winged Becard. On female, **whitish supraloral** and **dark wings with rufous edgings** distinguish it from female Rose-throated Becard. Uncommon on eastern slopes of Northern Highlands (but absent on Sierra Dipilto-Jalapa); above 4,000 ft (1,200 m). Reaches southernmost distribution in Nicaragua. Occurs in cloud forest, highland pine and pine-oak forest, and forest edge. Single or in pairs, forages from mid-canopy to canopy; joins mixed-species flocks. Whistles a clear and high-pitched *súp súp-wee*.

Rose-throated Becard (Cabezón Gorgirrosado)   *Pachyramphus aglaiae*
6.5 in (17 cm). Only Nicaraguan becard with **square tail** (not graduated). Male is mostly gray, with noticeable **black crown** (pale rose throat is seldom visible). Uniform rufous wings distinguish female Rose-throated from female Gray-collared Becard and black crown separates it from Cinnamon Becard. Two subspecies occur. *Latirostris* subspecies: Male is paler gray; common breeding resident in Pacific lowlands and foothills; to 2,300 ft (700 m). *Hypophaeus* subspecies: Male is darker gray; uncommon winter resident (roughly Sept to April) in Northern Highlands and southern Caribbean; to 4,400 ft (1,350 m). Winter resident birds probably join resident population in Pacific lowlands and foothills. Found in dry forest, gallery forest, secondary forest, forest edge, open woodland, and open areas with scattered trees. Forages, alone or in pairs, from mid-canopy to canopy, gleans insects from vegetation and eats small fruit. Inconspicuous, often silent, and easily overlooked. Sometimes displays a complex repertoire of vocalizations, including a sharply descending *peeur* whistle (0.5 second in length) and a variety of squeaky chips and twittering sounds.

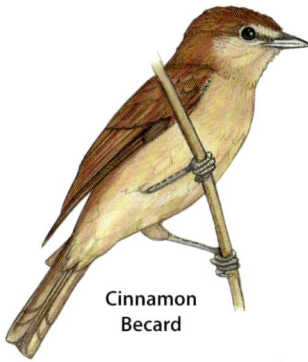

**Cinnamon
Becard**

male

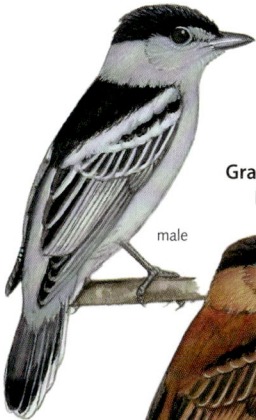

**Gray-collared
Becard**

female

**White-winged
Becard**

male

female

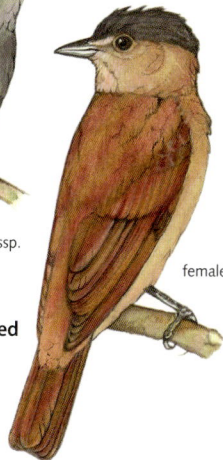

male
latirostris ssp.

**Rose-throated
Becard**

female

An extremely diverse family of frugivorous, canopy dwelling birds. Contingas occur only in the Neotropics. Of the 6 species that occur in the country, only the Rufous Piha does not exhibit extreme sexual dimorphism. Males have vibrant colors or ornamental feathers and skin flaps, while females are more cryptic; often the best way to identify a drab female is to look for her colorful mate. Because of their reliance on fruiting trees, several members of this family perform seasonal or altitudinal migrations to search for fruit. Migration patterns within Nicaragua are poorly understood, but information about such migrations is essential to developing conservation programs.

Purple-throated Fruitcrow (Frutero Gorgirrojo) *Querula purpurata*
11 in (28 cm). If not seen clearly, best distinguished by **husky shape** and **stout, gray-blue bill**. Male's **purplish-red throat** is diagnostic, but not visible in poor light. Rare to uncommon in southeastern Caribbean lowlands and foothills; primarily known from Río San Juan WR and Indio Maiz BR, but also reaches Punta Gorda NR; to 1,300 ft (400 m). Reaches northernmost distribution in Nicaragua. Occurs in humid lowland forest, in mid-canopy and canopy. Very social; found in groups of 6 or more roaming forest or perched together on the same branch. Eats fruit and large insects, and descends to lower levels and forest edge to visit fruiting trees. Joins flocks of other large frugivorous birds like toucans, oropendolas, and cotingas. Loudly whistles a worried *huáp*, sometimes repeated quickly though sporadically. Into main song, sometimes interjects a smooth ascending *wuuUP*.

Bare-necked Umbrellabird *Cephalopterus glabricollis*
(Pájaro Sombrilla Centroamericano)
M 17 in (43 cm); F 14.5 in (37 cm). **Large size** and **long crest** (shorter on female) make it unmistakable. Very rare, with records from Nov to Feb. Birds from breeding populations in Costa Rican highlands migrate short distances to Nicaragua. Only found south of Lake Nicaragua and Indio Maíz BR; to 1,300 ft (400 m). Reaches northernmost distribution in Nicaragua. Prefers humid lowland forest, swamps, old second growth, abandoned cacao plantations, and fruiting trees in semi-open areas. From understory to mid-canopy, forages for fruit, insects, lizards, and frogs. Usually alone, but migrating groups are sometimes seen flying over rivers and open areas. Most likely silent while in Nicaragua, when it is away from breeding distribution.

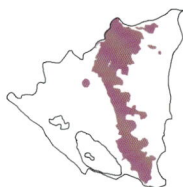

Lovely Cotinga (Cotinga Linda) *Cotinga amabilis*
7.5 in (19 cm). Flamboyant male is unmistakable; female could be confused with a dove, but **white scaling on brown upperparts** and **spotted underparts** are distinctive. Rare to locally common in foothills of Saslaya NP and very rare in Caribbean lowlands; from 1,600 to 4,000 ft (500 to 1,200 m), but may descend to 3,300 ft (100 m). Two historical records suggest it was once rare in the Northern Highlands (Peñas Blancas NR, June 1908 and June 1909). Probably performs seasonal altitudinal migration corresponding to availability of fruit (not yet documented). Inhabits humid lowland forest, forest edge, and nearby second growth with emergent trees. Moves alone within canopy; also descends to lower canopy to find fruiting trees with berries, and on such forays it is sometimes seen in groups. Individuals perch in very high, emergent, bare branches. Usually silent and hard to detect, it is potentially more numerous than the few records suggest; can be detected by the soft clicking made by their wings in flight.

Purple-throated Fruitcrow

female

male

Bare-necked Umbrellabird

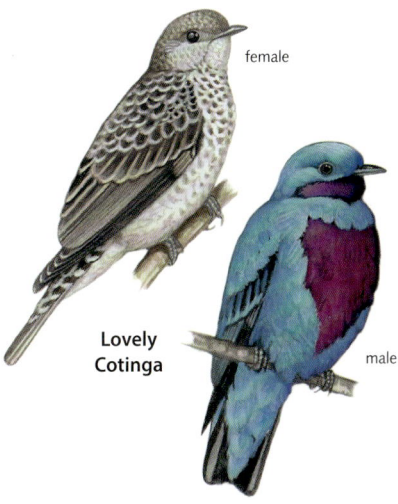

male

female

Lovely Cotinga

female

male

Lovely Cotingas typicaly perch on exposed branches above the canopy

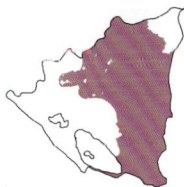

Rufous Piha (Pia Rojiza) — *Lipaugus unirufus*

9.5 in (24 cm). Almost identical to smaller Rufous Mourner (p. 276), but with **stouter bill** and **paler throat**; pale, indistinct eye ring further distinguishes it. Lack of dark wing coverts and tawny wing bars distinguish it from smaller Speckled Mourner. Locally common in foothills and highlands of Musún NR and Saslaya NP; locally common in Indio Maíz BR but uncommon in rest of Caribbean; to 3,300 ft (1,000 m). Very rare on eastern slopes of Northern Highlands; to 4,600 ft (1,400 m). Forages for insects and fruit in mid-canopy and canopy of cloud forest and humid lowland forest. Perches motionless for long periods of time; easily overlooked. Given its height in the canopy, sudden vocalizations made in response to forest noises are not easy to track down. Delivers an explosive *PEEAAH!* that can be startling in the context of a quiet forest. Also whistles a liquid, clear *peeur* (downslurred) and an erratic *wéeúr*.

Three-wattled Bellbird — *Procnias tricarunculatus*
(Pájaro Campana Centroamericano)

M 12 in (30 cm); F 10 in (25 cm). Male's chestnut-brown body, white head and breast, and **black wattles** make him unmistakable; on female, large size and **streaked underparts** on **lemon-yellow plumage** are distinctive. Very common in highlands of Saslaya NP, uncommon in Musún NR, and otherwise uncommon in Caribbean foothills and lowlands; uncommon on eastern slopes of Northern Highlands and northern region of Sierra Chontaleña; uncommon on Maderas Volcano, on Ometepe Island. Breeds in Northern Highlands and other high elevation locations within their distribution, from Feb to July; above 3,300 ft (1,000 m). After breeding season, it migrates and descends in elevation to the Caribbean lowlands and foothills; below 2,600 ft (800 m). Found in cloud forest, humid lowland forest, and occasionally in open areas with scattered tall trees. Male sings from an emergent bare branch over canopy and is difficult to locate from the forest floor; female moves in mid-canopy. Feeds on fruit, especially those of Lauraceae. Males unashamedly proclaim their near emblematic metallic *BOINK!*, which rises sharply in pitch on the second syllable. Also gives a less intense *bonk* and sweeter, high-pitched whistles such as *fa-siá*. **VU**

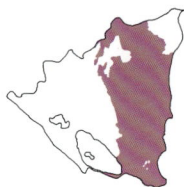

Snowy Cotinga (Cotinga Nevada) — *Carpodectes nitidus*

8.5 in (21 cm). On male, note white plumage and **dark eyes and bill**; unmistakable. Grayish female has **dusky wings** marked with **white wing bars** and **white edging on coverts and secondaries**; shorter tail and lack of bright color on head eliminate confusion with both tityras (p. 288). Uncommon in Caribbean lowlands and foothills; to 1,600 ft (500 m). Occurs in humid lowland forest, along streams, in secondary forest, and in forest remnants. Usually seen alone or in small groups perched high in the canopy (sometimes just a single male appears on an emergent bare branch) or in bigger groups feeding at a fruiting tree. Performs seasonal movements according to availability of fruit. Endemic to Central American Caribbean slope EBA.

Rufous Piha

male

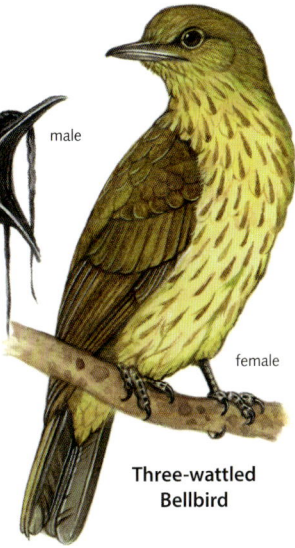

female

Three-wattled Bellbird

female

male

Snowy Cotinga

These small, plump birds, with short bills and tails, occur exclusively in the Neotropics. On males, black plumage contrasts strikingly with yellow, red, orange, blue, and other bright colors. Females and immature birds are drab, with olive-green or olive-yellow coloration. Male manakins are known for their extraordinary dancing skills, showcased during courtship displays at lek sites on the ground or small branches. These displays are performed to impress observing females. Finding a manakin lek in the forest understory is a quintessential Neotropical forest experience.

Long-tailed Manakin (Saltarín Toledo) *Chiroxiphia linearis*

4.5 in (11 cm); on M tail adds 5 in (13 cm); on F 1 in (3 cm). Only Nicaraguan manakin with **elongated central rectrices** (very long on male, short on female). Common in Pacific (especially in Cosigüina Volcano NR, San Cristóbal-Casita NR, Mombacho Volcano NR, and on Sierras Managua); rare to uncommon on Sierra Isabelia and Sierra Dariense of Northern Highlands; to 4,300 ft (1,300 m). Prefers dry forest, gallery forest, and tall second growth; in smaller numbers, also found in cloud forest. Alone or in small groups; moves in understory and mid-canopy in search of fruit. Two or 3 males collaboratively perform on small branches with hop-and-flutter flight courtship displays, simultaneously uttering a buzzy *za-WOW!* Best known for its melodic, arcing *to-lé-do* song (thus the Spanish common name), which is often followed by a plaintive and nasal *wyaah*. Also whistles a clear *féeo-fi*.

White-ruffed Manakin (Saltarín Gorgiblanco) *Corapipo altera*

4 in (10 cm). **Erectile white ruff** on **glossy blue-black body** makes male unmistakable. Female's **whitish-gray throat** disinguishes it from female Red-capped Manakin and female White-collared Manakin (both sympatric). Local and abundant in Saslaya NP and common in highlands of Musún NR (and probably in nearby Quirragua NR), but generally uncommon in Caribbean foothills and rare in lowlands; rare on eastern slopes of Northern Highlands (Jan to June); above 2,000 ft (600 m). After breeding season (probably July to Jan), descends from highlands to foothills; as low as 1,000 ft (300 m). Moves through understory and mid-canopy in search of fruit and small insects. Several males perform bounding, fluttering flight courtship displays at leks. Emits a variety of short *spreeá* twitters, often ascending before ending with a piercing tone.

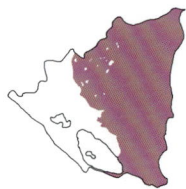

White-collared Manakin (Saltarín Cuelliblanco) *Manacus candei*

5 in (12 cm). Male is unmistakable. **Orange legs** distinguish female from other female manakins. Abundant in Caribbean lowlands and foothills; uncommon on eastern slopes of Sierra Amerrisque; common south of Lake Nicaragua (Guatuzos WR); to 2,600 ft (800 m), but found higher on Cerro Musún NR (and probably in nearby Quirragua NR), to 4,000 ft (1,200 m). Rare on eastern slopes of Northern Highlands; to 4,000 ft (1,200 m). Prefers dense tangled understory of humid lowland forest edge, secondary forest, and streamsides in woodlands. Moves in loosely formed groups that are detected by the slapping and buzzing mechanical noises made with its wings, especially while darting between vertical perches at leks. Calls include a soft and churring *meur*, descending in pitch.

Red-capped Manakin (Saltarín Cabecirrojo) *Ceratopipra mentalis*

4.5 in (11 cm). Striking male is unmistakable. **Yellowish thighs** and lack of pale gray throat distinguish female from female White-ruffed Manakin; female distinguished from female White-collared Manakin by gray legs (not orange). Common in Caribbean, but rare south of Lake Nicaragua (Guatuzos WR); uncommon on eastern slopes of Northern Highlands (absent from Sierra Dipilto-Jalapa), to 4,300 ft (1,300 m). Five individual records from Rivas Isthmus of southern Pacific (San Emilio, April 1905) suggest that a population there was extirpated because of habitat loss. Found in humid lowland forest and cloud forest, and also wanders into adjacent tall second growth and open areas. Moves within understory and mid-canopy to feed on fruit. Sings a distinctive *pi-pi-pi-seéuuu-CHIP* phrase, consisting of 3 introductory short notes, a drawn out middle note that quickly rises before trailing off in intensity, and a final harsh note.

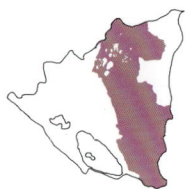

Gray-headed Piprites (Saltarín Cabecigrís) *Piprites griseiceps*

4.5 in (11 cm). Nondescript, with **bold, white eye ring** on gray head; coloration and rounded-head appearance is reminiscent of female manakins, but tail is longer. Rare in Caribbean lowlands and foothills; to 2,000 ft (600 m). Joins mixed-species flocks in understory and mid-canopy of humid lowland forest and forest edge. Song is a mix of short, rich whistles and rolling notes: *whit wit wit wi-dah-ridiridiridi-reer* (entire song lasts 2 seconds). Call is a single *whit!* note. Endemic to Central American Caribbean slope EBA.

Long-tailed Manakin

female

male

immature male

Two male Long-tailed Manakins perform a courtship display at a lek, with a female looking on.

White-ruffed Manakin

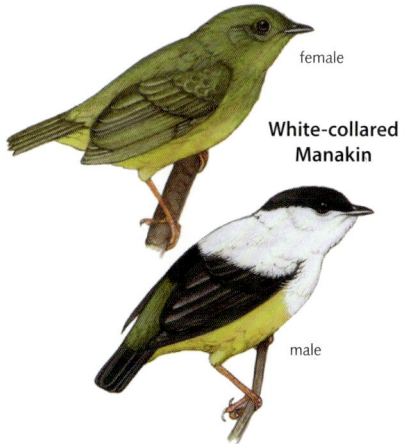

female

male

White-collared Manakin

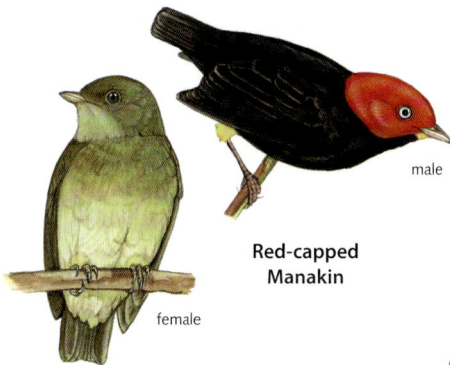

female

male

Red-capped Manakin

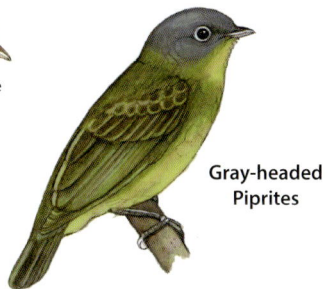

female

male

Gray-headed Piprites

Vireos are easily confused with New World warblers (family Parulidae). They have stout, hook-tipped bills, while warblers have delicate bills without a hooked tip; they also forage more slowly than do warblers and are generally more robust. Vireos are primarily found within forest, where they search for insects and some fruit; they are prone to join mixed-species foraging flocks. Song and calls are persistent, but not known to be melodious. Of the 15 species in Nicaragua, 8 are migratory (7 from North America and 1 from South America).

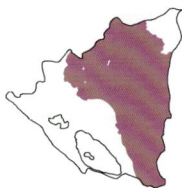

Tawny-crowned Greenlet (Verdillo Leonado)   *Tunchiornis ochraceiceps*

4.5 in (11 cm). **Pale white iris** rules out female Plain Antvireo (p. 244); **lack of wing bars** distinguishes it from female Streak-crowned Antvireo (p. 244). Locally common on eastern slopes of Northern Highlands; in the Caribbean, common in foothills (common in foothills and highlands of Saslaya NP) and uncommon in lowlands; to 4,400 ft (1,350 m). Found in humid lowland forest and cloud forest. Roams in dense understory in search of insects and berries. Found in pairs, small groups, and mixed-species flocks composed of antwrens, antvireos, ant-tanagers, and other small passerines. Whistles a high-pitched and monotone *fweep*, with a clear serene quality; this is repeated at length every 1 to 3 seconds. Gives an agitated and nasal *da-da-da…*; during pauses in the series, sometimes adds in a soft *wit*.

Lesser Greenlet (Verdillo Menudo)   *Pachysylvia decurtata*

4 in (10 cm). Very similar to Nashville Warbler (p. 366), but has pale olive-green throat and breast (not yellow) and stout vireo bill. Combination of **white eye ring**, short tail, and relatively large head distinguishes it from any similar New World warblers. Common countrywide (slightly less numerous in Pacific); to 4,900 ft (1,500 m). Occurs in humid lowland forest and edge, cloud forest and edge, dry forest, shade coffee plantations, and tall second growth. In pairs, small groups, and mixed-species flocks, forages very actively for insects and berries, from mid-canopy to canopy. Song consists of two sweetly whistled phrases that alternate sporadically: *fee-áw-pur* (in a down-up-down pattern) and *fít-per-áw* (in an up-down-up pattern); there is noticeable spacing between the phrases. Calls with a buzzy *twí*, singly or in duplets and triplets.

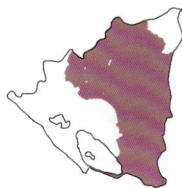

Green Shrike-Vireo (Vireón Esmeralda)   *Vireolanius pulchellus*

5.5 in (14 cm). Similar to Blue-crowned Chlorophonia (only small range overlap), p. 340, but note larger size; **stout, long, hook-tipped bill**; and **yellow throat**. Immature is drabber, lacks blue crown, and has a distinctive yellow superciliary and malar stripe. Rare on eastern slopes of Northern Highlands; common in Caribbean foothills and highlands but less numerous in lowlands; generally to 3,300 ft (1,000 m) but occasionally wanders up to 4,300 ft (1,300 m). Dwells in humid lowland forest and, occasionally, in cloud forest and edge. Forages very high in canopy (more often heard than seen), usually in pairs, for large insects, caterpillars, berries, and seeds; sometimes accompanies mixed-species flocks down to mid-canopy. Song is a harmoniously whistled *pídá-pídá-pídá* (1 second in length and repeated every 1–2 seconds). Calls include a sharp chatter (*chi-chi-chi…*) and a harsh, scratchy *dreh* note.

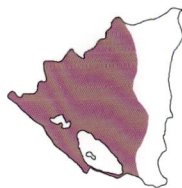

Rufous-browed Peppershrike (Vireón Cejirrufo)   *Cyclarhis gujanensis*

6.5 in (16 cm). Stout hook-tipped bill and **broad, rufous superciliary** (extends onto forehead) make it unmistakable. Common countrywide; above 1,000 ft (300 m), but uncommon in lowlands. Found in secondary forest, second growth, woodland, dry forest, mangroves, humid lowland forest, and cloud forest edge. Generally forages in pairs but sometimes alone, for insects and caterpillars taken from the outer foliage of the crowns of trees. Employs an impressive repertoire of vocalizations, including a variety of fast, oscillating, warbling songs, all with a bubbly quality.

Tawny-crowned
Greenlet

Lesser
Greenlet

Green
Shrike-Vireo

Rufous-browed
Peppershrike

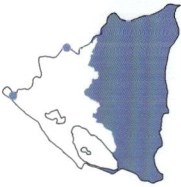

White-eyed Vireo (Vireo Ojiblanco)   *Vireo griseus*
5 in (13 cm). Combination of **white iris** and **yellow spectacle** is distinctive, also note conspicuous white wing bars. Immature has brown to grayish iris; bold yellow spectacles and **yellow flanks** distinguish immature from smaller Mangrove Vireo. Winter resident (mid-Oct to April). Rare in Caribbean. Accidental in Pacific and in Northern Highlands (El Corozo, Nueva Segovia Dept., Jan 1955; Hato Nuevo RSP, Chinandega Dept., May 2015). Prefers scrub, semi-open areas, and disturbed areas. Easily overlooked because of secretive behavior. Sometimes vocalizes with jumbled phrases, including a harsh *chik-der-vee*.

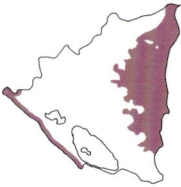

Mangrove Vireo (Vireo de los Manglares)   *Vireo pallens*
4.5 in (11 cm). Two subspecies occur in Nicaragua, both with **prominent lores** connecting to **narrow upper eye crescents**; also note **two faint, whitish wing bars**. Similar to larger White-eyed Vireo, but compare iris color and head markings. *Pallens* subspecies: Duller overall, gray to amber iris, and most often with pale lemon lores (although some individuals have *semiflavus*-like coloration). Locally common on Pacific coast; restricted to mangroves. *Semiflavus* subspecies: Yellower overall, amber iris, and bold yellow lores. Locally common in Caribbean lowlands. Found in mangroves, scrub, second growth, and forest edge. They forage alone or in pairs for insects, from understory to mid-canopy. Sings a dry, buzzy, twangy 4- or 5-noted *twee-twee-twee-twee* (1 second in length, with each note upslurred). Calls include a short, sharp, downslurred *tweo*; a scolding chatter that mixes a raspy *chí-chí-chí…* with a dry *ba-da-da-da…* (reminiscent of a car alarm).

Bell's Vireo (Vireo Pálido)   *Vireo bellii*
5 in (12 cm). Drab. **Faint white superciliary** and **broken eye ring** form a weak spectacle; superciliary is less prominent than on larger Warbling Vireo (p. 302). **White wing bars are faint** (upper one indistinct) and tertials have white edgings. Very warbler-like in appearance but bill is stout and hook-tipped. Accidental winter resident, known from a single record in Northern Highlands (Matagalpa Dept., April 1917). Recent records in southwestern Honduras (bordering Nicaragua) suggest it might occur in Gulf of Fonseca area (Chinandega Dept.). Searches for insects and spiders in dense understory of young second growth, scrub, and thorn forest. Calls with an agitated, raspy, nasal *whA!*, delivered quickly and sporadically. **NT**

Yellow-throated Vireo (Vireo Pechiamarillo)   *Vireo flavifrons*
5.5 in (14 cm). Only vireo with combination of **yellow breast** and **white underparts**. Also note **white wing bars** and **yellow spectacles**. Common winter resident countrywide (Oct to April); to 5,600 ft (1,700 m). Visits a variety of forest types, forest edge, coffee plantations, and tall second growth. Forages in mid-canopy and canopy; joins mixed-species flocks composed of warblers, tanagers, and greenlets. When gleaning insects, searches thoroughly in one spot before moving on to next search site. Gives a somewhat aggressive cackle: *chi-chi-chi…* . During boreal spring months, sings a variety of slurred, squeaky phrases, with 1–3 seconds between phrases.

Blue-headed Vireo (Vireo de Anteojos)   *Vireo solitarius*
5.5 in (14 cm). Note **bold white spectacles; blue-gray head, olive upperparts**; and white wing bars. Very similar to smaller Plumbeous Vireo (formerly considered conspecific), but head, back, and spectacles are richer in coloration and show more contrast. Winter resident (Sept to April). Common in the Northern Highlands; above 2,600 ft (800 m). Very rare in Pacific and Caribbean. Found in a variety of forest habitats and edges. Usually alone; joins mixed flocks of warblers, tanagers, and other vireos foraging in mid-canopy and canopy for insects and small fruit.

Plumbeous Vireo (Vireo Plomizo)   *Vireo plumbeus*
5 in (12 cm). Note **white spectacles, gray head, gray-olive upperparts, yellowish flanks**, and white wing bars. Similar to larger Blue-headed Vireo (formerly considered conspecific), but head, back, and spectacles are drabber and show less contrast. Uncommon in Northern Highlands; above 3,600 ft (1,100 m). Reaches southernmost distribution in Nicaragua. Found in highland pine and pine-oak forest and forest edge. Forages from mid-canopy to canopy; joins mixed-species flocks of warblers and tanagers. Sings a series of sweetly slurred phrases, with some buzz notes: includes a down-up *feu-wEE*, an up-down *fít-beuw*, and an arcing *zeeU*.

White-eyed
Vireo

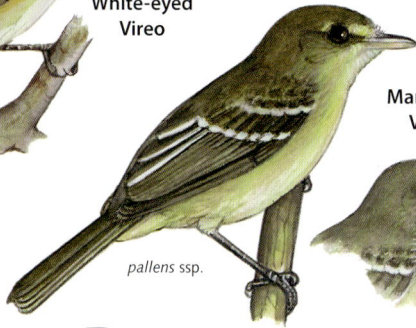

Mangrove
Vireo

pallens ssp.

semiflavus ssp.

Bell's Vireo

Yellow-throated
Vireo

Blue-headed
Vireo

Plumbeous
Vireo

Philadelphia Vireo (Vireo Canadiense) *Vireo philadelphicus*

5 in (13 cm). Only vireo with **combination of white superciliary** and **yellow throat**. Further distinguished from Warbling Vireo by dark lore and bolder eye line. Very similar to smaller Tennessee Warbler (p. 366), but stout vireo bill and more sluggish foraging movement rule it out. Winter resident (mid-Sept to April). Common in Pacific and Northern Highlands; rare in Caribbean; to 4,600 ft (1,400 m). Occurs in dry forest, woodland, cloud forest edge, shade coffee plantations, tall second growth, and thorn forest. Usually solitary, but also joins mixed-species flocks or groups of Tennessee Warblers; forages from mid-canopy to canopy for insects, fruit, and seeds.

Warbling Vireo (Vireo Grisáceo) *Vireo gilvus*

5.5 in (14 cm). On **pale face**, note **white superciliary, pale lore**, and **faint dark eye line**. Nominate subspecies (*gilvus*) is overall darker and slightly larger with thicker bill than subspecies *swainsoni*. Distinguished from smaller Philadelphia Vireo by pale lore (not dark) and white throat (not yellow). Lacks black border on superciliaries (as seen on larger Red-eyed Vireo); also lacks wing bars (as seen on smaller Bell's Vireo, p. 300). Stout vireo bill and more sluggish foraging movement rule out Tennessee Warbler (p. 366). Winter resident (mid-Sept to April). Common in Northern Highlands and uncommon in Pacific; to 5,000 ft (1,550 m). Rare in Caribbean lowlands. Prefers thorn forest, highland pine and pine-oak forest, dry forest, second growth, and shade coffee plantations. Forages in canopy for insects and berries, alone or in mixed-species flocks of warblers and other vireos. Calls with a labored *whyáh* (sounds like a nasal whine).

Brown-capped Vireo (Vireo Montañero) *Vireo leucophrys*

4.5 in (12 cm). Very small. **Brown crown and nape** distinguish it from Warbling Vireo and Philadelphia Vireo. White superciliary distinctly contrasts with **dark eye line**. Stout vireo bill rules out Tennessee Warbler (p. 366). Rare in Northern Highlands; above 4,300 ft (1,300 m). Found in cloud forest and edge, secondary forest, and shaded coffee plantations. In canopy, forages alone or in pairs for insects and berries, also joins mixed-species flocks. When foraging, sometimes hangs upside down to inspect curled leaves. Sings a fast-paced, jumbled series of alternating high and low notes (1–2 seconds in length).

Red-eyed Vireo (Vireo Ojirrojo) *Vireo olivaceus*

5.5 in (14 cm). **White superciliary bordered by black** is distinctive. Head pattern is bolder than on Yellow-green Vireo; additionally, yellow on sides of neck, flanks, and undertail coverts is paler, and dark bill is shorter. Red iris is not always seen. Common passage migrant countrywide (mid-Aug to Nov and March to mid-May); to 5,600 ft (1,700 m). Forages in mid-canopy and canopy for insects and fruit in a variety of forest habitats, woodland and edge, and gardens with trees. Joins mixed-species flocks. Generally silent during migration, but occasionally gives a buzzy, nasal *míyah* call.

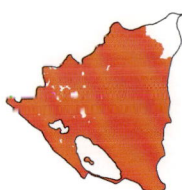

Yellow-green Vireo (Vireo Cabeciqrís) *Vireo flavoviridis*

5.5 in (14 cm). Shows **bright yellow on sides of neck, flanks, and undertail coverts**; further distinguished from Red-eyed Vireo by longer, paler bill and less contrast on head pattern. Note **yellow wing edgings on flight feathers**. Breeding migrant (March to Oct). Abundant in Pacific lowlands and foothills (especially in Sierras Managua and Los Pueblos Plateau). Uncommon in Northern Highlands; to 4,400 ft (1,350 m). Rare in Caribbean. Prefers secondary forest, dry forest, gallery forest, forest edge, and second growth near cloud forest and humid lowland forest; also can be found in wooded urban gardens. Forages in canopy for insects and berries. Sings at length a compilation of sweet, cheery, slurred phrases, each characterized by quick rises and falls in pitch (1–2 second pause between phrases). Calls with a wheezy and downslurred *wyeea!*

Philadelphia
Vireo

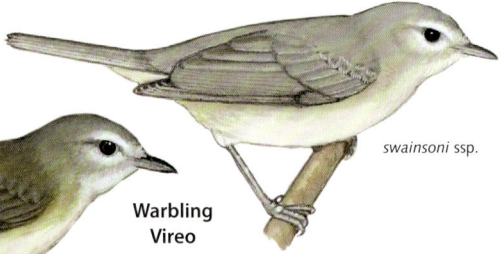

Warbling
Vireo

swainsoni ssp.

gilvus ssp.

Brown-capped
Vireo

Red-eyed
Vireo

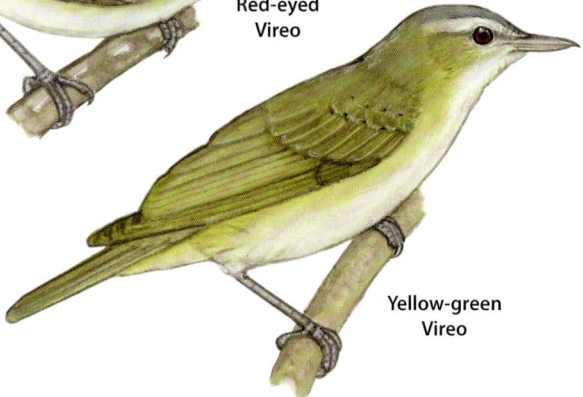

Yellow-green
Vireo

The members of this cosmopolitan family are known for their social behavior, high intelligence, and noisy vocalizations. They have strong bills and legs. Corvids eat a wide variety of food, including other birds' eggs and nestlings. They are most often seen moving in small flocks. Of the 7 species in Nicaragua, 5 reach their southernmost distribution in the country.

White-throated Magpie-Jay (Urraca Copetona) — *Calocitta formosa*

19 in (48 cm). Size and plumage are striking. **Black crest** adorns the head and **black "necklace"** interrupts all-white underparts. **Long, graduated tail** is unmistakable. Abundant in Pacific; uncommon in dry intermontane valleys of Northern Highlands and south through Sierra Chontaleña, to the eastern and southern borders of Lake Nicaragua; to 4,600 ft (1,400 m). Generally favors dry forest and edge, woodland, mangroves, open areas with scattered trees, thorn forest, and scrub; when it is found in humid areas, it prefers open agricultural land with scattered trees. Conspicuously forages in small noisy groups for bird eggs and nestlings, fruit, seeds, insects, and trash piles. Emits a variety of unmusical and querulous vocalizations, including a grating *ARGH!*, a gravelly churr, and a high pitched, strained *peeó*.

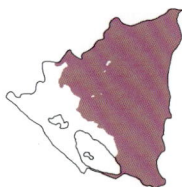

Brown Jay (Urraca Parda) — *Psilorhinus morio*

16 in (40 cm). The only jay with bicolored plumage; note **dark brown upperparts** and white underparts (brown on head extends below to breast). On immature, note yellow bill and eye ring. In flight, also note white-tipped outer rectrices. Common on eastern slopes of Northern Highlands; to 5,600 ft (1,700 m). Uncommon in northern region of Sierra Chontaleña, on eastern slopes of Sierra Amerrisque, and in Caribbean lowlands and foothills; to 3,000 ft (900 m). Only jay found in most of the Caribbean. In flocks of 10 or more, roams open and semi-open areas in search of fruit, grains, large insects, lizards, frogs, and bird eggs and nestlings. Screams a loud, strained *PIYAH*. Speed of delivery varies; often repeated at length.

Green Jay (Urraca Verde) — *Cyanocorax yncas*

11 in (27 cm). A gorgeous bird. Vagrant. Only 2 records in Nicaragua, both in the Mosquitia, near Coco River (Jan and Feb 2000). Prefers forest edge and plantations. Forages in understory and mid-canopy; pairs or small flocks feed on fruit, grains, and small vertebrates. Delivers a metallic, rapid-fire *TSIT-TSIT-TSIT-TSIT*; also vocalizes with raspy, agitated screams.

White-throated Magpie-Jay

immature

adult

Brown Jay

Green Jay

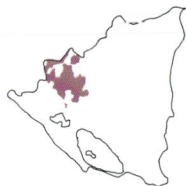

Bushy-crested Jay (Urraca Pechinegra)   *Cyanocorax melanocyaneus*

12 in (31 cm). Note **yellow iris, black head**, and blue on rest of body (with exception of legs); may appear black in poor light. Immature has yellow bill and dark iris. No similar jay within its range. Common to locally abundant in Northern Highlands; above 3,000 ft (900 m). Occurs in cloud forest and edge, highland pine and pine-oak forest, second growth, and shaded coffee plantations. Moves in groups of 10 or more at all levels of vegetation and on ground. Forages for insects, seeds, figs, and other fruit. Opportunistically joins mixed-species flocks to follow army ant swarms. Commonly makes a variety of harsh, raspy vocalizations, including a repeated *ERH ERH ERH…*, often very raucous; a double-noted *choc-tá*; and other sweeter, chattering vocalizations. Reaches southernmost distribution in northern Nicaragua. Endemic to North Central American highlands EBA.

Steller's Jay (Urraca de los Pinares)   *Cyanocitta stelleri*

12 in (30 cm). **Horizontal crest** is prominent; **white eye crescents** mark an otherwise entirely blue body. Locally common in Northern Highlands, as far south as Sierra Isabelia (most highly concentrated in Sierra Dipilto-Jalapa); above 4,000 ft (1,200 m). There are several historical records from Yalí Volcano NR, where once common but now very rare. Reaches southernmost distribution in Nicaragua. Restricted to highland pine and pine-oak forest, where it primarily moves in canopy but also descends to ground. Found in pairs or groups; sometimes joins mixed-species flocks. Feeds on insects, seeds, berries, small vertebrates, and bird eggs and nestlings. Communicates with agitated, scratchy repetitions (of *neeáh* or *yeáh*); also makes a distinctly different *da-ríe da-ríe*. Its forlorn *BEOW* scream faintly resembles that of a Red-tailed Hawk.

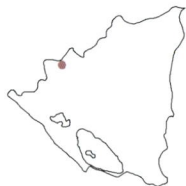

Unicolored Jay (Urraca Unicolor)   *Aphelocoma unicolor*

13 in (33 cm). Only jay with **uniform purplish-blue plumage**; immature is dusky-brown, with yellow bill. Locally uncommon on Cerro Mogotón in Dipilto-Jalapa NR; above 5,400 ft (1,650 m). Reaches southernmost distribution in Nicaragua. Occurs in cloud forest and edge and highland pine and pine-oak forest. Forages omnivorously in pairs or small groups in canopy (making it very difficult to see). Easier to locate by vocalization: Produces raspy, sporadic combinations of *ruáh* and *rí* calls, often changing in inflection and intensity and sometimes turning into drawn-out squeals or short grunts.

Common Raven (Cuervo Común)   *Corvus corax*

24 in (61 cm). **Large** and **all black**, with **long, massive bill** and **long, wedge-shaped tail**. In flight, note long narrow wings and finger-like primaries. Often soars. Rare but locally uncommon in Northern Highlands; accidental in Pacific (a single record on Telica Volcano, Sept 2016); generally above 4,000 feet (1,200 m) but may descend in search of food, as demonstrated by an exceptional record in western Caribbean lowlands at 600 ft (200 m). Reaches southernmost New World distribution in Nicaragua. Found in highland pine and pine-oak forest, dry forest, and thorn forest; prefers habitat with nearby cliffs used for nesting. Found alone or in pairs; feeds on carrion, grains, small animals, fruit, eggs, and at trash piles. In flight, makes a throaty yet subdued *rawh* call.

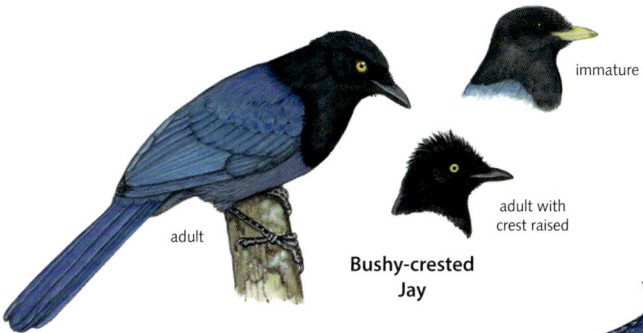

immature

adult

adult with
crest raised

**Bushy-crested
Jay**

Steller's Jay

immature

adult

**Unicolored
Jay**

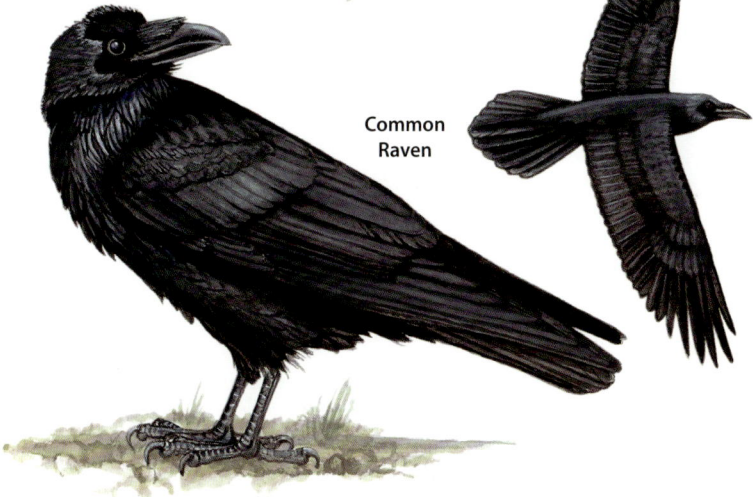

**Common
Raven**

These cosmopolitan aerial experts glide over open areas in search of flying insects. The majority of the species that occur in Nicaragua are migrants from North America (only 4 species breed in Nicaragua). Multiple species of swallow sometimes join together to form massive flocks, at which time careful attention to detail is necessary to sift through the species. Although they are superficially similar to swifts, swallows differ in plumage, body form, and flight pattern. Their plumages are mainly bicolored and often show colorful greens and blues, or combinations of red and orange hues. Flight pattern is smooth, with more gliding than in swifts; unlike swifts, swallows can be seen perched on branches, fence posts, and wires.

Purple Martin (Avión Púrpura) — *Progne subis*

8 in (20 cm). **Uniformly dark** male has **glossy purple and black plumage**. On female, upperparts lack the glossy purple on the forecrown, hindcrown, and nape; female is further distinguished from Gray-breasted Martin by underparts that are uniformly white, on which gray mottling varies in intensity. In flight, forked tail is usually visible at some point during their maneuvering. Passage migrant (Aug to Oct and Feb to April). Uncommon throughout Caribbean; rare in Pacific; very rare throughout Northern Highlands; generally found in lowlands and foothills. Migrates through the Caribbean lowlands in large flocks of up to several hundred. In flight, calls with rich, bubbly *twí* and *twer* notes.

Gray-breasted Martin (Avión Pechigrís) — *Progne chalybea*

6.5 in (16 cm). Both male and female have **purplish-blue and black upperparts** contrasting with whitish-gray underparts. It is distinguished from the female Purple Martin by uniform color from forecrown to nape and **gray breast band**. Common countrywide; to 4,600 ft (1,400 m). Forages over open areas, forest edge, and second-growth; often found in urban areas. Calls include a soft, trilled *chiua*, a clear *chi*, and a buzzy *zeep!*

Bank Swallow (Avión Zapador) — *Riparia riparia*

5 in (13 cm). **Broad, brown breast band** on **white throat** is diagnostic. Brown stripe on the middle of the breast is of variable intensity. (The SA migrant Brown-chested Martin has been recorded in the Guatuzo Plains in Costa Rica; Bank Swallow is smaller and with white extending up and behind the ear coverts.) Winter resident (mid-Aug to May); uncommon in Pacific lowlands, Guatuzo Plains, and San Juan River headwaters; very rare in rest of Caribbean. Winter resident often found near water. Passage migrant (mid-Aug to Oct and April to May); common in Pacific, Northern Highlands, Guatuzo Plains, and San Juan River headwaters; rare in the rest of the Caribbean. Known to migrate with Cliff Swallows and Barn Swallows. In flight, repeats muffled, electric *vrí* and *vree* notes.

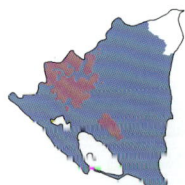

Northern Rough-winged Swallow (Golondrina Alirrasposa Norteña) — *Stelgidopteryx serripennis*

5 in (13 cm). Drab. Brown upperparts contrast with pale underparts. **Gray throat** and **brown rump** distinguish it from Southern Rough-winged Swallow. Both breeding resident and NA migrant populations occur in Nicaragua, but the distribution of each is not precisely understood. Breeding resident birds are uncommon and most likely limited to Northern Highlands (from 2,300 to 5,200 ft [700 to 1,600 m]) and Sierra Amerrisque (generally above 1,600 ft [500 m], but small numbers do occur at lower elevations in the southern Pacific during breeding season). Winter resident (Sept to April), uncommon countrywide; to 5,200 ft (1,600 m). Prefers to forage over water and nearby open areas; associates with mixed-species swallow flocks. In flight, calls with a combination of liquid, burry *zwert* and *zwít* notes.

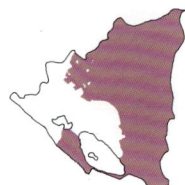

Southern Rough-winged Swallow (Golondrina Alirrasposa Sureña) — *Stelgidopteryx ruficollis*

5 in (13 cm). Drab. Brown upperparts contrast with pale underparts. **Cinnamon throat** and **gray rump** distinguish it from Northern Rough-winged Swallow. Common in Caribbean. Performs a post-breeding dispersal (Sept to March), when it is uncommon in Pacific (mainly south of Lake Managua) and on eastern slopes of the Northern Highlands; to 4,100 ft (1,250 m). Found foraging in open areas, generally close to water. In flight, calls with a combination of liquid, burry *zoo* and *zwee* notes.

Purple Martin

male

female

Gray-breasted
Martin

Bank Swallow

Northern
Rough-winged
Swallow

Southern
Rough-winged
Swallow

Tree Swallow (Golondrina Bicolor)   *Tachycineta bicolor*
6 in (15 cm). Bicolored pattern created by contrast between upperparts and underparts is similar to that on Blue-and-White Swallow, but note **white vent and undertail coverts**. Lack of white on rump distinguishes it from Mangrove Swallow and Violet-green Swallow. Diffuse gray breast band on immature is much fainter than that on the Bank Swallow. Winter resident (Oct to April); rare countrywide, but irruptive in some years; records to 4,100 ft (1,250 m) but could occur higher. Sometimes present in low numbers within mixed-swallow foraging flocks; during irruptive years, it forms same-species flocks of up to several hundred birds. Generally forages in open areas close to water. Gives a staccato *chik* note in flight, which, when large numbers vocalize, creates a liquid sound.

Mangrove Swallow (Golondrina Rabiblanca)   *Tachycineta albilinea*
4.5 in (11 cm). **Entirely white rump** distinguishes it from all other blue or green swallows. **Thin, white superciliary** that stops above mid-eye also sets it apart. Immature lacks blue-green color but has the diagnostic white rump. Common countrywide, but less likely in northern region of Northern Highlands; to 3,300 ft (1,000 m). Almost exclusively found near bodies of water or adjacent open areas, where it cruises for insects low over the water and conspicuously perches on debris jutting out of the water. In flight, calls with a soft, liquid *whít* or a buzzy *zeep!*

Blue-and-white Swallow (Golondrina Blanquiazul)   *Pygochelidon cyanoleuca*
5 in (13 cm). **Black vent and undertail coverts** distinguish it from all blue or green swallows. Status is unknown. Most likely occurs as a vagrant, with birds arriving from South America (*patagonica* subspecies) as they migrate north (May to Sept). On *patagonica* subspecies, note white extending onto black vent and undertail coverts. The CA breeding resident (*cyanoleuca* subspecies), which occurs in Costa Rica, has not been confirmed in Nicaragua.

Violet-green Swallow (Golondrina Cariblanca)   *Tachycineta thalassina*
5 in (13 cm). **White underparts extend onto the face and behind the eye**, unlike on similar blue or green swallows. **Dark central stripe on white rump** is also unique. Very rare winter resident (Nov to Feb), but irruptive in some years; records at Playitas-Moyuá-Tecomapa lagoons, Sébaco Valley, Casita Volcano, and throughout Northern Highlands. Forages over grasslands, forest edge, and marshes; joins mixed-swallow flocks but is often seen in large flocks of its own. In flight, calls with a screechy, erratic *choo* note.

Tree Swallow

Mangrove
Swallow

Blue-and-white
Swallow
(*cyanoleuca* ssp.)

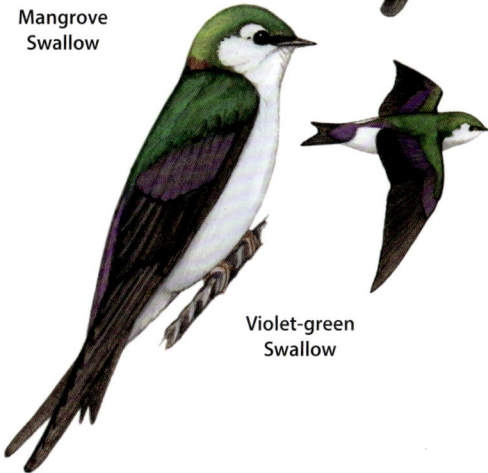

Violet-green
Swallow

Cliff Swallow (Golondrina Gorginegra)   *Petrochelidon pyrrhonota*
5.5 in (14 cm). Similar to the less common Cave Swallow, but distinguished by **chestnut on throat and sides of head**, **black throat patch**, and paler rump and flanks. *Pyrrhonota* subspecies (more likely) has **buffy forehead**, but beware the possibility of *melanogaster* subspecies, whose forehead is the same color as that on Cave Swallow. Rare winter resident (Aug to early May) in Pacific, Guatuzo Plains, and San Juan River headwaters. Passage migrant (Aug to early Nov and mid-Feb to early May); abundant in Pacific, Guatuzo Plains, and San Juan River; common on western slopes of Northern Highlands, and uncommon throughout rest of country; to 4,900 ft (1,500 m). Often migrates in flocks of several hundred that include Barn Swallows and Bank Swallows; winter residents are often found with flocks of Barn Swallows. In flight, repeats a rich *zee* and a buzzy, downslurred *zeeú!*

Cave Swallow (Golondrina Pueblera)   *Petrochelidon fulva*
5.5 in (14 cm). Similar to more common Cliff Swallow, but distinguished by the **cinnamon throat and sides of head** (which noticeably contrast with dark crown), lack of black throat patch, and darker rump and flanks. Cinnamon forehead further distinguishes it, but beware the possibility of Cliff Swallow (*melanogaster* subspecies), which has same forehead color. Winter resident (mid-Nov to mid-Feb); very rare in Pacific lowlands and extreme southwestern Caribbean, but possible into foothills and highlands. Joins mixed-species swallow flocks to forage in open areas. In flight, calls with an electric, laser-like *whulT* and a harsh, strained *CHU*.

Barn Swallow (Golondrina Común)   *Hirundo rustica*
5.5 in (14 cm). Most often seen of the migratory swallows. **Deeply forked tail** is unique among the swallows, although length of tail can vary with age and sex. **Chestnut forehead** contrasts with blue upperparts and **chestnut throat** contrasts with **cinnamon underparts**. Winter resident (Sept to April); abundant countrywide, with very large numbers possible during passage months (Sept to Oct and March to April); to 4,600 ft (1,400 m). Forages in a wide variety of open areas, including agricultural fields, pastures, urban zones, and over water. In flight, gives liquid *whit* notes.

Cliff Swallow
(*pyrrhonota* ssp.)

Cave Swallow

Barn Swallow

TREECREEPERS Certhiidae

These specialized insectivorous tree climbers occur primarily in the Northern Hemisphere but also in Sub-Saharan Africa; a single species represents the family in the Americas. Cryptically colored upperparts allow treecreepers to camouflage themselves on tree trunks. Although behavior is similar to that of some Furnarid woodcreepers, distinguishing between the 2 groups is generally not a problem.

Brown Creeper (Agateador Americano)   *Certhia americana*
5 in (13 cm). Distinguished from all Furnarid woodcreepers by small size, **crisp white underparts**, and **pale stripe across primaries** (seen in flight). Bill is slender and decurved; white superciliary is easily noticed on brown streaked head. Uncommon on Sierra Dipilto-Jalapa and very rare throughout rest of Northern Highlands, as far south as Matagalpa; from 3,300 to 5,900 ft (1,000 to 1,800 m). Reaches southernmost distribution in Nicaragua. Found solely in highland pine and pine-oak forest, where it forages for insects hiding under bark. Searches in a predictable pattern, climbing up a trunk in spiraling fashion before dropping down to the base of a nearby trunk to start again. Usually solitary but sometimes joins mixed-species flocks composed of various species of New Word warblers. Whistles a reedy, clear 5-note song: *seet-seee-suét-soo-suét* (1.5 seconds in length). Call is a reedy, quivering *tseep*.

WRENS Troglodytidae

This is an exclusively New World family, with the sole exception of the Eurasian Wren. The plumage is brown and other earthy hues, and they have barred wings and tails and long, slender bills. Although inconspicuous and shy, they are great songsters (sometimes singing in duets) and are frequently detected that way. An agitated wren will often briefly enter into view—its tail cocked in characteristic fashion—then drop back into dense cover. Essentially insectivorous, they forage in forest understory and tangles and thickets for insects, spiders, and other arthropods.

House Wren (Chochín Casero)   *Troglodytes aedon*
4.5 in (11 cm). Distinguished from Rufous-browed Wren by **indistinct pale superciliary** and habitat (House Wren in disturbed habitat and Rufous-browed in cloud forest and highland pine and pine-oak forest and edge). Head and back are unmarked, unlike Sedge Wren (streaked). Abundant in Northern Highlands; common in Caribbean; generally rare in Pacific lowlands, but locally uncommon in foothills and highlands. Prefers disturbed habitat; found in open areas, coffee and cacao plantations, urban areas, and forest edge. Gleans insects and spiders from foliage. Very active and vocal. Characteristically sings a *tu-Eaw-wawawa…*, consisting of sharp, fast notes (1 second in total length), often followed by a low-pitched trill; also makes a variety of curt, shrill notes combined with pleasant trills.

Rufous-browed Wren (Chochín Cejirrufo)   *Troglodytes rufociliatus*
4 in (10 cm). Very small. **Rufous superciliary** over **dark eye line, rich cinnamon underparts on throat and breast**, and **heavily barred flanks** all distinguish it from House Wren. Rare in Northern Highlands; above 4,100 ft (1,250 m). Occurs in cloud forest and highland pine and pine-oak forest and edge. Forages at all levels within forest, searching for insects on moss- and epiphyte-laden branches. Song is a jumbled mix of short, ethereal notes and light trills (2 seconds in total length). Reaches southernmost distribution in Nicaragua. Endemic to North Central American highlands EBA

Sedge Wren (Chochín Sabanero)   *Cistothorus platensis*
4.5 in (11 cm). No other Nicaraguan wren has **streaked crown and back**; similar House Wren lacks streaking. Very rare. Only 2 records: 1 in Northern Highlands (Jinotega Dept., 1929) and 1 from the Mosquitia (Feb, 1962). Inhabits flooded fields and marshes, where it forages for insects. Makes nomadic movements in response to seasonal changes in water levels. Flies only short distances when flushed before quickly disappearing into vegetation. Easily overlooked because of secretive behavior. Song combines buzzy, abrupt, repeated notes (*ju-ju-ju…*) and a dry trill.

Brown
Creeper

House
Wren

Sedge
Wren

Rufous-browed
Wren

Rock Wren (Charralero de las Rocas)   *Salpinctes obsoletus*
5.5 in (14 cm). Only Nicaraguan wren with **coarsely barred and spotted throat, breast, and upper belly** and **barred flanks, vent, and undertail coverts**. Lacks cinnamon belly of larger Band-backed Wren. Locally uncommon on Sierra Maribios; above 2,300 ft (700 m). Locally common on Sierra Isabelia (Jinotega and Estelí Depts.); above 3,400 ft (1,200 m). Occurs in semi-arid, rocky grassland with scattered shrubs. In pairs, probes on ground for insects and spiders. Vocalizations have a variety of double-noted phrases, including a reedy *ti-wEE*; a lower-pitched *pí-urh*; and a forceful and abrupt *chu-chu-chu....* .

Band-backed Wren (Saltapiñuela Barreteada)   *Campylorhynchus zonatus*
8 in (20 cm). **Extensively banded back** and **heavily spotted breast** are distinctive; **tawny from belly to undertail coverts**. Common to locally abundant in Northern Highlands, uncommon in eastern foothills; above 1,950 ft (600 m). Very rare in southern Caribbean lowlands along San Juan River. Prefers highland pine and pine-oak forest, light gaps and edge of cloud forest, woodland, open areas with scattered trees, and second growth. Forages at all levels of vegetation in noisy groups of up to 10; occasionally joins mixed-species flocks. Utters a series of dry, rhythmic cackling notes that can grow to a raucous sound when several group members are vocalizing.

Rufous-naped Wren (Saltapiñuela Nuquirrufa)   *Campylorhynchus rufinucha*
6.5 in (17 cm). Combination of **all-white underparts** and **rufous nape** is distinctive. **Graduated tail shows broad white tips**. Abundant in Pacific; uncommon in dry intermontane valleys of Northern Highlands and south into Sierra Chontaleña; to 4.600 ft (1,400 m). Prefers arid habitat such as thorn forest, scrub, woodland, open areas with scattered trees, second growth, mangroves, and urban areas. Actively, and with much commotion, moves in pairs or family groups at all levels of vegetation to forage for insects and spiders. Delivers a series of mocking, slurred, or forceful whistles, often intermixed with a choppy *wut-er-wEER*. Calls with a raspy bark (*rhaw!*) that transforms into a rattle when delivered quickly.

Rock Wren

Band-backed Wren

Rufous-naped Wren

A family group of Rufous-naped Wrens in a village.

Carolina Wren (Charralero Cejiblanco)   *Thryothorus ludovicianus*
5 in (13 cm). Bold white superciliary contrasts with broad dark eye line. **Cinnamon-buffy underparts from breast to vent** and **barred undertail coverts** distinguish it from Cabanis's Wren. Locally uncommon in San Cristóbal-Casita NR; very rare in eastern foothills of Pacific, with 1 historical record (south of Ciudad Darío, Matagalpa Dept., March 1917); from 1,500 to 3,300 ft (450 to 1,000 m). Reaches southernmost distribution in Nicaragua. Found in dry forest, thorn forest, and scrub. Often overlooked. Forages on ground or in understory for insects, spiders, and fruit. Usually in pairs. Male sings year-round; throughout the day, makes a rolling, rich double-noted *chu-áw* (repeated 5–6 times). Call is a reedy, bubbly *tuhwee*.

Rufous-and-white Wren (Charralero Rufiblanco)   *Thryophilus rufalbus*
6 in (15 cm). Note warm rufous upperparts, white underparts, **dark malar stripe**, and **barred undertail coverts**. Unmarked flanks distinguish it from duller Banded Wren. Common in Pacific. Uncommon in Northern Highlands, south through Sierra Chontaleña, and southwestern Caribbean; to 4,600 ft (1,400 m). In Pacific, prefers dry forest and gallery forest; in Northern Highlands and Caribbean, found in cloud forest, forest remnants, and second growth; in Caribbean, also found in humid lowland forest edge. In pairs, forages in understory and on ground for insects and other invertebrates. Unique song starts with 2–4 slow, hollow toot-whistles, trailed by a series of short, fast whistles, and finishing with a sharply upward-inflected note. Calls with a harsh and lively *SVET SVET*.

Banded Wren (Charralero Fajeado)   *Thryophilus pleurostictus*
5.5 in (14 cm). **Heavily black barred flanks** on white underparts are diagnostic, and distinguish it from Rufous-and-white Wren. Common in Pacific, rare to uncommon in dry intermontane valleys and western slopes of Northern Highlands; to 4,000 ft (1,200 m). Found in thorn forest, scrub, and dry forest. Forages in pairs or small groups for insects and spiders on ground and in understory. Furtive; more often heard than seen. Sings loud, clear, repeated whistles, interjected with trills and buzzes; combines these sounds in various ways to produce several melodic songs.

Cabanis's Wren (Charralero de Cabanis)   *Cantorchilus modestus*
5.5 in (14 cm). Bold white superciliary contrasts with broad dark eye line. **Rufous-brown upperparts** and **cinnamon-buffy underparts, from belly to vent**, distinguish it from Canebrake Wren. **Unbarred undertail coverts** separate it from Carolina Wren and Rufous-and-white Wren. Common countrywide; to 6,600 ft (2,000 m). Frequents a variety of forest habitats and disturbed areas, where it forages in understory. Furtive; more often heard than seen. Sings a fast, shrill, ascending *dit-dit-dirít* and a lower-pitched, slurred *dawEUR*. Calls with a thin *twit twit*, sometimes followed by a quavering whistle.

Canebrake Wren (Charralero de Zeledón)   *Cantorchilus zeledoni*
6 in (15 cm). Very plain. Combination of **grayish head and upperparts, underparts with cinnamon-buffy only on undertail coverts**, and pale grayish flanks distinguishes it from Cabanis's Wren. Common in southern Caribbean lowlands (mainly in Guatuzo Plains and along San Juan River and tributaries) and uncommon at San Miguelito Wetlands and surrounding area; to 300 ft (100 m). Prefers bamboo thickets along rivers, flooded grassland, overgrown pasture, and in second growth; avoids primary forest. Sings 2 songs: a thin, shrill *it-uít* or *it-uít FWEE* and a bubbly 3-noted *chu-chu-ree*, rapidly repeated and oscillating up and down in pitch; pairs often duet the 2 songs simultaneously. Reaches northernmost distribution in Nicaragua. Endemic to Central American Caribbean slope EBA.

Carolina Wren

immature

adult

Rufous-and-white
Wren

immature

adult

Banded
Wren

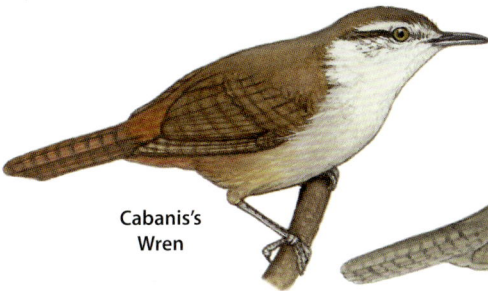

Cabanis's
Wren

Canebrake
Wren

Spot-breasted Wren (Charralero Pechimoteado)   *Pheugopedius maculipectus*
5 in (13 cm). Heavily **spotted throat and breast** and unbarred wings rule out Stripe-breasted Wren. Common on eastern slopes of Northern Highlands; to 4,600 ft (1,400 m). Common in Caribbean and rare on northern region and eastern slopes of Sierra Chontaleña; to 3,300 ft (1,000 m). Occurs in edge of cloud forest and humid lowland forest, secondary forest, and dense plantations. In pairs, forages in understory tangles and mid-canopy for insects and spiders. Sings a 5-noted *fwee-fwi-fa-fue-fuah*, consisting of clear, sweet whistles given with an oscillating pitch pattern (1.5 seconds in total length); final note sometimes upslurred, sometimes downslurred. Delivers both a typical monotone rattling trill and a buzzy, ascending trill.

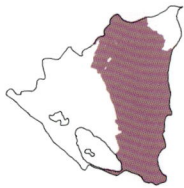

Stripe-breasted Wren (Charralero Pechirrayado)   *Cantorchilus thoracicus*
5 in (12 cm). **Streaked breast** and **heavily barred wings** distinguish it from Spot-breasted Wren. Uncommon in Caribbean lowlands and foothills; to 2,600 ft (800 m). Prefers humid lowland forest, forest edge, forest remnants, swamps, second growth, and streamside vegetation. In pairs or small groups, forages for insects and spiders in dense vine tangles of understory and mid-canopy. Sings several 3-noted phrases, each repeated multiple times before switching to a different song type. Repertoire includes a tiny *fwít-éur-ur* (low-high-low pitch pattern); a *fee-feur-ur* (high-low-high pitch pattern); and an ascending *wi-wer-dít*; often sing duets together. Also repeats a monotone *FWIT*.

Bay Wren (Charralero Cabecinegro)   *Cantorchilus nigricapillus*
5.5 in (14 cm). **White facial markings** stand out on **black head**. Prominent **white throat** excludes Black-throated Wren (extensive black throat). Common in Caribbean lowlands, west to Guatuzos WR; uncommon in Caribbean foothills and in San Miguelito Wetlands; to 2,300 ft (700 m). Reaches northernmost distribution in Nicaragua. Found in humid lowland forest edge, forest light gaps, second growth, secondary forest, and thickets next to streams and rivers. Gleans insects from foliage in understory. Although inquisitive, remains hidden in dense tangles, where it is seen only briefly. Territorial. Very vocal; has a large repertoire, but most often heard making a rapid series of loud, twittering whistles (varying greatly in pitch), usually including 3 rapid repetitions of a single note intermixed with jumbled phrases.

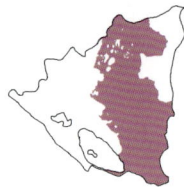

Black-throated Wren (Charralero Gorginegro)   *Pheugopedius atrogularis*
6 in (15 cm). No other Nicaraguan wren has **black throat and upper breast** (Bay Wren has white throat and ear patch). May be confused with Chestnut-backed Antbird (p. 246), but note absence of blue orbital skin. Common in southern Caribbean, in lowlands and foothills; very rare to rare in northern Caribbean, in lowlands and foothills (records along Grande de Matagalpa River, Escondido River, in Wawashan NR, and in eastern foothills of Sierra Amerrisque); to 2,000 ft (600 m). Found in humid lowland forest and edge, secondary forest, dense second growth, and streamside thickets. Forages for insects and spiders in dense tangles and thickets of understory. Usually in pairs; extremely furtive. Best way to locate is through its song. Sings a distinctive *fFEo-WHI-WHI-WHI-WHI*; starts with a powerful, double-noted, up-down whistle that is followed by 4 rapid-fire staccato notes. May only give portions of its song or intermix it with other notes, including short trills. Endemic to Central American Caribbean slope EBA.

Spot-breasted Wren

adult

immature

Stripe-breasted Wren

Bay Wren

Black-throated Wren

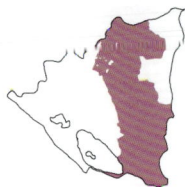

Song Wren (Chochín Cariazul)   Cyphorhinus phaeocephalus
5 in (13 cm). Oddly antbird-like. **Barring on wings and tail** distinguish it from female Bare-crowned Antbird (p. 246). Common in Saslaya NP and at other locations in Caribbean foothills; very rare in Caribbean lowlands; to 2,600 ft (800 m). On the floor of humid lowland forest, roams about in small family groups, tossing leaves in search of spiders and other invertebrates. Opportunistically joins mixed-species flocks that cross its territory, including those following army ant swarms. Easily detected by voice. Beautifully whistles a relaxed, ascending *fu-fi-fí* or a pure, double-noted *fit-fer*; also mixes in a dry, wooden *wet-wit-er* clucking.

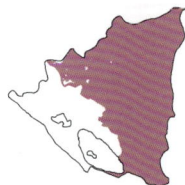

White-breasted Wood-Wren (Chochín Pechiblanco)   Henicorhina leucosticta
4.5 in (11 cm). Note **white throat and breast** and **cinnamon flanks**. Gray-breasted Wood-Wren (which occurs at higher elevations) has gray breast and rufous-brown flanks. Immature has grayish breast and cannot be positively identified if not in presence of adult. Common in Northern Highlands and Caribbean; to 4,600 ft (1,400 m).   Inhabits humid lowland forest, cloud forest, forest edge with dense vegetation, forest light gaps, and old second growth. In pairs or small family groups, moves about understory. More often heard than seen; very vocal, with a diverse repertoire of musical, cheery phrases. Whistled songs include a descending *fwee-fi-fu* with an abrupt final note; an ascending, melodic *fu-fi-fee* with an elongated final note; and a *feeu-fu-fia* (a high-low-high pitch pattern), with the first note noticeably downslurred. Calls with a short and sharp *pweet*. Song pattern is typically less complex and shorter than that of Gray-breasted Wood-Wren.

Gray-breasted Wood-Wren (Chochín Pechigrís)   Henicorhina leucophrys
4.5 in (11 cm). Set apart from very similar White-breasted Wood-Wren by **gray breast**; also note **rufous-brown flanks** (not cinnamon). Very rare in Northern Highlands (more likely in Sierra Dipilto-Jalapa); above 4,600 ft (1,400 m). Occurs in cloud forest, highland pine and pine-oak forest, and adjacent dense second growth. Forages in pairs or family groups in understory and on ground. Sings a jumbled phrase of 5–6 sweet, crisp whistles, with last either upslurred or downslurred; song has a rolling quality. Alone or in duet, often sings incessantly. Song pattern is typically more complex and longer than that of White-breasted Wood-Wren.

Nightingale Wren (Chochín Ruiseñor)   Microcerculus philomela
4.5 in (11 cm). **Dark brown** overall with **black scaling**; also note **pale gray throat** and upper breast and **short tail** (looks entirely dark in shaded understory). Uncommon on eastern slopes of Northern Highlands and in Caribbean foothills and highlands; from 1,000 to 4,600 ft (300 to 1,400 m). Locally common in Saslaya NP and uncommon in Musún NR (probably in nearby Quirragua NR). Inhabits humid lowland forest and cloud forest. Alone, picks insects off ground; opportunistically follows army ant swarms to feed on the arthropods they stir up. Dark color, behavior, and habitat make this species difficult to detect, but song is unique; consists of a series of short, monotone whistles delivered in an unmistakable up-down cadence.

Song Wren

White-breasted
Wood-Wren

Gray-breasted
Wood-Wren

Nightingale
Wren

These small, insectivorous birds occur only in the Americas. Gnatwrens are mostly brown and have long bills; they mainly inhabit the understory of forest and woodland. Gnatcatchers often cock their long, narrow tail. All members of the family forage tirelessly for insects found on foliage.

Tawny-faced Gnatwren (Cazajején Carirrufo)   *Microbates cinereiventris*
4 in (10 cm). Combination of **tawny face**, black malar stripe, and **white throat** with **necklace of black streaks** is distinctive. Has shorter tail and shorter, broader bill than does Long-billed Gnatwren. Lack of barring on wings and tail rules out most wrens. Rare in southern Caribbean lowlands and foothills and locally uncommon in Indio Maíz BR; to 1,300 ft (400 m). Reaches northernmost distribution in Nicaragua. Occurs in humid lowland forest; pairs or small family groups roam the understory and on ground probing leaf litter for insects. Opportunistically attends army ants swarms and joins mixed-species flocks of antbirds, antvireos, and tanagers. Sings a single and clear, downslurred *feauu*, repeated with pause of several seconds between whistles. Calls with a nasal, downslurred buzz, a raspy chuckle, and an up-down up-down *zwee-zur zwee-zur.*

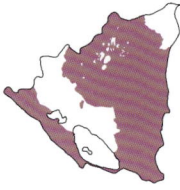

Long-billed Gnatwren (Cazajején Picudo)   *Ramphocaenus melanurus*
5 in (12 cm). **Very long, slender bill** and **long, graduated white-tipped tail** distinguish it from Tawny-faced Gnatwren, and all wrens. Locally uncommon in Pacific (prefers more humid areas); to 4,000 ft (1,200 m). Rare to uncommon on eastern slopes in Northern Highlands; to 4,300 ft (1,300 m). Locally very rare in northern region of Sierra Chontaleña; above 3,300 ft (1,000 m). Uncommon in Caribbean lowlands and foothills; generally to 2,000 ft (600 m), but to 4,000 ft (1,200 m) in Musún NR. Favors humid lowland forest edge, forest light gaps, secondary forest, second growth, and overgrown plantations. Furtive; very hard to see as it forages, in pairs, in dense vine tangles of understory and, less frequently, in mid-canopy. Rapidly sings a string of whistled notes; slightly rises in pitch at beginning before leveling off or slightly falling (2 seconds in length). Also communicates using a raspy chatter.

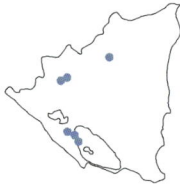

Blue-gray Gnatcatcher (Perlita Grisácea)   *Polioptila caerulea*
4.5 in (11 cm). Only *Polioptila* gnatcatcher with **white eye ring** and **pale gray-blue crown**; approaching spring migration, male may show narrow black superciliary. Accidental winter resident (Sept to March). Three recent records in Pacific (Mombacho Volcano NR, Feb 2004; Zapatera NP, Nov 2015; Sierras Managua, March 2017); 3 records in Northern Highlands (Yali Volcano NR, Sept 2015; Moropotente NR, Nov and Dec 2016); and a single record in Sasaya NP (March 2016); to 4,600 ft (1,400 m). Favors highland pine and pine-oak forest, but in the Pacific lowlands it is found in a variety of forest habitats close to swamps and other bodies of water. Moves agitatedly while foraging in canopy, occasionally down to understory. Call is a thin, buzzy *zwee*.

White-lored Gnatcatcher (Perlita Cabecinegra)   *Polioptila albiloris*
4 in (10 cm). Breeding male (April to July) is unmistakable, with black crown extending to eye. (Despite its name, the White-lored breeding male has dark lores.) Nonbreeding plumage is very similar to that of Tropical Gnatcatcher (with little range overlap), but note **narrow white superciliary** and **black eye line (gray on female)**. Note broad white edging on tertials (forming a patch). Common in Pacific. Uncommon on western slopes of Northern Highlands; to 4,600 ft (1,400 m). Accidental on western slopes of Sierra Chontaleña. Found in dry forest edge, thorn forest, scrub, second growth, highland pine and pine-oak forest, and cloud forest edge. Very active and nervous; usually in pairs, forages from mid-canopy to canopy. Song consists of 15–17 rapid-fire, downslurred *spí* whistles (entire song lasts 2 seconds), which sometimes gradually descend in pitch towards middle of song. Call is a nasal, plaintive *whea*.

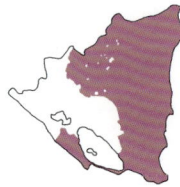

Tropical Gnatcatcher (Perlita Tropical)   *Polioptila plumbea*
4 in (10 cm). **Broad white superciliary and lore** distinguish it from very similar White-lored Gnatcatcher; also note less extensive white edging on tertials. Common in Caribbean lowlands and foothills; to 2,600 ft (800 m). Uncommon on eastern slopes of Northern Highlands; to 4,400 ft (1,350 m). Locally rare to uncommon in Pacific, at Los Pueblos Plateau, Mecatepe-Río Manares NR, and southeastern spur of Rivas Isthmus; to 2,000 ft (600 m). Forages actively, usually in pairs, in canopy of humid lowland forest edge, stream edge, secondary forest, and gallery forest. Song consists of 18–22 rapid-fire, upslurred *spí* whistles (song lasts 2 seconds), which descend in pitch and speed up toward latter half of song; similar to that of White-lored Gnatcatcher, but shriller and slightly faster paced. Call is a nasal and agitated *wheep*.

Tawny-faced Gnatwren

Long-billed Gnatwren

breeding male

Blue-gray Gnatcatcher

nonbreeding male/female

breeding male

White-lored Gnatcatcher

nonbreeding male

female

male

Tropical Gnatcatcher

female

DIPPERS Cinclidae

The 5 species (3 in the Americas) of this small family live near clear rushing rivers and streams. They are the only passerines that display truly aquatic behaviors. Their ability to swim and dive allows them to feed on aquatic invertebrates, but also makes them susceptible to water pollution.

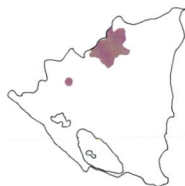

American Dipper (Mirlo Acuático Americano) *Cinclus mexicanus*

7.5 in (19 cm). Has plump **slate-gray body, short tail, and long legs**. Locally common on turbulent mountain streams of Saslaya NP and other mountain streams throughout Bosawas Biosphere Reserve; above 1,000 ft (300 m). Formally occurred in the Northern Highlands, but there are no recent records for it there. Found in fast flowing, unpolluted mountain streams, where it wades streamside, perches on wet rocks, or swims and dives into turbulent water to feed on aquatic invertebrates and small fish. Characteristically flies low and direct over streams as it moves upstream and downstream to cover its territory. Song consists of a long, lively series of phrases continuously delivered in sets of chirps, buzzes, and clicks.

THRUSHES and ALLIES Turdidae

A large family with near-global distribution, reaching higher diversity in temperate latitudes. Representatives of the family in Nicaragua are medium-sized, stocky, and have long legs; their body is held in a nearly horizontal posture. Most have drab plumage, though some species have brightly colored bills, legs, and orbital rings. Sexes are mostly alike; immatures have buffy or rufous spots. Breeding resident species sometimes perform altitudinal migrations in search of fruit, supplementing a mainly insectivorous diet.

Eastern Bluebird (Celeste Oriental) *Sialia sialis*

7 in (17 cm). Unmistakable. Note **bright blue head and upperparts** and **cinnamon-rufous throat, breast, and flanks**. Female is drabber, with gray head and back; immature has diffuse blue coloration but shows profuse white spotting. Nicaragua has 2 subspecies, geographically separated from each other, though they are very similar in appearance. *Meridionalis* subspecies: Common in Northern Highlands, primarily on western slopes (not found on Sierra Dariense, although there are historical records in Yucul GRR [1907 and 1961]; locally rare in San Cristóbal-Casita NR; above 3,500 ft [1,000 m]). Dwells in highland pine and pine-oak forest with scarce understory. *Caribaea* subspecies: Common in the Mosquitia. Limited to lowland pine savanna. Reaches southernmost distribution in Nicaragua. Both subspecies perch conspicuously on exposed branches or wires and drop to ground to catch prey. Forages in pairs or small family groups for insects and other invertebrates. Delivers a combination of slurred, twittering, dulcet whistles, including an up-down *fuít-feud*.

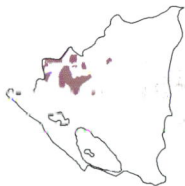

Slate-colored Solitaire (Solitario Gris) *Myadestes unicolor*

8 in (20 cm). Note **slate-gray** coloration, **broken white eye ring**, and **dark lore**. Unmistakable. Immature scalloped with dark brown and buffy markings. Common in Northern Highlands, Musún NR, and probably Quirragua NR, above 3,300 ft (1,000 m). Locally common in Saslaya NP; above 2,600 ft (800 m). Reaches southernmost distribution in Nicaragua. Inconspicuously perches for long periods in mid-canopy and canopy of cloud forest. Easily overlooked because of shy behavior, cryptic coloration, and poor light in interior forest. Easiest to locate by its beautiful and frequently heard song. Leisurely sings varied combinations of lilting, metallic notes interspersed with elongated screeching notes and trills. Scolds with a raspy *weyat*.

American
Dipper

immature

female

Eastern
Bluebird

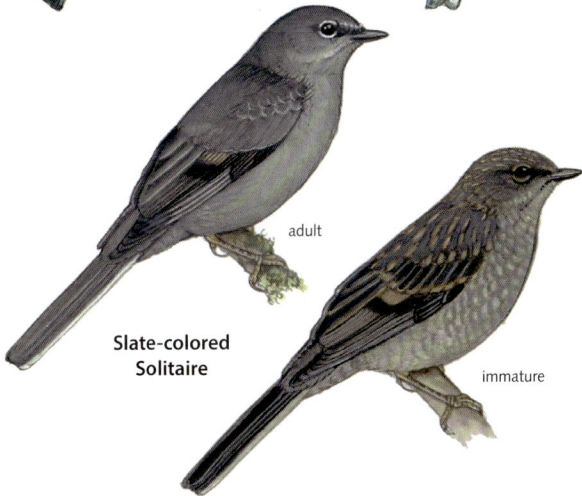

male

adult

Slate-colored
Solitaire

immature

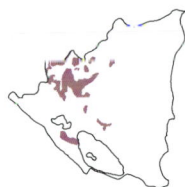

Orange-billed Nightingale-Thrush
Catharus aurantiirostris

(Zorzal Piquinaranja)

6.5 in (16 cm). **Orange bill, orbital ring, and legs** distinguish it from Ruddy-capped Nightingale-Thrush; **rufous-brown crown and upperparts** distinguish it from Black-headed Nightingale-Thrush. Distributed in disjunct populations in foothills and highlands across the country. Common in Northern Highlands; above 3,300 ft (1,000 m). Local in Pacific (common on Sierras Managua and Los Pueblos Plateau; uncommon on Mombacho Volcano NR). Rare to uncommon on Sierra Chontaleña and very rare in Saslaya NP. Found in cloud forest edge, highland pine and pine-oak forest, woodlands with dense understory, second growth, coffee plantations, slightly humid dry forest, and thorn forest. Forages on ground and in understory for insects; visits mid-canopy and canopy to feed on fruiting trees. Produces a variety of songs composed of high-pitched, unmelodic notes, among them a 3-noted *teu-tu-tí*, with the abrupt, final note rising dramatically in pitch; a *see-u see-u* (with a high-low high-low pitch pattern); and songs incorporating reedy trills.

Ruddy-capped Nightingale-Thrush (Zorzal Gorrirrojizo)
Catharus frantzii

7 in (17 cm). **Brown upperparts** and **rufous crown** distinguish it from Black-headed Nightingale-Thrush. **Bicolored bill** (black upper mandible and orange lower mandible) and lack of orange orbital ring and legs distinguish it from Orange-billed Nightingale-Thrush. Locally common in Caribbean highlands (Musún NR, probably Quirragua NR, and Saslaya NP); rare to uncommon in Northern Highlands, but locally common on Sierra Dipilto-Jalapa and in Kilambé NR; above 3,900 ft (1,200 m). Found in cloud forest and highland pine and pine-oak forest. Alone or in pairs, forages on ground and understory for insects, other invertebrates, and berries; occasionally attends army ant swarms. Given the fact that it is shy and elusive and lives in dense vegetation, perhaps best detected by song. Song is a languid jumble of ethereal whistles (does not employ trills used by Slate-colored Solitaire). Call is a weak, slightly upslurred *thee*.

Black-headed Nightingale-Thrush
Catharus mexicanus

(Zorzal Cabecinegro)

6.5 in (16 cm). **Orange bill, orbital ring, and legs** and dark crown distinguish it from Ruddy-capped Nightingale-Thrush. **Black crown and nape** (female with black only on crown) and **olive-brown upperparts** distinguish it from Orange-billed Nightingale-Thrush. Common in foothills and highlands of northern Caribbean, but abundant in Saslaya NP. Common on eastern slopes of Northern Highlands; uncommon in Musún NR and probably Quirragua NR; typically above 3,300 ft (1,000 m). Locally uncommon on Maderas Volcano (Ometepe Island); above 2,000 ft (600 m). Inhabits cloud forest, humid lowland forest, and adjacent mature second growth. Forages on ground and in understory for insects, other invertebrates, and fruit. Difficult to see in poor light conditions and is often best located by voice. Sings 2 high-pitched and liquid songs: one is 6-noted with noticeable spikes in pitch on the second and final note; the other is a 4-noted *swee-si-si-suu* that finishes with a tremulous quality. Call is a whistled, upslurred *swee*.

Spotted Nightingale-Thrush (Zorzal Pechiamarillo)
Catharus dryas

7 in (17 cm). Unmistakable. Note **orange bill and orbital ring, black head**, and **yellow breast with dusky spots**. Locally uncommon in Tepesomoto-Pataste NR; above 4,600 ft (1,400 m). Occurs in cloud forest; forages on ground or in understory for insects, invertebrates, and small fruit. Shy and inconspicuous, it is difficult to spot within the dense vegetation that it inhabits; often best located by voice. Beautiful song consists of single or double-noted ethereal, fluted notes, with several seconds of spacing between each note. Also calls with a nasal *weyah*-út.

Orange-billed
Nightingale-Thrush

Ruddy-capped
Nightingale-Thrush

Black-headed
Nightingale-Thrush

Spotted
Nightingale-Thrush

Veery (Zorzal Dorsicanelo)   *Catharus fuscescens*
7 in (17 cm). **Warm reddish-brown upperparts** and **tawny-washed breast** with
indistinct spots distinguish it from Gray-cheeked Thrush and Swainson's Thrush. Passage
migrant (Sep to Oct and March to April). Rare to uncommon in Northern Highlands and
Caribbean (more likely along coast); to 4,000 ft (1,200 m). Favors humid lowland forest
and second growth with dense understory, where it forages on the ground for fruit and
insects. Inconspicuous and shy, it is easily overlooked. Call is a gravelly *breet*.

Gray-cheeked Thrush (Zorzal Carigrís)   *Catharus minimus*
7 in (17 cm). **Gray on cheeks** not always distinguishable from color of head. Has more
distinct spotting on breast than does Veery; upperparts are drab, gray-brown (lacking
reddish hues on Veery). Inconspicuous eye ring (sometimes enlarged behind) and lack
of buffy lore distinguish it from Swainson's Thrush. Passage migrant (mid-Sept to Nov
and late March to mid-May). Uncommon in most of country; perhaps more common in
Caribbean lowlands. Not found in northwest Pacific. Skulks on ground and in understory
of forest edge, tall second growth, and thickets. Shy and probably more numerous than
the few records suggest. Call is a sharp *wee*-úr.

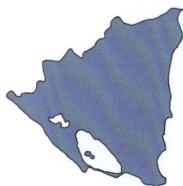

Swainson's Thrush (Zorzal Ustulado)   *Catharus ustulatus*
7 in (17 cm). **Olive-brown upperparts** and distinct spotting on breast distinguish it from
Veery. **Buffy wash on face** and **buffy spectacles** distinguish it from very similar Gray-
cheeked Thrush. Winter resident (mid-Sept to mid-May). Common countrywide, at all
elevations. Becomes abundant, mostly in Northern Highlands, when passage migrants
move through the country. Forages in understory of forest and forest edge for fruit
and insects. Begins singing in March; delivers a jumbled, ethereal song that trails off in
intensity at the end as it rises in pitch. Call is a brisk *whep!*

Wood Thrush (Zorzal Grande)   *Hylocichla mustelina*
7.5 in (19 cm). Note large size, **bold white eye ring**, and **rufous crown and nape**.
Rufous upperparts and **heavy black spots on white underparts** distinguish it from
all migrant *Catharus* thrushes. Winter resident (Oct to April). Common to abundant in
Northern Highlands and Caribbean; in Pacific, rare in lowlands, uncommon in foothills
and highlands. Found in forest, forest edge, coffee plantations adjacent to forest, tall
second growth, and scrub. Forages on ground and in understory for insects and small
invertebrates. Calls with a rapid *whit!-whit!-whit!*…, and a short, dry chuckle. **NT**

Veery

Gray-cheeked
Thrush

Swainson's
Thrush

Wood Thrush

Black Thrush (Sensontle Negro)   *Turdus infuscatus*

8.5 in (22 cm). Male unmistakable, with **black plumage** punctuated by **yellow eye ring, bill, and legs**. **Yellow legs** of brown female and immature help distinguish it from similar Mountain Thrush and Clay-colored Thrush. Very rare and local in Dipilto Jalapa NR; above 4,000 ft (1,200 m). Favors cloud forest, highland pine forest, and forest edge. Forages in mid-canopy and canopy, but descends to ground in forest light gaps. Song is a long-winded series of alternating phrases, with noticeable pauses between phrases; phrases include single or repeated dulcet, buzzy whistles. Call is a clucking *ruh! ruh! ruh!*

Mountain Thrush (Sensontle Montañés)   *Turdus plebejus*

10 in (25 cm). Dark olive-brown overall, with paler underparts. Undertail coverts show dark, buffy scaling. **Blackish legs** distinguish it from female and immature Black Thrush; **black bill** further distinguishes it from Clay-colored Thrush. Common in highlands of Saslaya NP; uncommon in highlands of Musún NR and probably also in Quirragua NR; locally uncommon in Northern Highlands (Tepesomoto-Pataste NR, Dipilto-Jalapa NR, Miraflor-Moropotente NR, Kilambé NR, and Arenal NR); above 4,000 ft (1,200 m). Found in cloud forest, forest edge, and forest light gaps. Forages high in canopy. Location in canopy, dark coloration, and indistinct song make it difficult to detect. Repeats at length an anxious *tue-eet tue-eet tue-eet…* (4 repetitions delivered per second); pitch pattern of each double note is low-high. Call is a clucking *rah! rah! rah!*

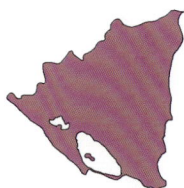

Clay-colored Thrush (Sensontle Pardo)   *Turdus grayi*

10 in (25 cm). **Yellowish bill** and warm brown coloration distinguish it from Mountain Thrush; **grayish legs** further distinguish it from female and immature Black Thrush. Abundant countrywide at all elevations. Occurs in open areas with scattered trees, woodland, gardens, plantations, dry forest and edge, tree-lined roadsides, and second growth. Conspicuously forages alone or in pairs from ground to mid-canopy; flocks often attend prolifically fruiting trees. Characteristically flicks tail up and down. Song is a long series that alternates whistled-notes with guttural clucks and shrill sounds; maintains a rhythmic and even tempo. Call is a rising and forceful *peeurYA!*

White-throated Thrush (Sensontle Gorgiblanco)   *Turdus assimilis*

9 in (23 cm). Note **black streaks on white throat, white collar**, black to dark brown upperparts, and **gray underparts**. Common in foothills and highlands of northern Caribbean (Musún NR, probably Quirragua NR, and Saslaya NP); locally uncommon on Maderas Volcano (Ometepe Island); locally uncommon in Northern Highlands (absent from Sierra Dipilto-Jalapa); above 1,300 ft (400 m). Accidental in Caribbean lowlands. Occurs in cloud forest and humid lowland forest, where it feeds on insects and fruit within mid-canopy and canopy. Performs elevational migration from highlands to Caribbean foothills after breeding. Sings a long series of alternating phrases, including single repeated whistles, double-noted up-down whistles, slurred whistles, short chattering, and trills; does not keep a rhythmic, even tempo. Call is a rising, soft *peeur.*

female

Black Thrush

male

**Mountain
Thrush**

immature

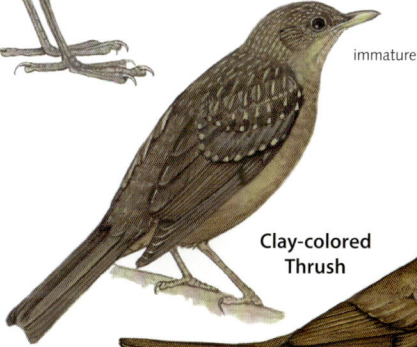

**Clay-colored
Thrush**

adult

**White-throated
Thrush**

MOCKINGBIRDS   Mimidae

The members of this New World family have long tails that are often cocked up. Mockingbirds are known for mimicking the vocalizations of other birds. Their diet includes arthropods and some fruit. The distribution and abundance of each of Nicaragua's 3 species are quite distinct. The Northern Mockingbird is abundant—indeed ubiquitous—and occurs in open areas; the Blue-and-white Mockingbird is very rare and occurs at specific locations within the country; and the Gray Catbird is a North American migrant whose abundance varies from location to location.

Blue-and-white Mockingbird (Mulato Pechiblanco)   *Melanotis hypoleucus*
10.5 in (27 cm). Sleek and unmistakable. **Deep blue upperparts** strikingly contrast with **white underparts**; also note **black mask**. Very rare and local in Miraflor NR and Tisey-Estanzuela NR (expected in Tepesomoto-Pataste NR and Dipilto-Jalapa NR); above 4,000 ft (1,200 m). Favors dense vegetative cover within second growth, scrub, highland pine and pine-oak forest edge, and harvested corn fields. Forages mainly on ground for insects, small invertebrates, and fruit. Makes a variety of short liquid or sharp whistles, soft barks, nasal grunts, and short trills; delivered with a choppy cadence or sometimes with even spacing and several seconds between phrases. Reaches southernmost distribution in Nicaragua. Endemic to North Central American highlands EBA.

Gray Catbird (Maullador Gris)   *Dumetella carolinensis*
8 in (20 cm). Combination of **black crown, rufous undertail coverts**, and slate-gray plumage is unique. Winter resident (mid-Oct to mid-May). Common in Caribbean; uncommon in Northern Highlands; and rare in Pacific (only as a passage migrant). Skulks in dense understory, preferring thickets, vine tangles, shrubbery, and forest edge; visits canopy of fruiting trees. Calls with a short, wheezy *meaw*.

Tropical Mockingbird (Sensontle Tropical)   *Mimus gilvus*
10 in (25 cm). Has a slender body and long legs. Note **pale gray upperparts** and **white underparts**. Dusky wings have **faint white wing bars** and **white-tipped tail** is long and broad (often cocked up). Immature is duller. Common in Pacific lowlands and foothills; to 2,000 ft (600 m). Common in dry intermontane valleys of Northern Highlands; to 4,600 ft (1,400 m). Favors open areas, grassland with scattered trees, agricultural fields, scrub, and urban areas. Conspicuously perches in exposed places; alone forages on ground and in surrounding vegetation for small invertebrates and fruit. Gives a long series of repeated and varying phrases that include clear whistles, squeaky whistles, trills, and chuckles.

WAXWINGS   Bombycillidae

This small family has just 3 species. The name is derived from the red, waxy tips on their secondary feathers. Bills are small and stubby, hinting at their fruit mashing tendencies. Frugivory also partly contributes to their gregarious behavior, as they gather in the hundreds at fruit-laden trees and often move in large numbers in search of the same.

Cedar Waxwing (Ampelís Americano)   *Bombycilla cedrorum*
7.5 in (19 cm). Has a sleek, silky appearance. Note **rich brown head, black mask**, and **crest projecting off the back of the head**. Square tail showcases a **broad yellow terminal band**. Winter resident (Jan to mid-May); only present in irruptive years. Rare in Northern Highlands; very rare in Pacific foothills and highlands, possible in lowlands; above 2,100 ft (650 m). Accidental in Caribbean. Typically forages for berries in large flocks within highland pine and pine-oak forest, cloud forest, and shaded coffee plantations. Vocalizes with very high-pitched, thin *seeeu* whistles.

Blue-and-white Mockingbird

adult

immature

Gray Catbird

Tropical Mockingbird

Cedar Waxwing

MUNIAS   Estrildidae

This Old World family also includes waxbills and a variety of finchlike allies that inhabit warm areas of Africa, Australasia, and the South Pacific islands. They are gregarious in nature and known to be colonial nesters. One species has recently colonized areas within Nicaragua, most likely descendants of birds that escaped from cages.

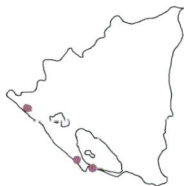

Tricolored Munia (Munia Tricolor)   *Lonchura malacca*
5 in (13 cm). Note **stout, conical bill** on this small-bodied bird. Black head and breast, chestnut upperparts, and white underparts give the male its unmistakable tricolor look. On female, larger size and 2-toned coloration help distinguish it from brown female seedeaters. Native to Southeast Asia; escaped caged birds appear to have locally colonized locations in Pacific lowlands close to Chinandega, Rivas, and Cárdenas. Prefers agricultural fields and grasslands, where flocks forage within the vegetation. Makes tinny, squeaky utterances.

OLD WORLD SPARROWS   Passeridae

This Old World family is represented by a single species in Central America. The House Sparrow was introduced to New York in 1850, and by the 1980s it had arrived as far south as Panama. It is stockier and has a thicker bill than the native New World Sparrows.

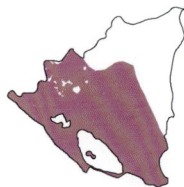

House Sparrow (Gorrión Común)   *Passer domesticus*
6 in (15 cm). On male, note **gray crown, chestnut upperparts,** and **black bill, throat, and breast**; on nonbreeding male, however, coloration is muted. Drab, nondescript female is stockier than drab New World Sparrows; best way to identify female is often the presence of the distinctive male. Locally abundant throughout Pacific and in Northern Highlands; to 4,300 ft (1,300 m). Locally common throughout much of Caribbean. Colonizes urban areas and agricultural fields, where small flocks opportunistically feed on ground for seeds, fruit, and all manner of food scraps. Sings a rapid series of liquid, sliding notes. Whistles a rich *chíu* call.

A group of Tricolored Munias foraging in tall grass.

Tricolored Munia

adult

immature

female

House Sparrow

male

Birds of Nicaragua / **337**

This is a cosmopolitan family. Nicaragua is home to 11 species, all of which are sexually dimorphic and often quite colorful. Finches typically have conical bills specialized for feeding on seeds. Euphonias and chlorophonias (formerly placed in family Thraupidae) are primarily frugivorous but also feed on insects. Euphonias often present identification challenges within the family itself, as many of these colorful birds resemble each other. On the "blue-and-yellow" males (4 species), it is helpful to note the extent of yellow on the head, throat, and vent; on the respective female (primarily green-yellowish), note head color and also plumage pattern on the underparts.

Red Crossbill (Piquituerto Común)   *Loxia curvirostra*

6.5 in (16 cm). **Bill crossed at the tips** is diagnostic. Dusky wings and tail contrast with dull red plumage (of male) and ochre-yellow (of female). Locally abundant to uncommon in Northern Highlands; above 3,900 ft (1,200 m). Locally uncommon in the Mosquitia. Accidental record on summit of San Cristóbal Volcano (April 1891). Reaches southernmost distribution in Nicaragua. Restricted to highland pine and pine-oak forest and lowland pine savanna. Exhibits nomadic behavior and erratic seasonal movements in response to pine cone crops. Feeds and travels in flocks, which can make noisy, sweet chatter (often given in flight); crossed bill specializes in extracting pine seeds from cones. Calls with a bubbly *pip! pip! pip!* (given singly or repeated at length) and a softer, double-noted *jee-jeú*.

Black-headed Siskin (Verdecillo Cabecinegro)   *Spinus notatus*

4.5 in (11 cm). Adult male is unmistakable. **Yellow wing bar** and **broad yellow patch at base of primaries and secondaries** are distinctive in all plumages. Locally common in Northern Highlands; above 3,600 ft (1,100 m). Locally common in the Mosquitia. Reaches southernmost distribution in Nicaragua. Found in highland pine and pine-oak forest, lowland pine savanna, and occasionally in cloud forest edge. Small flocks of up to 20 individuals feed on seeds at all levels, including on ground. Song is a fast repetition of warbling notes, most with metallic undertones. Calls include a nasal *jea!* and a soft, downslurred *peeuu*.

Lesser Goldfinch (Verdecillo Menor)   *Spinus psaltria*

4 in (10 cm). **White patch on base of primaries** is distinctive in all plumages. Locally common to uncommon in Northern Highlands and northern region of Sierra Chontaleña; rare in San Cristóbal-Casita NR; above 2,000 ft (600 m). Accidental in Caribbean foothills. Found in highland pine and pine-oak forest, scrub, thickets, grassland, and open areas with scattered trees. In small flocks, forages for seeds on the ground, but also at all canopy levels; exhibits nomadic seasonal movements in search of food. Sings a compilation of fast repeated phrases chaotically pieced together, including whistles, yelps, barks, chatter, and trills. Calls include a sharp, clear *fEEuu* whistle.

Red Crossbill

female

male

female

male

Black-headed Siskin

Lesser Goldfinch

male

female

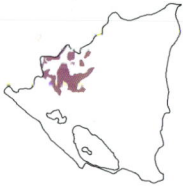

Blue-crowned Chlorophonia   *Chlorophonia occipitalis*
(Clorofonia Coroniazul)

5 in (13 cm). **Turquoise blue restricted to crown** distinguishes it from both sexes of Golden-browed Chlorophonia (with turquoise blue extending onto nape). Male lacks yellow on head (male Golden-browed has yellow forecrown and superciliary). Uncommon to locally common in Northern Highlands; above 3,300 ft (1,000 m). Reaches southernmost distribution in Nicaragua. Inhabits cloud forest and edge, and is also seen in shaded coffee plantations. In pairs or small flocks, forages in canopy of fruiting trees. Green coloration and life high in the canopy mean that this species is more often heard than it is seen. Delivers series of slow, low-pitched, soft yips and yelps. Gives a short and twangy thrush-like *pacá* call and a drawn-out, mournful *feeea* whistle. Also vocalizes with a nasal *rhe!*

Golden-browed Chlorophonia   *Chlorophonia callophrys*
(Clorofonia Cejidorada)

5 in (13 cm). On male, **bright yellow forecrown and superciliary** distinguish it from male Blue-crowned Chlorophonia. Female is almost identical to female Blue-crowned Chlorophonia, but **turquoise blue crown extends to cover the nape** (if this field mark is not seen, females cannot safely be identified). Very rare; 3 recent records exist from Northern Highland locations in Jinotega and Matagalpa departments (Feb 2002; April 2003; March 2015); above 3,900 ft (1,200 m). In pairs or small flocks, found in cloud forest and edge, where it forages in canopy for berries, figs, and insects; joins flocks of Blue-crowned Chlorophonia. Song is a steady, repeated, monotone, mournful whistle: *puuuuh* (repeated every 4 seconds). Sometimes calls with a soft *pup* whistle or a nasal *erh!* accompanied by scratchy notes.

Elegant Euphonia (Eufonia Capuchiazul)   *Euphonia elegantissima*

4.5 in (11 cm). Unmistakable. Note **turquoise blue hood** on both male and female. Uncommon in Northern Highlands; above 3,400 ft (1,050 m). Favors cloud forest, highland pine and pine-oak forest, forest edge, and semi-open areas with scattered trees. In pairs or small flocks, forages in canopy, but descends to lower levels to visit fruiting trees or clumps of mistletoe. Usually quiet and therefore easily overlooked. Song composed of thin, rolling, twittering notes. Calls with a clear *cheú* or muffled *chu*, as well as a nasal *erp!*

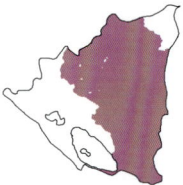

Olive-backed Euphonia (Eufonia Olivácea)   *Euphonia gouldi*

4 in (10 cm). Both sexes distinguished from other female euphonias by **rufous from belly to undertail coverts** (more extensive on male). Forecrown is yellow on male and rufous on female. Locally common on eastern slopes of Northern Highlands; to 4,300 ft (1,300 m). Locally common in northern region of Sierra Chontaleña and common in Caribbean; to 3,300 ft (1,000 m). Found in humid lowland forest, forested river edge, cloud forest, second growth, shaded plantations, and semi-open areas with scattered trees. In pairs or small groups, forages in canopy, but descends to visit fruiting trees; also visits fruit feeders. Joins mixed-species flocks of other euphonias, tanagers, honeycreepers, warblers, and greenlets. Sings with complex, jumbled twittering sounds. Calls include a nasal, downslurred *weeu* followed by a rapid *chichichichi* and a 3-part *chi-weeuu-chi*, the middle note consisting of a liquid trill.

**Blue-crowned
Chlorophonia**

male

female

male

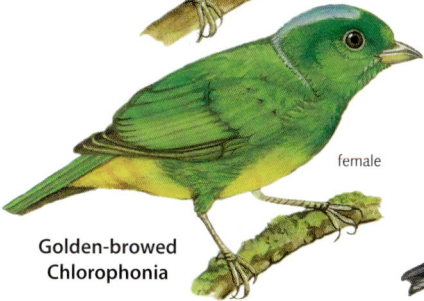

female

**Golden-browed
Chlorophonia**

male

**Elegant
Euphonia**

female

male

**Olive-backed
Euphonia**

female

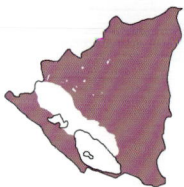

Yellow-throated Euphonia (Eufonia Gorgiamarilla)   *Euphonia hirundinacea*
4.5 in (11 cm). Male is only male euphonia that has a **yellow throat**. On female, whitish belly is similar to that on female White-vented Euphonia, but note yellow vent and undertail coverts (not white). Locally common on Sierras Managua, Mombacho Volcano NR, and Rivas Isthmus, but rare in other locations in the Pacific; common in Northern Highlands; uncommon in northern region and eastern slopes of Sierra Chontaleña; to 4,900 ft (1,500 m). In Caribbean, common in lowlands and foothills and uncommon in highlands; to 3,600 ft (1,100 m). In pairs or small groups, forages in canopy for fruit and insects; descends lower in canopy to seek out fruiting trees. Vocalizations are impressively varied. Song variations include a rapid series of abrupt, squeaky, and burry notes, with an unmusical effect. Calls include a *pit!-zeeuu-pit!* (2 staccato notes bookend a reedy trill in the middle); a burry *pip-zuéy*; a high-pitched double whistle (*swee-swee*); and an enthusiastic *chee-dee-dee*.

Scrub Euphonia (Eufonia Gorginegra)   *Euphonia affinis*
4 in (10 cm). Male is similar to other male euphonias with dark blue throats; distinguished from male White-vented Euphonia by **entirely yellow underparts**; distinguished from male Yellow-crowned Euphonia by restricted **yellow crown (barely reaches midcrown)** and **white inner web on outer rectrices** (creates a white patch at base of undertail on perched birds). Female distinguished from all other female euphonias by **gray hindcrown and nape**. Common in Pacific and uncommon in dry intermontane valleys of Northern Highlands; to 4,600 ft (1,400 m). Rare on Sierra Chontaleña and eastern borders of Lake Nicaragua, south to San Miguelito Wetlands. Accidental in Caribbean. Found in dry forest, thorn forest, scrub, second growth, semi-open areas with scattered trees, gallery forest, and fruiting trees in urban areas. In pairs or small groups, forages in canopy for berries (e.g., mistletoe and figs). Song is a series of weak, squeaky, chittering notes that oscillate in pitch, sometimes interjected with short buzzes. Calls with a melancholic *fwee-fwee* whistle, each note slightly rising, or a faster *fee-fee-fee* (similar to that of Yellow-crowned Euphonia).

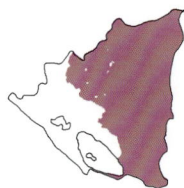

Yellow-crowned Euphonia (Eufonia Coronigualda)   *Euphonia luteicapilla*
4 in (10 cm). In Nicaragua, this is the only male euphonia with a **complete yellow crown**, which distinguishes it from Scrub Euphonia and White-vented Euphonia; further distinguished from Scrub Euphonia by lack of white on undertail. Female has green-olive upperparts, yellowish underparts, and no white on tail; she is further distinguished from other female euphonias by **thinner and more pointed bill**. Common in Caribbean lowlands and foothills, but occasionally ascends to 4,300 ft (1,300 m) on eastern slopes of Northern Highlands and on Sierra Chontaleña. Favors semi-open areas with scattered trees, scrub, second growth, and forest edge. In pairs or small groups, forages in canopy. Song is a jumbled series of chittering noises that terminate with a buzzy, descending, 3-note phrase: *zer-zee-ZEEP!* (2 seconds in total length); song repeated seemingly without end. Call is 2-3 ascending whistled notes: *fwee-fwee-fwee* (similar to Scrub Euphonia).

White-vented Euphonia (Eufonia Menuda)   *Euphonia minuta*
4 in (10 cm). **White vent and undertail coverts** on both sexes distinguish it from all other similar Nicaraguan euphonias. Rare to uncommon in Caribbean; to 3,300 ft (1,000 m). Found in humid lowland forest and forested river edge. Moves very high in canopy, alone, in pairs, or in mixed-species flocks of frugivorous birds; easily overlooked. Eats small fruit (e.g., mistletoe and figs). Sings a series of sporadic, squeaky notes, sometimes interspersed with sharp whistles, with unmusical effect. Calls include a thin *tsip* and lively *weep* or *peep* whistle.

Yellow-throated Euphonia

male

female

male

Scrub Euphonia

female

male

Yellow-crowned Euphonia

female

male

White-vented Euphonia

female

This family exhibits a high level of diversity in the Americas. Species in Nicaragua loosely fit into 3 groups: forest floor skulkers, grassland and open area dwellers, and understory and mid canopy specialists (genus *Chlorospingus*). All have the conical bill typical of the family. The secretive forest floor skulkers often only allow brief sightings within dense understory vegetation, but even a glance at the vibrant patterns on the head is often sufficient to make an id. Some of the grassland and open area species are equally challenging to see, as their dull hues allow them to blend in with their surroundings. Only 2 species are present as NA migrants.

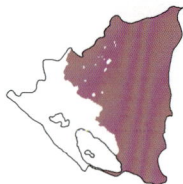

Orange-billed Sparrow (Pinzón Piquinaranja) *Arremon aurantiirostris*
6.5 in (17 cm). Note bold black-and-white pattern on head and diagnostic **bright orange bill**; immature keeps a black bill for almost a year. Common in Caribbean lowlands and foothills; to 2,600 ft (800 m). Rare on eastern slopes of Northern Highlands (absent on Sierra Dipilto-Jalapa); to 4,300 ft (1,300 m). Found in humid lowland forest and secondary forest. Hops and scratches on the ground in search of food; often in pairs or small family groups. Sings a fast series of sibilant, reedy notes that jump up and down in pitch. Call is an abrupt, scratchy *tit!*

Chestnut-capped Brush-Finch *Arremon brunneinucha*
(Saltón Gorgiblanco)
7.5 in (19 cm). The only forest floor specialist with combination of **chestnut crown, white throat**, and black mask. Common in Northern Highlands, and locally common in northern region of Sierra Chontaleña; uncommon and local in Mombacho Volcano NR; from 2,600 to 5,900 ft (800 to 1,800 m). Common on northern peaks in Caribbean (Quirragua NR, Musún NR, and Saslaya NP); above 1,800 ft (550 m). Found in cloud forest, highland pine forest, humid lowland forest, and adjacent edges. Turns over leaves on the ground in search of food; often seen in pairs. Sings a fast-paced series of 5–6 high-pitched, sibilant whistles, each exhibiting steep rises and falls in pitch (1 second in total length).

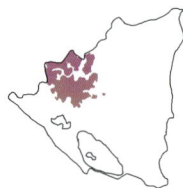

White-naped Brush-Finch (Saltón Gorgiamarillo) *Atlapetes albinucha*
7 in (18 cm). Note **white stripe on center of the crown** on black head and **bright yellow throat**. Common in Northern Highlands, Quiragua NR, and Musún NR; above 2,600 ft (800 m). Found in highland pine and pine-oak forest, cloud forest, forest edge, and secondary forest. Moves through understory with wary gestures; energetically forages in leaf litter on the ground for food. Song is slow series of repeated, high-pitched, varied notes that erratically rise and fall in pitch (1 note per second).

White-eared Ground-Sparrow (Pinzón Orejiblanco) *Melozone leucotis*
6.5 in (17 cm). **White lore, broken eye ring, and ear patch** create an unmistakable head pattern. Common on southern Segoviana Plateau, Sierra Isabelia, and Sierra Darense; above 3,600 ft (1,100 m). Uncommon and local on northern extent of Sierra Chontaleña; above 3,000 ft (900 m). Prefers dense understory or forest edge of cloud forest and adjacent secondary forest, as well as shaded coffee plantations. In search of food, flings debris from leaf litter with a hop-and-scratch method. Song begins with 2 squeaky, downslurred notes followed by 4–5 short, quickly repeated notes (1 second in total length).

Orange-billed
Sparrow

Chestnut-capped
Brush-Finch

White-naped
Brush-Finch

White-eared
Ground-Sparrow

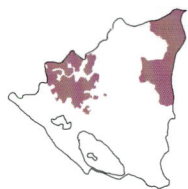

Rusty Sparrow (Sabanero Rojizo)   *Aimophila rufescens*
6.5 in (17 cm). Combination of **broad black lateral throat stripe** and **rufous stripes on the crown** is unique. Common in Northern Highlands, Quiragua NR, and Musún NR; locally rare on Sierra Chontaleña; from 2,600 to 5,200 ft (800 to 1,600 m). Uncommon in the Mosquitia, and accidental in rest of Caribbean lowlands. Found in second growth, scrub, forest edge, overgrown clearings, dense understory vegetation of highland pine and pine-oak forest, and lowland pine savanna. Forages on the ground for seeds and invertebrate prey. Makes a variety of rich, whistled songs, including a 3-noted *swí-suree-surree* delivered in a jerky fashion (0.5 second in length); a 4-noted *swítah-swítah-sí-siú* (1 second in length), with the final note noticeably more intense; and a 6-noted *suya-suya-swiyá-suh-suh* (1 seccond in length). Calls with a soft *tup*; also vocalizes with an often repeated churr.

Stripe-headed Sparrow (Sabanero Cabecilistado)   *Peucaea ruficauda*
7 in (18 cm). A striking sparrow, with a bold **black-and-white head** pattern and stout, bicolored bill. Common throughout Pacific and on western slopes of Northern Highlands; to 4,900 ft (1,500 m). Inhabits dry forest, thorn forest, scrub, and their associated edges, as well as arid grasslands and agricultural areas. In small flocks, forages on the ground for seeds. Song consists of a constant squeaky twittering. Calls include an electric *thwut!* and a thin *seet!*

Botteri's Sparrow (Sabanero Dorsilistado)   *Peucaea botterii*
6 in (15 cm). Nondescript, but combination of **pale supraloral**, **faint brown postocular stripe**, and **faint, short black malar stripe** helps distinguish it from similar sparrows. Gray underparts distinguish it from Grasshopper Sparrow (which has buffy underparts). Brown rump helps to distinguish it from Chipping Sparrow (with a gray rump). On heavily streaked immature, pale vent helps to distinguish it from immature Chipping Sparrow (with a white vent). *Spadiconigrescens* subspecies: Common in the Mosquitia, where it associates with palmetto thickets in lowland pine savanna. *Vulcanica* subspecies: Historical population on San Cristóbal Volcano, above 3,800 ft (1,150 m), thought to be extirpated by 1970 fire. Vocalizations are high-pitched and include a short trill. Call is a sharp, thin *tsít*.

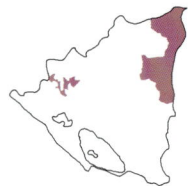

Chipping Sparrow (Sabanero Pechigrís)   *Spizella passerina*
5.5 in (14 cm). In breeding plumage, bright **rufous crown** is diagnostic. In all plumages, note 2 **white wing bars** and distinctive **black eye line** (with pale superciliary). Gray underparts distinguish it from Grasshopper Sparrow (which has buffy underparts). Gray rump helps to distinguish it from Botteri's Sparrow (with brown rump). On heavily streaked immature, white vent helps to distinguish it from immature Botteri's Sparrow (with pale vent). Abundant in the Mosquitia. Common on western slopes of Sierra Isabelia and Sierra Dariense in Northern Highlands; from 3,600 to 5,200 ft (1,100 to 1,600 m). Found in highland pine and pine-oak forest and lowland pine savanna. In small to large groups, forages in understory. Song is a chippy, high-pitched trill (2 seconds in length). Call is a sharp, thin *tsít*.

Grasshopper Sparrow (Sabanero Colicorto)   *Ammodramus savannarum*
5 in (13 cm). Three subspecies occur in Nicaragua, 1 breeding population and 2 NA migrants. All have typical **flat head, a yellow supraloral**, and a large bill. Buffy underparts distinguish it from Botteri's Sparrow and Chipping Sparrow (both with gray underparts). *Cracens* subspecies: Uncommon breeding resident in the Mosquitia, where it prefers lowland pine savanna. *Perpallidus* and *pratensis* subspecies: Winter resident (Nov to April). Rare in Pacific. Rare on western slopes of Northern Highlands; to 3,900 ft (1,200 m). Prefers grasslands and marshes. Song is a fast series of descending staccato notes (*cht-chrt-ch*) that are immediately followed by a buzzy, insect-like *bzzeeee*. Call is a short, buzzy *tzíp*.

Savannah Sparrow (Sabanero de Pradera)   *Passerculus sandwichensis*
5.5 in (14 cm). Combination of **yellow supraloral** (sometimes too faint to see) and **heavy black streaking on white breast and flanks** is diagnostic. **Black malar stripe** widens to a broad base on white throat; also note buffy ear patch. Accidental winter resident, with only 1 record, a medium-sized group at Tisma NR (Feb 2016). Prefers grasslands and marshes, where it searches on the ground for food. Occurs alone or in small groups. Calls with a staccato *chít*.

Rusty Sparrow

Stripe-headed
Sparrow

Botteri's
Sparrow

Chipping
Sparrow

nonbreeding

Grasshopper
Sparrow

Savannah
Sparrow

Rufous-collared Sparrow (Chíngolo) · *Zonotrichia capensis*

5.5 in (14 cm). No other passerine has the distinct **rufous nuchal collar**. Immature shares head pattern and white throat of adult, but lower underparts are heavily streaked. Rare in Northern Highlands; above 3,300 ft (1,000 m). Frequents grasslands, agricultural areas, and other open and disturbed areas. Forages on the ground for seeds and invertebrate prey. Typically whistles a 3-noted, melodic *swee-swee-sweeuuu* song, with the final note sliding downward in pitch. Calls with an abrupt *chau* note.

Olive Sparrow (Pinzón Aceituno) · *Arremonops rufivirgatus*

5.5 in (14 cm). Has **brown stripes on buffy head**; lower mandible of bicolored bill is pale. The olive upperparts and striped head appearance are very similar to that of Black-striped Sparrow (which has a gray head), but the 2 species are unlikely to overlap in distribution. Uncommon in Pacific lowlands and foothills, primarily south from Lake Managua; to 2,600 ft (800 m). Found on the ground in dry forest and gallery forest, where it picks through leaf litter in search of seeds and invertebrate prey. Whistles a high-pitched, 3-part song that starts with 2 abrupt notes, repeats a downslurred *tseeó* in the middle, and finishes with a rapid *tu-tu-tu…*; with each repetition, the song increases in pace. Calls with a soft, reedy *tih*.

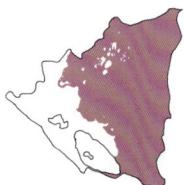

Black-striped Sparrow (Pinzón Cabecilistado) · *Arremonops conirostris*

6 in (16 cm). Has **black stripes on gray head**; the bill is entirely black. The olive upperparts and striped head appearance are very similar to that on Olive Sparrow (which has a buffy head), but unlikely to overlap in distribution. Common in Caribbean lowlands and foothills; to 2,000 ft (600 m). Rare on eastern slopes of Northern Highlands (absent on Sierra Dipilto-Jalapa) and locally rare on northern extent of Sierra Chontaleña; accidental in Pacific; to 4,600 ft (1,400 m). Found in humid lowland forest, cloud forest, and forest edge, as well as adjacent disturbed habitat with sufficient dense vegetation. Searches for invertebrate prey while hopping on the ground. Unique song is a series of whistles: Begins with the noticeably spaced *whit!, fer, KREAH* (each note distinctly different) and often transitions into a lengthy *cho! cho! cho!…*; song accelerates before ending abruptly. Also calls with a scratchy *tchep!*

Common Chlorospingus (Clorospingus Común) · *Chlorospingus flavopectus*

5.5 in (14 cm). Conspicuous **teardrop postocular spot** on slate-gray head is unique among the passerines. From below, white throat and belly stand out from yellow underparts. Abundant throughout Northern Highlands; above 3,400 ft (1,050 m). Abundant in Caribbean highlands at Quirragua NR, Musún NR, and Saslaya NP; above 2,300 ft (700 m). In cloud forest and forest edge, groups of 2 to 10 actively forage from understory to mid-canopy. It is often the nucleus species of large, diverse mixed-species flocks, which it facilitates by making high-pitched chittering contact calls. Non-musical song is an incessantly repeated harsh, high-pitched *tchep!*, delivered with a changing pace.

Rufous-collared
Sparrow

immature

adult

Olive
Sparrow

Black-striped
Sparrow

Common
Chlorospingus

Restricted to the Americas, these medium to very large birds have stout, long pointed bills, strong legs, and long pointed wings. Sexual dimorphism is common, with females often smaller and drabber than males. Of the 23 species in Nicaragua, 4 are NA migrants. Blackbirds, grackles, and cowbirds are terrestrial and feed on seeds and arthropods; orioles, caciques, and oropendolas are arboreal and feed on fruit, arthropods, and nectar.

Yellow-billed Cacique (Cacique Picoplata)   *Amblycercus holosericeus*

M 10 in (25 cm); F 8.5 in (22 cm). Combination of **pale-yellow bill** and contrasting black plumage is distinctive. Pale yellow iris distinguishes it from Scarlet-rumped Cacique (pale blue iris), a helpful feature when the scarlet rump of the latter species is not visible. Common in Caribbean. Locally common on Sierras Managua, but rare to uncommon in other regions of the Pacific; uncommon on eastern slopes of Northern Highlands; to 4,600 ft (1,400 m). Found in dry forest, cloud forest, humid lowland forest edge, secondary forest, second growth, river-edge vegetation, and bamboo thickets. Alone or in pairs, skulks in dense understory and tangles, where it is heard more often than it is seen. Excitedly repeats a whistled *fá-wEE* phrase, with whistles often varying in quality. Repeats a double-noted *weeu-wu*, and also delivers a variety of croaks and trills.

Scarlet-rumped Cacique (Cacique Rabirrojo)   *Cacicus uropygialis*

M 9.5 in (24 cm); F 8.5 in (22 cm). **Scarlet rump** is often concealed when perched. **Pale blue iris** further distinguishes it from Yellow-billed Cacique (pale yellow iris). Uncommon in Caribbean lowlands and foothills; to 2,300 ft (700 m). Inhabits humid lowland forest and edge, forest light gaps, tall second growth, and secondary forest. Groups forage in canopy for arthropods, small vertebrates, fruit, and seeds; accompanies oropendolas, fruitcrows, and nunbirds. Whistles a single, arcing *peeur* several times before repeating the same note in emphatic rapid-fire fashion.

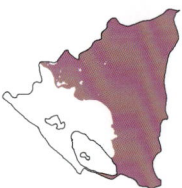

Chestnut-headed Oropendola (Oropéndola Alinegra)   *Psarocolius wagleri*

M 14 in (35 cm); F 11 in (27 cm). Note **chestnut head and neck and black wings and back**. Also note **pale white bill** (with frontal shield) and **pale blue iris**, which distinguish it from larger Montezuma Oropendola (with multi-colored bill and face). Common on eastern slopes of Northern Highlands (Feb to May); to 4,600 ft (1,400 m). Locally common in Caribbean. Dwells in both humid lowland forest and cloud forest. In small to large flocks, moves about in canopy feeding on arthropods, small vertebrates, fruit, and nectar. Breeds in colonies on forest edge and adjacent open areas, often in an isolated tree and sometimes with or nearby Montezuma Oropendola colonies; broods are sometimes parasitized by Yellow-billed Cacique and Giant Cowbird. Produces a variety of unique wooden or hollow vocalizations, as well as raspy *cha* call and a quivering electronic noise; vocalizations are subtler and more subdued than those of Montezuma Oropendola.

Montezuma Oropendola (Oropéndola Mayor)   *Psarocolius montezuma*

M 20 in (50 cm); F 15.5 in (39 cm). Combination of **black head and neck, bicolored bill,** and **light blue facial skin** distinguishes it from smaller Chestnut-headed Oropendola. Common in Caribbean lowlands and foothills. Locally common in Sierras Managua and Rivas Isthmus; uncommon in other regions of the Pacific; uncommon to locally common on eastern slopes of Northern Highlands; to 4,600 ft (1,400 m). Inhabits dry forest and humid lowland forest and edge, cloud forest, adjacent open areas, second growth, and shaded plantations. In small to large flocks, moves in canopy feeding on arthropods, small invertebrates, fruit, and nectar. Breeds in colonies on forest edge and adjacent open areas, often in an isolated tree; broods sometimes parasitized by Giant Cowbird. Displaying males emit a bizarre, escalating metallic chortle, often accompanied by a crackling noise; more raucous than that of Chestnut-headed Oropendola. Also barks a scratchy *rha!* and gives short yelps.

Yellow-billed
Cacique

Scarlet-rumped
Cacique

Chestnut-headed
Oropendola

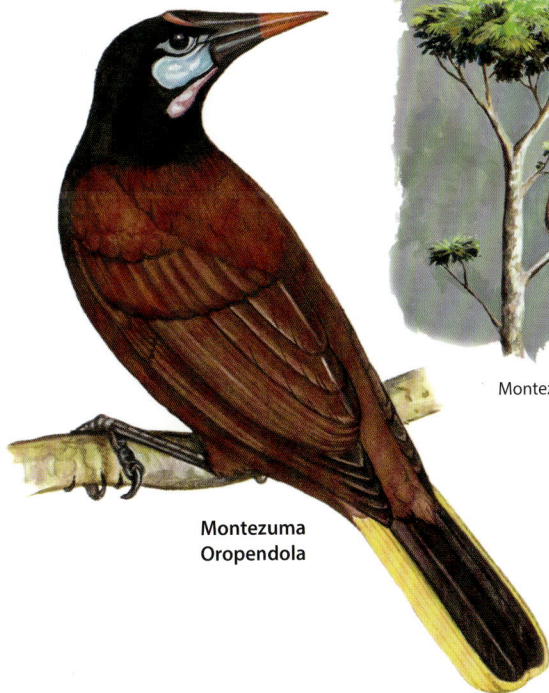

Montezuma Oropendola colony nest site.

Montezuma
Oropendola

Black-vented Oriole (Chichiltote Dorsinegro)   *Icterus wagleri*
8.5 in (21 cm). Only Nicaraguan oriole with **black vent and undertail coverts**. Very similar to Black-cowled Oriole but note that on Black-vented the yellow scapular patch connects to yellow underparts. Rare to uncommon in Northern Highlands; above 3,300 ft (1,000 m). Reaches southernmost distribution in Nicaragua. Occurs in highland pine and pine-oak forest, secondary forest, scrub, semi-open areas with scattered trees, and dry forest. In pairs, small groups, and (sometimes) flocks composed of other oriole species, forages for insects, larvae, and nectar. Whistles a sweet, slightly rising *feeep*; and yelps a nasal *WEYNT!*

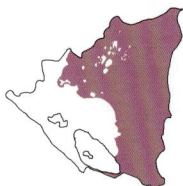

Black-cowled Oriole (Chichiltote Capuchinegro)   *Icterus prosthemelas*
8 in (20 cm). Very similar to Black-vented Oriole but **yellow scapular patch does not connect to yellow underparts**. On immature, olive back distinguishes it from larger Yellow-backed Oriole. Common in Caribbean lowlands and foothills; to 2,000 ft (600 m). Locally uncommon on eastern slopes of Northern Highlands; to 4,300 ft (1,300 m). Prefers humid lowland forest edge, open areas with scattered trees, cloud forest edge, secondary forest, shaded coffee plantations, and tall second growth. Alone, in pairs, or with other species of oriole, feeds in canopy on insects, fruit, and nectar; often seen in flowering trees. Delivers an eclectic series of whistles and metallic notes. Calls with a screeching, metallic *wheep*.

Yellow-backed Oriole (Chichiltote Dorsiamarillo)   *Icterus chrysater*
9 in (23 cm) Combination of **bright yellow back** and black wings and tail is distinctive (smaller immature Black-cowled Oriole has yellow-olive back). Common in Northern Highlands and locally rare on San Cristóbal-Casita NR; above 2,000 ft (600 m). Uncommon in the Mosquitia. Favors cloud forest edge, highland pine and pine-oak forest, lowland pine savanna, shaded coffee plantations, and secondary forest. In pairs, small groups, or with other oriole species, moves in canopy in search of insects, fruit, and nectar. Song is a series of slowly delivered, melodic whistles that oscillate up and down in pitch (5–7 seconds in total length). Also gives a squeaky, raspy *jeah*.

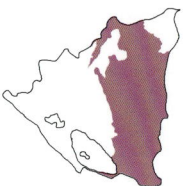

Yellow-tailed Oriole (Chichiltote Coliamarillo)   *Icterus mesomelas*
9 in (23 cm). **Yellow outer rectrices** are unique among orioles; undertail appears yellow when folded. Occurs in lowlands and foothills of Caribbean; locally common along San Juan River (but uncommon in other parts of southern Caribbean); rare in northern Caribbean; to 1,300 ft (400 m). Found in thickets and vine tangles in river edge, humid lowland forest edge, woodlands, swamps, and banana plantations; often near water. In pairs, small groups, or with other oriole species, forages in canopy for arthropods and fruit. Song is a cheerful, rolling repetition of *cheá-wer-herr-chítahuyah*, with some notes quite churry. Call is a low-pitched, muffled *cherp*.

Black-vented Oriole

adult

immature

adult

adult

immature

Black-cowled Oriole

immature

Yellow-backed Oriole

Yellow-tailed Oriole

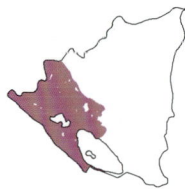

Streak-backed Oriole (Chichiltote Dorsilistado)   *Icterus pustulatus*

8 in (20 cm). **Streaked back** is diagnostic in all plumages, but note that it may appear solid black on some individuals. **Two white wing bars** (upper wing bar often includes yellow patch) and lack of spots on breast rule out Spot-breasted Oriole. Very similar to larger Altamira Oriole, but blue base on lower mandible is larger. In Pacific, common in lowlands and foothills; also common on western slopes and in dry intermontane valleys of Northern Highlands; to 5,200 ft (1,600 m). Inhabits dry forest, thorn forest, scrub, woodland, gallery forest, open areas with scattered trees, and second growth. In pairs, small groups, or with other oriole species, forages in canopy in insects, fruit, and nectar. Delivers a series of short, strained, descending whistles (typically repeated every 0.5–1 second). Call is an electronic-sounding *beaht*.

Spot-breasted Oriole (Chichiltote Maculado)   *Icterus pectoralis*

8.5 in (22 cm). **Black spots on sides of breast** (not always visible) and **lack of white wing bar** distinguish it from Streak-backed Oriole and Altamira Oriole. Combination of orange wing bar and **white patch on tertials and inner secondaries** is distinctive. Abundant in Pacific; common in dry intermontane valleys of Northern Highlands; to 5,200 ft (1,600 m). May be colonizing western portions of Caribbean, in areas where deforestation allows it to flourish. Prefers dry forest, secondary forests, thorn forest, scrub, woodland, second growth, gallery forest, and gardens with trees. Found alone, in pairs, or in groups with other oriole species; feeds on insects, fruit, nectar, and flowers. A beautiful songster; whistles a melodic tune of sweet, monotone notes that oscillate up and down in pitch. Call is a raspy *rhet* given multiple times.

Altamira Oriole (Chichiltote Mayor)   *Icterus gularis*

9.5 in (24 cm). Largest Nicaraguan oriole. Combination of **solid black back** and wing pattern (**orange upper wing bar** and **white lower wing bar**) is diagnostic; also, blue base on lower mandible is smaller than that on Streak-backed or Spot-breasted orioles. Uncommon in the Pacific. Common on western slopes and dry intermontane valleys of Northern Highlands; to 4,600 ft (1,400 m). Reaches southernmost distribution in Nicaragua. Prefers thorn forest, scrub, dry forest, second growth, and urban areas. In pairs, small groups, or with other oriole species, forages in the canopy for insects, flowers, nectar, and fruit. Whistles a slow series of sweet, abbreviated notes that gradually rise and then fall in pitch. Call is a nasal, raspy *rhe*.

Baltimore Oriole (Chichiltore Norteño)   *Icterus galbula*

7 in (18 cm). On male, combination of **entirely black head** and brilliant orange underparts makes male unmistakable; also note **orange outer tail rectrices**. Female has yellow-orange breast and undertail; **head and back are mottled** with black, orange, and olive; immature is similar but without mottling; coloration distinguishes both female and immature from female and immature Orchard Oriole (with yellow-olive color). Countrywide winter resident (Sept to early May). Common to abundant in Pacific; common in Northern Highlands and Caribbean; to 4,900 ft (1,500 m). Found in forest and edge, second growth, open areas with scattered trees, river edge vegetation, shaded coffee plantations, grasslands with scattered trees, gardens, and tree-lined roadsides. Forages primarily in canopy of flowering trees for nectar, insects, and fruit. Moves about in small groups and also joins mixed-species flocks. Whistles a compilation of soft, sweet *swee* and *su* notes, without apparent pattern. Calls with a raspy *reh*, often delivered in a fast chatter.

Orchard Oriole (Chichiltote Castaño)   *Icterus spurius*

6.5 in (17 cm). On male, combination of **chestnut and black** makes it unmistakable. On female and immature, olive back and yellow underparts distinguish it from female and immature Baltimore Oriole. Winter resident (Aug to April). Common in Pacific and Caribbean, and uncommon in Northern Highlands; to 4,600 ft (1,400 m). In addition to winter resident birds that visit the country, there is 1 record (Lake Apanás shoreline, Aug 2007) of a vagrant (*fuertesi* subspecies, resident of Mexico), on which the male shows ochre instead of chestnut. Found in woodlands, second growth, dry forest, river edge, lakeshores, marsh, grassland, tree-lined roadsides, and gardens. In small flocks and often with other oriole species, forages in canopy of flowering trees on nectar, pollen, insects, and fruit. Call is a raspy *chá*, given once, double-noted, or rattled off in fast succession.

immature

**Streak-backed
Oriole**

adult

adult with
black back

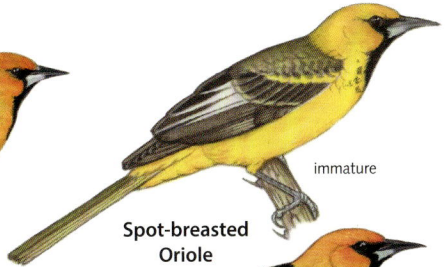

immature

**Spot-breasted
Oriole**

adult

**Altamira
Oriole**

immature

adult

**Baltimore
Oriole**

adult
female

immature

adult
male

adult male

immature
male

adult
female

**Orchard
Oriole**

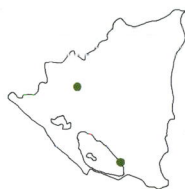

Yellow-headed Blackbird
(Tordo Cabeciamarillo)

Xanthocephalus xanthocephalus

9.5 in (24 cm). On male, **bright yellow head and breast** contrasting with the black body is diagnostic. On smaller female, yellow is duller and less extensive; also note brown body. Vagrant from NA, with only 2 known records (Jinotega Dept., Feb 2006; SE shore of Lake Nicaragua, April 2014). Occurs in agricultural fields, grassland, marshes with dense emergent vegetation, and rice fields. On ground, feeds on seeds and insects. Calls with a dry, harsh *chet*.

Bobolink (Tordo Arrocero)

Dolichonyx oryzivorus

6.5 in (16 cm). Note **sharp, pointed rectrices**. In breeding plumage (during boreal spring months), male is unmistakable. In nonbreeding plumage, male resembles female; on both, note **pale nape and lores** and pointed wings, which distinguish them from sparrows. Nonbreeding birds distinguished from larger female Red-winged Blackbird by **pinkish bill** that is shorter and stouter and **unstreaked breast and belly**; distinguished from Grasshopper Sparrow (p. 346) by larger size and **streaked flanks**. Passage migrant (late Aug to Oct and April to May). Rare in Caribbean and very rare in Pacific. Favors grassland and rice, sorghum, and other agricultural fields, where large flocks visit to feed on seeds and, occasionally, insects. In flight, gives a sweet *pink* contact call.

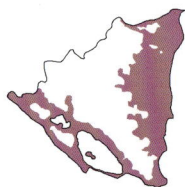

Red-winged Blackbird (Tordo Sargento)

Agelaius phoeniceus

8.5 in (22 cm). On adult male, **red-and-yellow shoulder** is distinctive (not always visible on perched birds). On female, **heavily streaked breast** distinguishes it from smaller female Bobolink and female Red-breasted Blackbird. Locally common to abundant in Pacific lowlands, but uncommon at Playitas-Moyúa-Tecomapa lagoons and in Sébaco Valley; locally common along San Juan River but uncommon in other parts of the Caribbean lowlands. Occurs at freshwater and brackish marshes and adjacent open areas, rice fields, and riparian thickets. Forages on ground and low vegetation for seeds and insects. Territorial during breeding season, but roosts and nests in colonies. Distinct 3-note song starts with a whistle and ends with a drawn-out, twangy rasp. Call is a dry, soft *chet*.

Red-breasted Blackbird (Zacatero Pechirrojo)

Leistes militaris

7 in (18 cm). On male, flashy **red throat and breast** is diagnostic. Female has unique **pink wash on breast**, and is further distinguished from female and nonbreeding Bobolink by darker-hued upperparts; distinguished from female Red-winged Blackbird by unstreaked breast. Locally rare to uncommon in southern Caribbean lowlands, around Pearl Lagoon and along San Juan River; rare in other regions of the southern Caribbean lowlands (expected in the Guatuzo Plains); very rare on Cosigüina Peninsula (a single record of 2 individuals). Reaches northernmost distribution in Nicaragua; as birds continue to expand range from the south, this species is likely to increase in abundance in Nicaragua. Found in flooded grassy fields, rice fields, and palm plantations. Feeds on or near the ground on seeds, insects, and some fruit. Distinctively sings a *fi-fi-fi-fi-ZEEEEE*, sometimes not giving the opening whistles but only the concluding explosive *ZEEEEE* buzz.

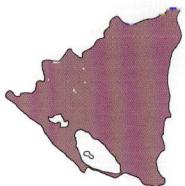

Eastern Meadowlark (Zacatero Común)

Sturnella magna

8.5 in (22 cm). **Black V-shaped band on bright yellow breast** is distinctive; in flight, note white outer rectrices. Locally common countrywide; to 5,200 ft (1,600 m). Found in grassland, agricultural fields, and, often, near water. Forages on ground or near ground for seeds, arthropods, and fruit. On the ground, the bird is so well camouflaged that it is hard to see; when singing from perch, however, characteristic field marks on underparts are easily seen. Sweetly sings a 3-noted *swee-swi-swilala*, with the third note noticeably dipping in pitch. Also delivers a short churr.

Yellow-headed Blackbird

male

female

breeding
male

nonbreeding
adult

Bobolink

male

female

**Red-winged
Blackbird**

**Red-breasted
Blackbird**

male

female

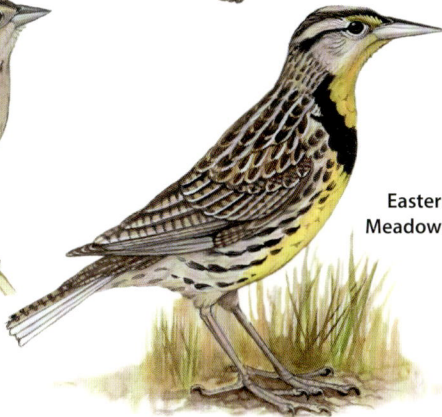

**Eastern
Meadowlark**

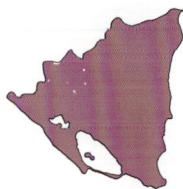

Bronzed Cowbird (Vaquero Ojirrojo)   *Molothrus aeneus*

8 in (20 cm). **Conical bill** (with straight culmen) and **lack of frontal shield** distinguish it from much larger Giant Cowbird. Distinguished from Melodious Blackbird by red iris (iris is brownish on immature). Abundant in Pacific; common in Northern Highlands; uncommon in southern and southeastern shores of Lake Nicaragua and along San Juan River headwaters; except for San Juan River, rare in rest of Caribbean, though it is extending range with the expansion of cattle ranches; to 5,200 ft (1,600 m). Frequents open areas, agricultural fields, woodland, roadsides, and urban areas. Gregarious; flies in small to very large flocks. Forages on ground for seeds and insects. A brood parasite on several small passerines. Gives a very thin, upslurred *weeep*, and a thin, downslurred whistle that morphs into a tinkling trill. Calls with a *ti-tu* or a *ti-tu-tu*.

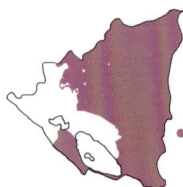

Giant Cowbird (Vaquero Grande)   *Molothrus oryzivorus*

M 13.5 in (34 cm); F 11.5 in (29 cm). **Long, stout bill** (with curved culmen) and **frontal shield** distinguish it from smaller Bronzed Cowbird. Locally uncommon in Northern Highlands and Caribbean, rare in Pacific; to 4,600 ft (1,400 m). Favors forest edge, adjacent woodland, and riverbanks. Alone or in small groups, forages on ground for insects and seeds; also forages in fruiting trees. Exhibits brood parasitism of oropendolas and caciques; easiest to find at oropendola colonies when those birds are laying eggs. Delivers a very thin, metallic *kwee-ti-ti*. Emits a loud and harsh *kuck!*

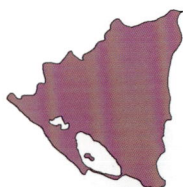

Melodious Blackbird (Cacique Piquinegro)   *Dives dives*

10.5 in (27 cm). The only Nicaraguan icterid that is **completely black**; black irises distinguish it from both cowbirds. Stout, pointed bill distinguishes it from Groove-billed Ani (p. 60). Common countrywide at all elevations. Occurs in woodland, forest edge, grassland, open areas with scattered trees, urban parks, tree-lined roadsides, and riparian thickets. Forages in pairs, either on ground for insects or in canopy of flowering trees for nectar. Song consists of eclectic whistles, including a lively, down-up *zEEu-wee*, a metallic *pít-WEE*, and a more subdued, downslurred *peeeo*. Calls continuously with a sharp *twit!*

Great-tailed Grackle (Zanate Grande)   *Quiscalus mexicanus*

M 13.5–18.5 in (34–47 cm); F 10.5–12.5 in (27–32 cm). Has a large bill and a very long, graduated tail. Very similar to smaller Nicaraguan Grackle, but note **proportionately longer tail** and **elongated body and neck**. Iridescent bluish sheen is seen in good light. In Nicaragua, female generally lacks contrasting superciliary; female further distinguished from female Nicaraguan Grackle by brown underparts that are richer. Abundant countrywide, but slightly less numerous in Caribbean; to 5,000 ft (1,550 m). Found in a variety of open habitats, including agricultural fields, coastal areas, wetlands, grasslands, and urban zones. Roosts in colonies at places close to humans. Feeds on seeds, insects, fruit, carrion, and trash. Sings with a variety of combined whistles and harsh notes; one commonly heard combination is *weelP! spi spi*, sung with a sharply grating quality. A commonly heard call, often given in flight, is a slightly strained *rea-reu*.

Nicaraguan Grackle (Zanate Nicaragüense)   *Quiscalus nicaraguensis*

M 11.5 in (29 cm); F 9.5 in (24 cm). Male often confused with male Great-tailed Grackle; keys to identification are habitat, colonial behavior, and vocalization. Noticeably **less glossy** than Great-tailed and tail is shorter. When singing, displayed tail creates a narrower triangle. **Female has pale buffy underparts** and **pale superciliary is sharply defined**. Very restricted distribution; locally uncommon from Lake Managua and south through Guatuzo Plains and San Juan River headwaters. Associated with emergent vegetation and shrubbery at freshwater marsh edge (particularly *Typha domingensis* and *Thalia geniculata*). Flocks forage on mudflats and flooded fields in search of seeds, snails, and insects. Great-tailed Grackle co-occurs in most of its habitat. Roosts and nests in colonies in marsh vegetation, never in urban areas or away from aquatic habitat. Male sings a sequence of 5 descending notes, screeching but ethereal in tone; first note is the longest and undergoes the greatest drop in pitch. Gives a variety of calls, including an upslurred, screeching *meeuw*; a short *mew*; and a raspy, forced *chu!* Endemic to Lake Nicaragua marshes EBA.

Two male Bronzed Cowbirds performing courtship displays for a group of females.

adult male

Bronzed Cowbird

immature

adult male

Giant Cowbird

immature

female

male

Great-tailed Grackle

Melodious Blackbird

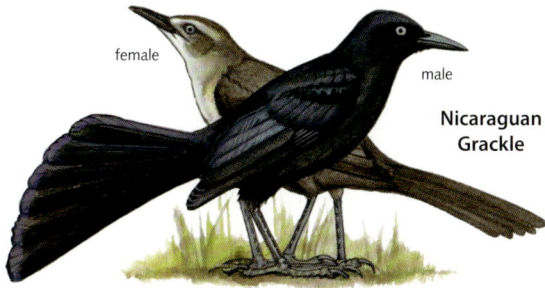

female

male

Nicaraguan Grackle

Although previously considered a New World warbler (Family Parulidae), the Yellow-breasted Chat was recently deemed to merit its own, newly described, family, of which it is the sole member. Note that the name *chat* is used for species within 4 separate families worldwide.

Yellow-breasted Chat (Reinita Grande)   *Icteria virens*
6 in (15 cm). Distinctive and large, with stout bill and long, rounded tail. **White spectacles and malar stripe** mark the head; **throat and breast are bright yellow**. From below, note white undertail coverts and belly. Compare to Fan-tailed Warbler (p. 384). Winter resident (mid-Sept to mid-April). Rare countrywide; to 4,300 ft (1,300 m). Prefers dense thickets, in which it skulks; mostly solitary. Calls include a raspy, twangy *chewt!*; a *chewt!-chewt!*, delivered with a scolding tone; and a nasal, plaintive *eearh*.

OLIVE WARBLER   Peucedramidae

The Olive Warbler is the sole member of its family! Although superficially similar to a New World warbler (Parulidae), it is just as closely related to the Thraupidae, Passerellidae, Cardinalidae, Icteridae, and Fringillidae. A specialist of coniferous forests, this insectivore reaches its southernmost distribution in Nicaragua.

Olive Warbler (Reinita Olivácea)   *Peucedramus taeniatus*
5 in (13 cm). Combination of **long, thin bill, black mask**, and head color (tawny on male, yellow on female) distinguishes it from any similar New World warbler; both male and female have 2 bold white wing bars and a deeply notched tail. In flight, note white outer rectrices and white patch at base of primaries. On female, faint black mask on yellow head and unstreaked underparts are similar to those on immature female Hermit Warbler (p. 378), but note notched tail. Locally uncommon on Sierra Dipilto-Jalapa and rare on Sierra Isabelia; from 3,900 to 5,900 ft (1,200 to 1,800 m). Reaches southernmost distribution in Nicaragua. Moves along branches in canopy of highland pine and pine-oak forest in search of insects; often flicks wings while foraging. Joins mixed-species flocks. Sings a rapid sequence of 9–11 strident notes (*FEE-FEE-FEE…*); sequence lasts 1 second. Calls with loud, falling *feeú* whistle.

The typical skulking behavior of a Yellow-breasted Chat.

Yellow-breasted Chat

Olive Warbler

female

male

The members of this New World family are colorful and diverse. Nicaragua has 48 species, 37 of which are migrants from North America. Habitat and foraging behavior are important keys to identification. Although patterns and coloration on the head are also useful for identification, sometimes one is only afforded views of the undertail, undertail coverts, and vent when birds are foraging in the trees. Adding to the difficulty of making a proper identification is the great variety in plumages, ranging from colorful breeding males to drab immature females.

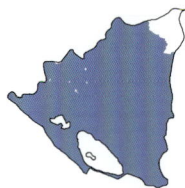

Ovenbird (Reinita Andarina)   *Seiurus aurocapilla*
5.5 in (14 cm). **Orange crown** (bordered by black lateral crown stripe) distinguishes it from Louisiana Waterthrush and Northern Waterthrush. **White eye ring** and short, thick bill further distinguish it. Winter resident (Oct to mid-May). Occurs countrywide. Common in Northern Highlands; uncommon in Caribbean; in the Pacific, it is uncommon on Sierras Managua, Mombacho Volcano, and Laguna Apoyo NR but rare in other areas; to 5,200 ft (1,600 m). Found in primary and secondary forest. Forages alone on ground, flipping leaf litter in search of prey. Walks steadily, bobbing its head as it does so; often cocks its tail. Call is a scratchy, harsh *chep!*

Worm-eating Warbler (Reinita Anteada)   *Helmitheros vermivorum*
5 in (13 cm). Note **black stripes on a buffy head** and olive upperparts. From below, note flesh-colored lower mandible; long, pointed bill; short, dark tail; and diagnostic **smudgy gray spots on the buffy undertail coverts**. Winter resident (Sept to April). Uncommon in Northern Highlands and Sierras Managua; rare throughout lowlands; to 4,300 ft (1,300 m). Often forages with mixed-species flocks in forest understory and mid-canopy, where it searches for caterpillars and other food in clumps of dead, hanging leaves. Call is a rich, excited *tsip!*

Louisiana Waterthrush (Reinita Acuática Cejiblanca)   *Parkesia motacilla*
5.5 in (14 cm). Almost identical to Northern Waterthrush, but larger and with a **longer bill**; white superciliary broadens behind the eye (superciliary is thinner on Northern and it does not broaden behind the eye); **unmarked white throat** is bordered by a broader malar stripe; has less streaking on underparts than does Northern; note buffy flanks. Winter resident (mid-July to mid-April). Uncommon countrywide; to 4,400 ft (1,350 m). Found on the banks of swift moving water (often under forest canopy). Slowly pumps rear of body as it walks in search of invertebrate prey. Call is a sharp *chip!*, very similar to that of Northern Waterthrush but more metallic.

Northern Waterthrush (Reinita Acuática Norteña)   *Parkesia noveboracensis*
5 in (13 cm). Almost identical to Louisiana Waterthrush, but smaller and with a **shorter bill**; buffy or white superciliary narrows behind the eye; **white throat with fine streaks** (unstreaked on Louisiana) and bordered by a narrower malar stripe; streaking on underparts is denser. Winter resident (Aug to mid-May). Common countrywide; to 4,900 ft (1,500 m). Rapidly pumps tail while walking on the ground in search of invertebrate prey; associated with a variety of aquatic habitats, but generally forages at standing water. Call is a sharp *chip!*, very similar to that of Louisiana Waterthrush but richer in tone.

Buff-rumped Warbler (Reinita Guardarribera)   *Myiothlypis fulvicauda*
5 in (13 cm). Note **buffy rump and base of tail**, brown upperparts, and brown mottling on pale breast. Common in Caribbean lowlands and foothills; to 2,600 ft (800 m). Rare on eastern slopes of Northern Highlands, but locally uncommon in Peñas Blancas NR; to 4,300 ft (1,300 m). Usually found in primary and secondary forests. Often seen hopping among rocks and debris at edge of rivers and streams. Habitually makes a pump-and-swing tail motion. Song is a repeated series of fast, strident notes that rise in pitch throughout the song and slow down at the end; song often begins with two noticeable introductory notes. Call is a lethargic, low-pitched *chrp*.

Ovenbird

Worm-eating
Warbler

Louisiana
Waterthrush

Northern
Waterthrush

Buff-rumped
Warbler

[Golden-winged Warbler and Blue-winged Warbler regularly hybridize, producing 2 hybrid forms: Brewster's Warbler and Lawrence's Warbler. Brewster's has gray and white plumage and lacks a black throat; Lawrence's has yellow plumage and a black throat. Lawrence's Warbler is the rarer of the 2 forms.]

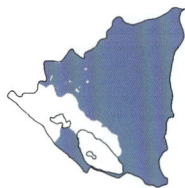

Golden-winged Warbler (Reinita Alidorada)   *Vermivora chrysoptera*

4.5 in (11 cm). Note **golden yellow wing patch** and crown on a predominately gray body. From below, also note black (on male) or gray (on female) throat and grayish-white underparts; all-white tail has thin black outer edges that taper to black-tipped outer rectrices. Winter resident (mid-Sept to mid-April). Common in Northern Highlands; generally uncommon throughout Caribbean but locally common along forested river edges); rare on Sierras Managua and Mombacho Volcano; to 4,900 ft (1,500 m). Found in primary and secondary forests and edges. Forages from understory to mid-canopy; often seen contorting body while searching dead leaf clumps for invertebrate prey. Joins mixed-species warbler flocks. Call is a sharp, thin *tsip!* **NT**

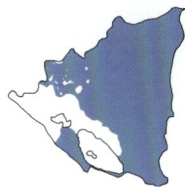

Blue-winged Warbler (Reinita Aliazul)   *Vermivora cyanoptera*

4.5 in (11 cm). Combination of **black eye line** and **two white wing bars** (less distinct on female) on **blue-gray wings** distinguishes it from other predominantly yellow passerines. From below, note contrast between white undertail coverts and yellow underparts; white on undertail tapers to a point because of black-tipped outer rectrices. Winter resident (mid-Sept to mid-April). Uncommon in Northern Highlands and Caribbean; rare on Sierras Managua and Mombacho Volcano and surrounding lowlands; to 4,600 ft (1,400 m). Found in a variety of forest types, from second growth to primary forest. Forages from understory to canopy, acrobatically prying open rolled leaves in search of invertebrate prey; often joins mixed-species flocks. Call is a sharp *chip!*

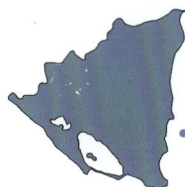

Prothonotary Warbler (Reinita Cebecidorada)   *Protonotaria citrea*

5 in (13 cm). Large bill and black eyes are prominent on **brilliant golden-yellow** plumage. **Unmarked blue-gray wings** and **white undertail coverts** help distinguish it from other yellow warblers. Winter resident (Aug to early April); rare to uncommon countrywide; to 4,900 ft (1,500 m). Found in swamps, mangroves, and dense vegetation bordering aquatic habitats. Alone forages on the ground and in understory. Calls with an unhurried *chip*.

female

Golden-winged Warbler

male

male

Blue-winged Warbler

female

Brewster's Warbler
(hybrid)

Lawrence's Warbler
(hybrid)

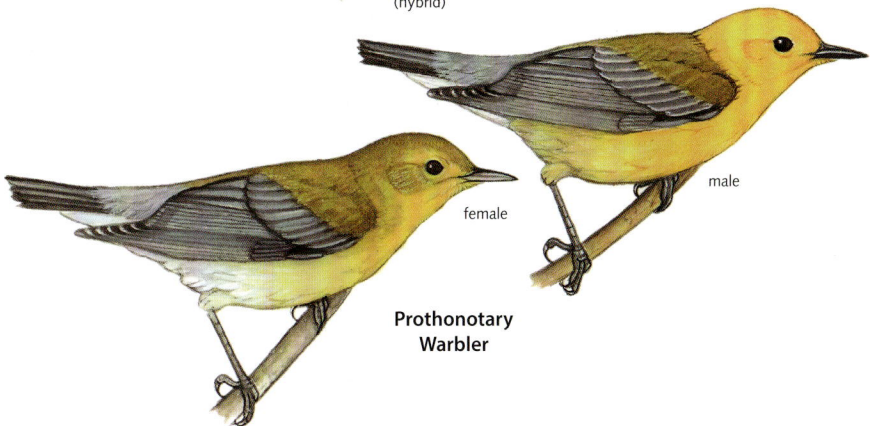

male

female

Prothonotary Warbler

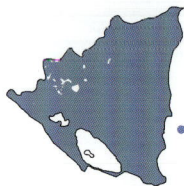

Northern Parula (Parula Norteña) — *Setophaga americana*

4 in (10 cm). This and Tropical Parula are the smallest of the Nicaraguan warblers. Note yellow-green back surrounded by blue upperparts. **Broken white eye ring**, two conspicuous wing bars, and **black and orange band on breast (on male)** distinguish it from Tropical Parula. Female and immature are duller but with same basic plumage pattern. In flight, note white underparts (from belly to undertail) and yellow throat that blends into yellow lower mandible. Rare winter resident (mid-Oct to April). Occurs countrywide; to 4,300 ft (1,300 m). Acrobatically forages at mid-canopy and canopy of forests, forest edge, and second growth. Call is a rich *chi!*

Tropical Parula (Parula Tropical) — *Setophaga pitiayumi*

4 in (10 cm). This and Northern Parula are the smallest of the Nicaraguan warblers. Has a yellow-green back surrounded by blue upperparts. **Black mask**, one whitish wing bar (sometimes not present), **orange breast (on male)**, and yellow underparts distinguish it from Northern Parula. Female and immature are duller but with same basic plumage pattern. In flight, note white underparts (from lower belly to undertail) and yellow throat that blends into yellow lower mandible. Common in Northern Highlands; from 2,600 to 5,200 ft (800 to 1,600 m). Locally common in Rivas Isthmus and on Ometepe Island; there is a historical record from Mombacho Volcano. Locally uncommon on northern extent of Sierra Chontaleña; above 1,600 ft (500 m). Rare in Caribbean lowlands and foothills, but locally uncommon in foothills and highlands of Quirragua NR, Musún NR, and Saslaya NP. Actively forages, sometimes hovering, at mid-canopy and canopy of forest and forest edge; joins mixed-species flocks. Short song transitions from a high-pitched warble to a short trill before abruptly ending with a downward inflected note. Call is an abbreviated *tip!*

Crescent-chested Warbler (Parula Cejiblanca) — *Oreothlypis superciliosa*

4.5 in (11 cm). **Chestnut crescent on yellow upper breast** is diagnostic, though it is sometimes faint or absent. **Broad, white superciliary** and unmarked wings distinguish it from Northern and Tropical Parula. From below, note contrasting yellow and white underparts and a dark undertail. Rare in Northern Highlands; above 3,300 ft (1,000 m). Reaches southernmost distribution in Nicaragua. Forages at mid-canopy and canopy of cloud forest and highland pine and pine-oak forest; sometimes found in mixed-species flocks. Song is a soft, insect-like trill (1 second in length). Call is a sharp, high-pitched *tsit!*

Tennessee Warbler (Reinita Verduzca) — *Oreothlypis peregrina*

4.5 in (11 cm). All plumages have at least a faint **superciliary** and a **thin, black eye-line**. Also note short tail, long primary projection, and short, fine-tipped bill. From below, note white undertail coverts and unstreaked underparts. Winter resident (Sept to April). Abundant in the Pacific and Northern Highlands; common in the Caribbean. Found at all levels in forest, second growth, coffee plantations, and gardens. Forages on nectar and berry pulp and actively searches for invertebrates. Calls with a sharp *tseet!*

Nashville Warbler (Reinita Cabecigrís) — *Oreothlypis ruficapilla*

4.5 in (11 cm). **Gray half-hood** and **yellow throat** distinguishes it from Lesser Greenlet (with whitish throat), p. 398; also note **white eye ring** (chestnut crown patch is mostly hidden). Female is slightly drabber, and should be compared to female and immature Mourning Warbler (p. 368). From below, note white vent contrasting with unstreaked yellow underparts and slender, dark tail. Accidental winter resident; only 1 record, from Tisey-Estanzuela NR (Feb 2016). Most likely found actively foraging in understory and mid-canopy of forest edge and second growth. Call is a thin, metallic *chewp!*

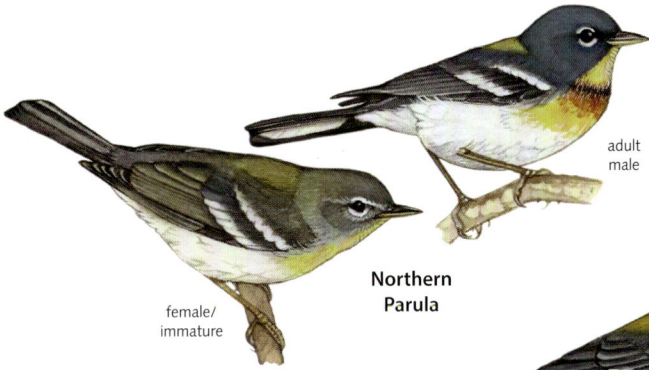

adult
male

female/
immature

**Northern
Parula**

male

**Tropical
Parula**

**Crescent-chested
Warbler**

immature

**Tennessee
Warbler**

adult
female

breeding
male

male

**Nashville
Warbler**

Connecticut Warbler (Reinita Ojianillada)   *Oporornis agilis*
5.5 in (14 cm). In all plumages, **thick, complete eye ring** distinguishes it from other similar looking *Geothlypis* warblers. On male, gray hood is paler than hood of MacGillivray's and Mourning Warbler (p. 368); male lacks the black patch on the breast. Female is slightly drabber than male; immature has drab olive-brown upperparts (including hood) and yellow underparts, with a buffy throat. From below, yellow undertail coverts reach ¾ the length of the tail. An accidental NA passage migrant (Sept to Oct and March to April); only 1 record (Peñas Blancas NR, Oct 2017). Forages by walking on the ground in dense second growth.

MacGillivray's Warbler (Reinita Ojeruda)   *Geothlypis tolmiei*
5 in (13 cm). On all plumages, note **thick, broken white eye ring**; larger Connecticut Warbler has complete eye ring. On male, black lores connect across forehead. Undertail coverts only reaching middle of tail further distinguishes it from Mourning Warbling (with undertail coverts reaching ¾ the length of tail). Winter resident (mid-Sept to April). Uncommon in Northern Highlands; above 2,500 ft (750 m). Rare in Pacific foothills. Forages on ground and understory of forest edge, shade coffee plantations, grasslands, and thickets in second growth. Calls with a harsh *chewt!*

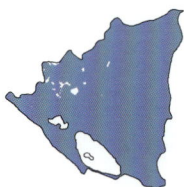

Mourning Warbler (Reinita Enlutada)   *Geothlypis philadelphia*
5 in (13 cm). On male, combination of **unmarked slate-gray head** and **black patch on breast** is distinct; some individuals are drabber. Female and immature sometimes have yellow on the throat, making them similar to Nashville Warbler (p. 366). On female and immature, eye ring can be absent, broken, or complete, but it is always thin if present; male rarely has the thin eye ring. Undertail coverts reaching ¾ the length of the tail help distinguish it from MacGillivray's Warbler (with undertail coverts only reaching middle of tail). Connecticut Warbler always has bold white eye ring. Winter resident (Sept to early May). Rare countrywide; to 4,300 ft (1,300 m). Skulks on ground in dense vegetation of forest, forest edge, and thickets in second growth. Calls is a scratchy *twit!*

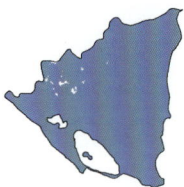

Kentucky Warbler (Reinita Cachetinegra)   *Geothlypis formosa*
5 in (13 cm). **Yellow spectacle** is diagnostic; black on crown and face is reduced on female; crown and face show olive on immature. From below, note bright yellow underparts and long undertail coverts that extend close to the tip of short tail. Winter resident (mid-Aug to April). Common on eastern slopes of Northern Highlands; generally uncommon in Caribbean (but common in Saslaya NP); very rare in Pacific and on western slopes of Northern Highlands; to 4,800 ft (1,450 m). In forest and forest edge, skulks on ground and in understory; also ventures into thickets in adjacent second growth. Call is a low-pitched *chup!*

Connecticut
Warbler

immature

male

MacGillivray's
Warbler

female

male

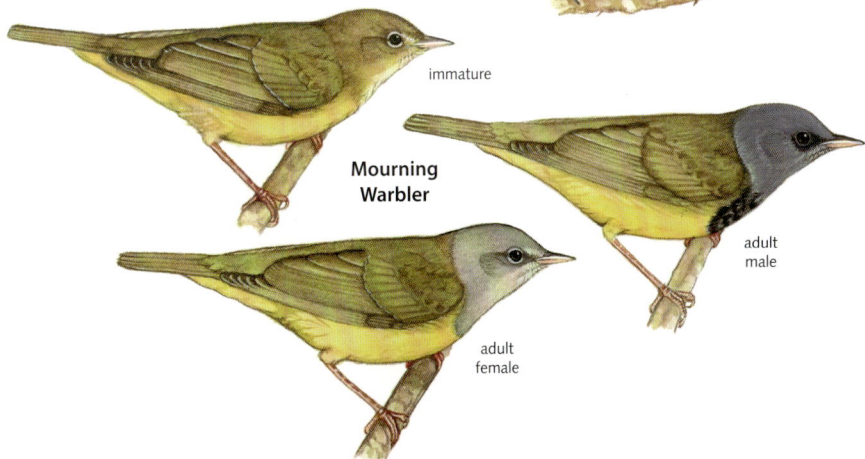

immature

Mourning
Warbler

adult
male

adult
female

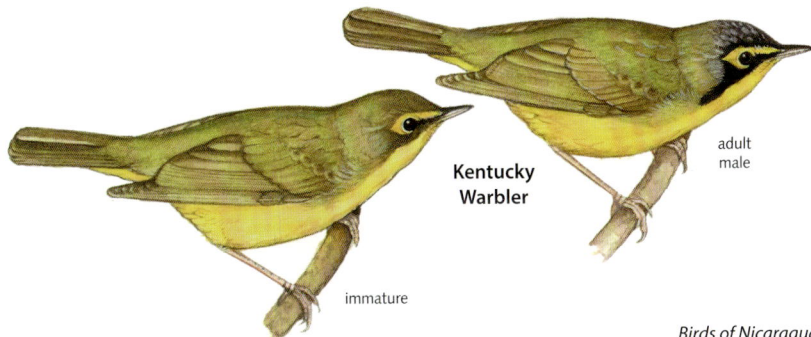

Kentucky
Warbler

adult
male

immature

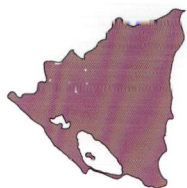

Gray-crowned Yellowthroat
(Enmascarado Coronigris)
Geothlypis poliocephala

5.5 in (14 cm). On male, note **small, black mask** and **gray crown**, both of which are reduced on female. Female underparts are entirely yellow. Compared to Olive-crowned and Common Yellowthroat, stout, bicolored bill is larger and graduated tail is longer. Common countrywide; to 5,400 ft (1,650 m). Inhabits highland pine forest and pine-oak forest, lowland pine savanna, grasslands, second growth, agricultural fields, and borders of freshwater marshes. Erratically flicks its cocked tail side-to-side while on exposed, low-vegetation perch. Sings a melodic, complex warble (3-4 seconds in length), often with a single soft, buzzy note added toward the end. Calls with a rich *chip chip chip*; a raspy *jeeree*; and a sweet, descending *tee-ji-ert*.

Olive-crowned Yellowthroat (Enmascarado Carinegro)
Geothlypis semiflava

5 in (13 cm). On male, **broad, black mask** directly borders olive crown and nape. Female has very faint yellow superciliary, olive head, and all-yellow underparts. Bill is larger than that of Common Yellowthroat. Present throughout Caribbean (rare in the north, common in the south). Prefers wet grasslands and vegetation that borders rivers and freshwater marshes. Sings a melodic, 2-phased warble; starts moderately paced before speeding up in the second half; sometimes ends with a short, dull trill. Call is a nasal, raspy downslurred *jeer*.

Common Yellowthroat (Enmascarado Norteño)
Geothlypis trichas

5 in (13 cm). On male, **broad, black mask is bordered by a narrow white and gray band**. Female has thin broken eye ring; the forehead sometimes has a slight rufous tinge; underparts are all yellow except for pale belly (no other *Geothlypis* warbler has a pale belly). Bill is smaller than that of Olive-crowned Yellowthroat; all plumages are slightly browner than those of Gray-crowned and Olive-crowned Yellowthroat. Winter resident (mid-Oct to mid-April). Common on eastern slopes of Northern Highlands and in the Caribbean; locally uncommon on western slopes of Northern Highlands and in Pacific; to 4,600 ft (1,400 m). Furtive; inhabits grasslands, marshes, and dense vegetation bordering water. Calls with a burry *chep!*

Hooded Warbler (Reinita Encapuchada)
Setophaga citrina

5.5 in (14 cm). In all plumages, note bright yellow face and dark lore; on male, **black hood** is diagnostic; on female, hood is greatly reduced. When flicking its tail open, flashes white inner web of outer rectrices (no other olive and yellow warblers have white on the inner web of the outer rectrices). From below, note white tail with yellow undertail coverts that reach half the length of the tail. Winter resident (mid-Oct to April). Uncommon on eastern slopes of Northern Highlands and in the Caribbean; locally rare on Sierras Managua and in adjacent lowlands; rare in the Pacific and on western slopes of Northern Highlands; to 4,300 ft (1,300 m). Forages in understory of forest, forest edge, and adjacent second growth; hovers to pick prey from foliage or sallies to capture them in midair. Call is a sharp *tchip!*

Gray-crowned Yellowthroat

male

female

A male Gray-crowned Yellowthroat singing in tall grass.

Olive-crowned Yellowthroat

male

female

Common Yellowthroat

female

male

Hooded Warbler

female/ immature

male

Cape May Warbler (Reinita Atigrada) *Setophaga tigrina*

4.5 in (11 cm). Breeding male is unmistakable; has **rufous ear patch**. In all plumages, note yellow or olive rump (also on Magnolia Warbler and Yellow-rumped Warbler); yellow-green wing edgings; at least faint wing bars; and streaking on underparts (of varying intensity); most birds show some yellow on the face. On adult female, black streaks on mostly yellow underparts distinguish it from Yellow-rumped Warbler (with mostly white underparts). From below, note white undertail coverts and black outer edges and black tip on a white tail. Very rare winter resident (mid-Oct to mid-April), countrywide. Feeds on nectar; attracted to flowering trees, where it is known to accompany groups of Tennessee Warblers and sometimes exhibit aggressive behavior. Call is a sharp, high-pitched *tsit!*

Magnolia Warbler (Reinita Colifajeada) *Setophaga magnolia*

5 in (13 cm). On breeding male, combination of **black mask, white superciliary**, and **black patch on breast** is diagnostic. In all plumages, note yellow rump (also on Cape May and Yellow-rumped Warbler); wing bars (sometimes faint); and streaking on underparts (of varying intensity). On adult male and female, black streaks on bright yellow underparts distinguish it from Yellow-rumped Warbler (which has mostly white underparts). In flight, **broad white band on middle of uppertail** (except for central rectrices) is diagnostic. From below, note white base and black tip on tail and white undertail coverts and lower belly. Winter resident (mid-Sept to early May). Common in Northern Highlands and Caribbean and uncommon in Pacific; to 4,600 ft (1,400 m). Frequents a wide variety of habitats. In forests, actively forages at understory and mid-canopy, where it sometimes joins mixed-species flocks. Call is a raspy *cheap!*

Yellow-rumped Warbler (Reinita Rabiamarilla) *Setophaga coronata*

5.5 in (14 cm). Nicaragua as 2 subspecies; both are NA migrants. Both subspecies have **yellow rumps** (as do Cape May Warbler and Magnolia Warbler), **yellow upper flanks**, and a yellow crown patch. They prefer open to semi-open habitats. *Coronata* subspecies (Myrtle Warbler): Black streaks on mostly white underparts distinguish it from Cape May Warbler and Magnolia Warbler (which have mostly yellow underparts). Winter resident (Oct to March). Uncommon in Northern Highlands and Caribbean; locally rare in Pacific; to 4,600 ft (1,400 m). Call is a slightly muffled *chip*. *Auduboni* subspecies (Audubon's Warbler): On breeding male, yellow throat distinguishes it from breeding male Myrtle Warbler. Very rare winter resident. Occurs in Northern Highlands. Call is a rich, lively *twit!*

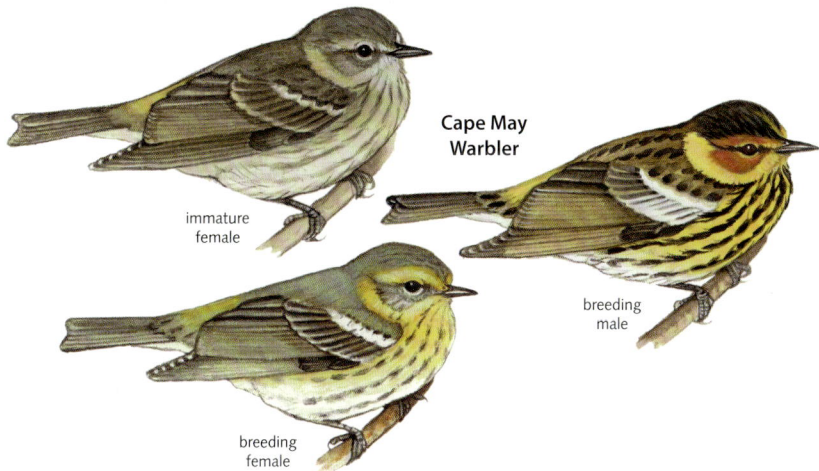

**Cape May
Warbler**

immature
female

breeding
male

breeding
female

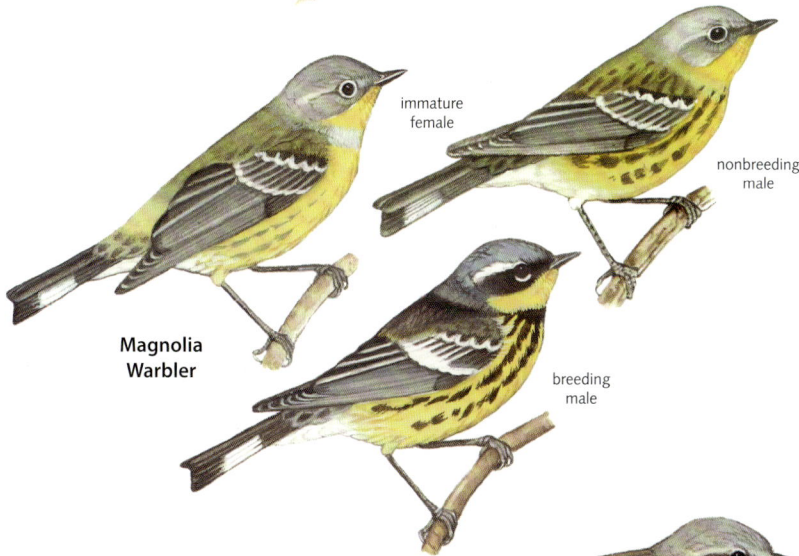

immature
female

nonbreeding
male

**Magnolia
Warbler**

breeding
male

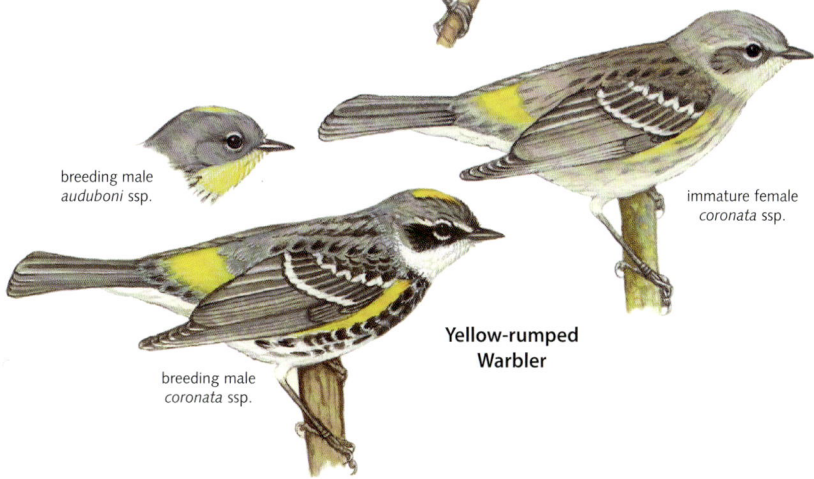

breeding male
auduboni ssp.

immature female
coronata ssp.

breeding male
coronata ssp.

**Yellow-rumped
Warbler**

Cerulean Warbler (Reinita Cerúlea)   *Setophaga cerulea*

4.5 in (11 cm). In all plumages, has **blue to blue-green upperparts** and **2 bold white wing bars**. Sky blue male is unmistakable. Female and immature have broad white or pale superciliary and yellow washed underparts; lack of pale stripes on back and lack of pale central crown stripe help distinguish them from immature female Blackburnian Warbler. From below, note white undertail with black-tipped outer rectrices; also note white subterminal band on fanned tail. Passage migrant (Aug to mid-Oct and March to mid-April). Uncommon in Northern Highlands and Caribbean and rare in Pacific; to 4,600 ft (1,400 m). Forages at mid-canopy and canopy; occasionally hovers as it picks insects from foliage. Sometimes forms small groups and also joins mixed-species flocks. Call is a high-pitched *chip*. **VU**

Blackburnian Warbler (Reinita Gorginaranja)   *Setophaga fusca*

5 in (13 cm). On breeding male, **bright orange coloration** is unmistakable. In all plumages, note **2 pale stripes on back** (which are diagnostic); a **dark triangular mask**; and streaked flanks (of varying intensity). From below, note white tail with black outer edges and white undertail coverts that blend into yellow-tinged vent. If present, pale central stripe on forecrown can help distinguish immature female from immature female Cerulean Warbler (which lacks central crown stripe). Also compare to Yellow-throated Warbler. Passage migrant (mid-Aug to mid-Oct and March to early May). Common in Northern Highlands and rare in lowlands throughout the country. Accidental winter resident. Found in primary and secondary forests and at forest edge. Forages at mid-canopy and canopy; joins mixed-species flocks. Call is a rich *chip!*

Yellow-throated Warbler (Reinita Gorgiamarilla)   *Setophaga dominica*

5 in (13 cm). Note large bill and **contrast between bright yellow throat and black mask**. **Vertical white patch** behind the black mask is absent on Grace's Warbler. From below, note white tail with narrow black outer edges. Compare to Blackburnian Warbler. Winter resident (mid-Oct to mid-March). Common in Northern Highlands and the Mosquitia; to 5,200 ft (1,600 m). Very rare throughout rest of Caribbean. Moves along branches inspecting bark for prey; prefers highland pine forest and pine-oak forest, cloud forest, and lowland pine savanna. Calls with a sharp *chip!*

Grace's Warbler (Reinita Pinera)   *Setophaga graciae*

5 in (13 cm). Has **bright yellow throat and mostly gray face**; also note small bill. **Yellow superciliary and lower eye crescent** further distinguish it from Yellow-throated Warbler. From below, note white tail with narrow black outer edges, white underparts that extend to streaked flanks, and yellow breast. There are two disjunct subspecies in Nicaragua. *Remota* subspecies: Common in highland pine forest and pine-oak forest (but wanders to cloud forest) in Northern highlands; above 3,300 ft (1,000 m). Uncommon on Casita Volcano; above 2,000 ft (600 m). *Decora* subspecies: Common in lowland pine savanna of the Mosquitia; to 1,000 ft (300 m). Both subspecies typically forage at mid-canopy to canopy and will join mixed-species flocks. Sing a series of repeated, distinct *tchew* notes (1.5–2 seconds in total length); second half of song is higher pitched and sometimes accelerates. Calls with a rich *chewt*.

immature
female

**Cerulean
Warbler**

breeding
male

immature
female

**Blackburnian
Warbler**

breeding
male

**Yellow-throated
Warbler**

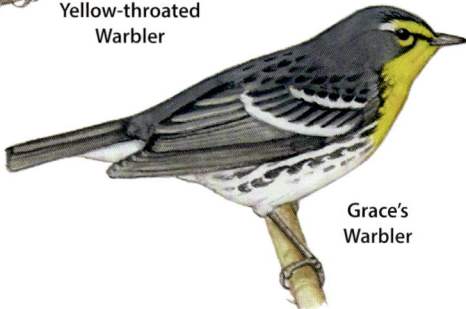

**Grace's
Warbler**

Bay-breasted Warbler (Reinita Castaña)   *Setophaga castanea*

5.5 in (13 cm). On breeding male, **chestnut crown and throat** is diagnostic. In all plumages, note streaking on mantle, back, and rump (of varying intensity); white wing bars (sometimes faint); and chestnut on flanks (of varying intensity). On immature, dark feet help distinguish it from very similar immature Blackpoll Warbler. Any confusion with nonbreeding or immature Chestnut-sided Warbler is precluded by white wing bars and lack of white eye ring. Passage migrant (mid-Sept to Oct and mid-March to April); common in Northern Highlands, and Caribbean; locally rare on Sierras Managua; to 4,600 ft (1,400 m). Also a winter resident (mid-Sept to April); rare along San Juan River. Found in forest, forest edge, and second growth. Leisurely forages at mid-canopy and canopy; sometimes forms large groups and often joins mixed-species flocks. Calls with a sharp *cheat!*

Chestnut-sided Warbler (Reinita Flanquicastaña)   *Setophaga pensylvanica*

5 in (13 cm). On breeding male, note y**ellow crown** and **chestnut flanks**; breeding female shows less chestnut. On nonbreeding adults and immature, note yellow-green upperparts, **white eye ring**, and two yellow wing bars. Any confusion with nonbreeding or immature Bay-breasted Warbler is precluded by yellowish wing bars and white eye ring. From below, note white tail with black outer edges. Winter resident (Sept to early May). Common throughout Northern Highlands and Caribbean; rare in most regions of the Pacific but locally common on Sierras Managua; to 4,900 ft (1,500 m). Forages at all levels, picking prey from foliage. Often seen with a moderately cocked tail. Call is a slightly muffled, rising *cheap!*

Blackpoll Warbler (Reinita Rayada)   *Setophaga striata*

5 in (13 cm). On breeding male, **black crown, white lower half of face, and black malar stripe** distinguish it from Black-and-white Warbler. In all plumages, shows streaking on mantle and flanks, even if faint; all but breeding male have black eye line and broken white eye ring. Yellow feet help to separate immature from very similar immature Bay-breasted Warbler. From below, note white undertail coverts that extend half the distance of the tail. Accidental passage migrant (mid-Oct to April); all 3 records are from the Caribbean. Calls with a sharp *chip!*

Black-and-white Warbler (Reinita Trepadora)   *Mniotilta varia*

5 in (13 cm). Distinguished from breeding male Blackpoll Warbler by **striped crown** and distinctive **white tertial edgings**. From below, note **large black spots on white undertail coverts**. Male has a black throat; female has a white throat. Winter resident (mid-Aug to April). Common countrywide; to 4,900 ft (1,500 m). Found virtually in all forest types and stages, as well as in mangroves and scrub. Individuals often join mixed-species flocks. **Creeps along trunks and branches** while probing for invertebrate prey, sometimes upside down. Call is a plain *chit!* Song (not always given in Nicaragua) is a rhythmic, high-pitched *sweeta-sweeta-sweeta…* (2 seconds in length).

Bay-breasted
Warbler

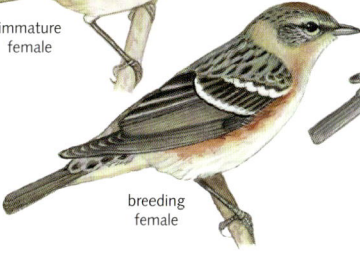

immature
female

breeding
female

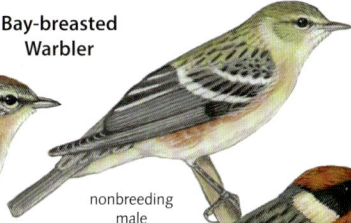

nonbreeding
male

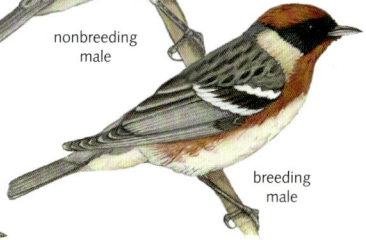

breeding
male

immature
female

Chestnut-sided
Warbler

breeding
male

immature

breeding
female

Blackpoll
Warbler

breeding
male

female

male

Black-and-white
Warbler

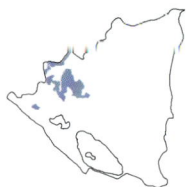

Townsend's Warbler (Reinita Bicolor)   *Setophaga townsendi*

5 in (13 cm). On male, note **black mask and yellow lower eye crescent**; all plumages have **yellow breast** and **black streaks** (of varying intensity) **on yellow flanks**. From below, note yellow breast and—on some birds—a few black streaks on undertail coverts. Winter resident (Sept to mid-April). Uncommon to locally common throughout Northern Highlands and on Casita Volcano; above 3,300 ft (1,000 m). Prefers highland pine forest and pine-oak forest, but can occur in cloud forest and forest edges. Actively forages at mid-canopy and canopy and often joins mixed-species flocks. Call is a slightly metallic *tsip!*

Hermit Warbler (Reinita Cabecigualda)   *Setophaga occidentalis*

5 in (13 cm). Male is distinctive, with **bright yellow head** and black nape. Female and immature have faint olive ear patch; **yellow lore** and **mostly unstreaked flanks** help to distinguish them. From below, note white underparts with variable amounts of black on the throat. Winter resident (mid-Aug to early April). Locally uncommon in Northern Highlands; above 3,300 ft (1,000 m). Rare on Casita Volcano; above 2,000 ft (600 m). Prefers highland pine forest and pine-oak forest, but can occur in cloud forest and forest edges. Actively forages at mid-canopy and canopy and often joins mixed-species flocks. Call is a weak, slightly raspy *tsip.*

Golden-cheeked Warbler (Reinita Cachetidorada)   *Setophaga chrysoparia*

5 in (13 cm). Male is distinctive, with **narrow black eye line**, yellow face, and black crown and back (Black-throated Green male has olive ear mask on yellow face and olive crown and back); female and immature are drabber, with olive crown and back. From below, note varying amounts of black on the throat; a few black streaks sometimes show on undertail coverts. Lack of yellow on vent and (sometimes) presence of streaking on back distinguish female and immature from female and immature Black-throated Green Warbler (which has yellow on vent and unstreaked back). Winter resident (Sept to mid-April). Locally uncommon in Northern Highlands; above 3,900 ft (1,200 m). Found solely in highland pine forest and pine-oak forest (and adjacent cloud forest). Actively forages at mid-canopy and canopy and often joins mixed-species flocks. Call is a sharp *tchep!* **EN**

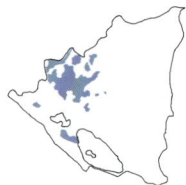

Black-throated Green Warbler (Reinita Gorginegra)   *Setophaga virens*

5 in (13 cm). On all plumages, note **diffuse olive mask** on yellow face and olive upperparts. From below, note black throat (amount of black varies) and **tinges of yellow on breast and vent**. Yellow on vent, olive mask, and unstreaked back help distinguish it from all plumages of Golden-cheeked Warbler (which lacks yellow on vent and olive mask, but may have streaked back). Winter resident (Sept to April). Abundant in Northern Highlands, in northern region of Sierra Chontaleña, and in Quirragua NR and Musún NR; above 3,000 ft (900 m). Locally uncommon on Casita Volcano and locally very rare on Sierras Managua; above 1,300 ft (400 m). Accidental passage migrant throughout lowlands and foothills. Prefers highland pine forest and pine-oak forest, cloud forest, and forest edges; in passage, uses a wide variety of habitats. Forages at mid-canopy and canopy with mixed-species flocks. Call is a thin *tsip!*

male

immature
female

**Townsend's
Warbler**

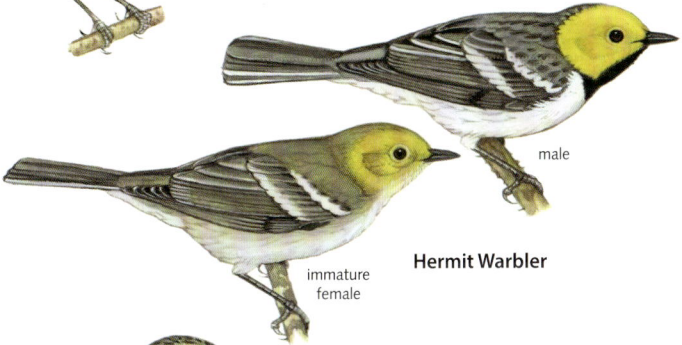

male

immature
female

Hermit Warbler

female

immature
female

male

**Golden-cheeked
Warbler**

immature
female

male

**Black-throated
Green Warbler**

Palm Warbler (Reinita Coronirrufa)   *Setophaga palmarum*
5 in (13 cm). Combination of **yellow rump and uppertail coverts** and **yellow undertail coverts** is unique among the warblers. Winter resident (mid-Oct to mid-April). Rare in the Mosquitia and on Caribbean coast; accidental in Northern Highlands. Found in lowland pine savanna, grasslands, and marshes. While foraging on ground and in understory, continually **pumps tail**. Calls with a sharp *chep!*

Prairie Warbler (Reinita Galana)   *Setophaga discolor*
4.5 in (11 cm). On male, note **yellow lower eye crescent bordered by a black stripe**, black eye line, and **black patch on the sides of upper breast**; marks on female are duller; immature is much duller, with gray on the face. **Rufous mantle patch** is diagnostic if seen. From below, note long, white tail with black outer edges and pale undertail coverts that blend into the yellow underparts. Winter resident (Aug to May). Very rare in the Mosquitia and Caribbean coast. Found in lowland pine savanna, mangroves, grasslands, and second growth. Actively forages from understory to mid-canopy; frequently pumps tail. Calls with a weak *tsip!*

Yellow Warbler (Reinita Amarilla)   *Setophaga petechia*
5 in (13 cm). There are two subspecies in Nicaragua. *Erithachorides* subspecies: On male, **chestnut head** is unmistakable; female often has some rufous on the head. Breeding resident. Locally common in mangroves on Pacific coast. Song begins a with a series of warbles and ends with cascading notes (1.5 seconds in length). Calls with a rich *cha! Aestiva* subspecies: Often brilliant yellow, but sometimes pale yellow. **Red streaking** on yellow underparts varies in intensity with sex and age. Note black eye and stout bill on unmarked face. Winter resident (late Aug to early May). Abundant in Pacific; common in Caribbean; generally uncommon in Northern Highlands but abundant in dry intermontane valleys; to 5,200 ft (1,600 m). Forages actively, sometimes in mixed-species flocks, in a wide variety of habitats, especially in disturbed areas. Call is a rich *chip!*

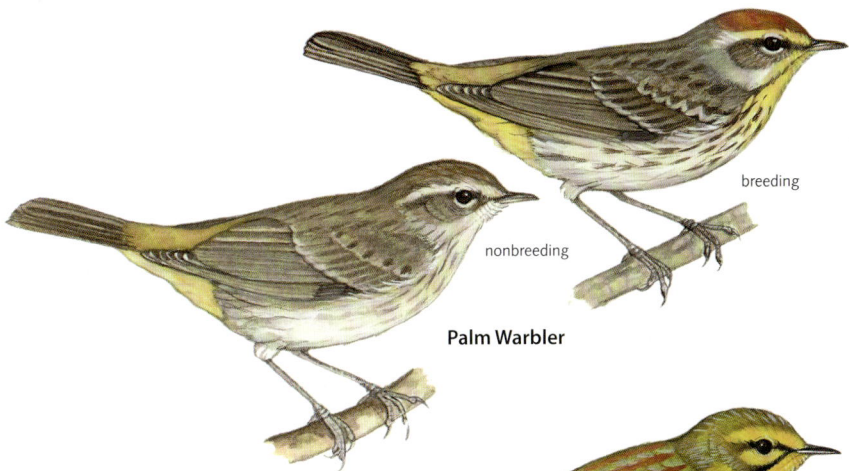

nonbreeding

breeding

Palm Warbler

male

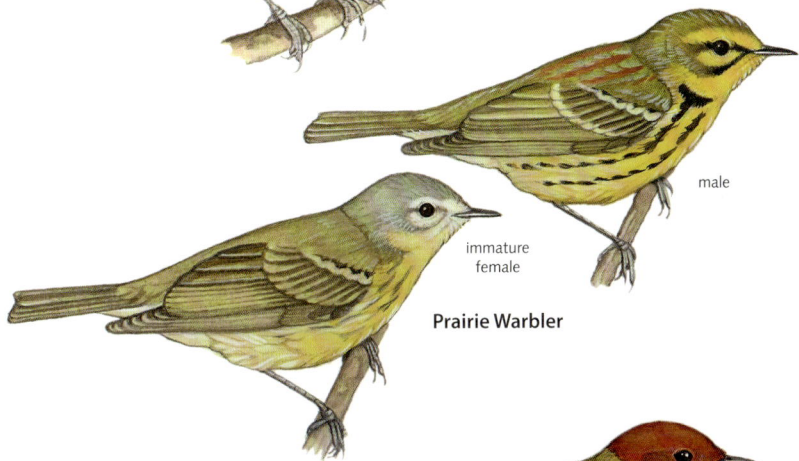

immature
female

Prairie Warbler

male
erithachorides ssp.

**Yellow
Warbler**

adult male
aestiva ssp.

female/immature
aestiva ssp.

Black-throated Blue Warbler (Reinita Azulinegra)   *Setophaga caerulescens*
5 in (13 cm). On adult male and female, distinct **white patch at base of primaries** is diagnostic. Male is unmistakable; female and immature have **thin whitish superciliary** and lower eye crescent on blue-green tinged head. From below, male has short, white undertail coverts that rarely cover the black base of tail; on female and immature, note gray tail and buffy underparts. Very rare winter resident (late Oct to mid-April); only records are between 3,900 and 4,600 ft (1,200 and 1,400 m) in Northern Highlands. Forages from understory to mid-canopy of forest and forest edge; joins mixed-species flocks. Call is an energetic *tchup!*

Canada Warbler (Reinita Pechirrayada)   *Cardellina canadensis*
5 in (13 cm). On breeding male, combination of **heavy black streaking on breast** and **bicolored spectacles** is distinctive. All plumages have mostly gray upperparts. From below, note combination of white undertail coverts, dark tail, and yellow underparts; the combination is diagnostic. Passage migrant (Sept to early Nov and mid-March to mid-May). Common in Northern Highlands and in Caribbean; rare throughout Pacific; to 4,900 ft (1,500 m). Found in a variety of forest habitats. Actively forages in understory; frequently seen with its tail cocked. Calls with a rich *chip!*

Wilson's Warbler (Reinita Gorrinegra)   *Cardellina pusilla*
4.5 in (11 cm). On male, **black on midcrown and hindcrown** is distinctive; on female and immature, note yellow forecrown, superciliary, and lore and olive crown and ear patch. From below, note long, dark tail and yellow underparts. Winter resident (Sept to early May). Abundant in Northern Highlands; uncommon on northern extent of Sierra Chontaleña; uncommon in the Pacific; rare in the Caribbean; to 5,900 ft (1,800 m). Found in forest, forest edge, shaded coffee plantations, and second growth. Forages actively, sometimes hovering, in the understory; often joins mixed-species flocks. Call is an energetic *chip!*

American Redstart (Candelita Norteña)   *Setophaga ruticilla*
5 in (13 cm). Adult male has **orange wing patch, upper flanks, and base of outer rectrices**; female is similar but with yellow replacing orange. From below, also note broad black terminal band on tail. Winter resident (mid-Aug to early May). Common countrywide; to 5,900 ft (1,800 m). Found in forests at mid-canopy and canopy; drops to lower levels at forest edge and second growth. Sallies for flying insects and hovers to pick prey off of foliage; characteristically fans tail while foraging. Often joins mixed-species flocks. Call is a rich *chip!*

female

male

**Black-throated
Blue Warbler**

female/
immature

male

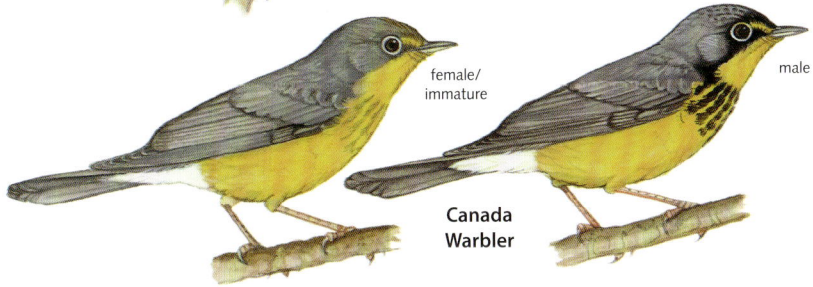

**Canada
Warbler**

female/
immature

**Wilson's
Warbler**

male

immature
male

adult
male

adult
female

**American
Redstart**

Fan-tailed Warbler (Reinita Alzacola) *Basileuterus lachrymosus*
6 in (15 cm). Note **yellow crown patch**, **broken white eye ring**, and **white supraloral spot**. The **white-tipped tail** further distinguishes it from Yellow-breasted Chat (with no white tips on tail), p. 360. Immature has dusky upperparts, pale yellow underparts, and 2 narrow white wing bars. Locally common in Pacific foothills and highlands and on western slopes of Northern Highlands, reaching northern region of Sierra Chontaleña. Reaches southernmost distribution in Nicaragua. Walks on ground of primary and secondary forests, where it is known to join army ant swarms; sporadically hops with a twisting motion and frequently fans its tail. Song is a sweet warble, with the final note sometimes exaggeratedly slurred. Call is a high-pitched, downslurred *seeuh*.

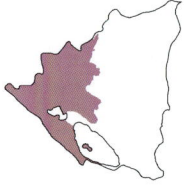

Rufous-capped Warbler (Reinita Cabecicastaña) *Basileuterus rufifrons*
5 in (13 cm). Combination of **rufous crown and ear patch** and white superciliary is diagnostic. From below, note dark tail and all-yellow underparts. Common in the Pacific, where it is less likely at lowland elevations; common in Northern Highlands; to 5,900 ft (1,800 m). Accidental in western portions of Caribbean. Found in forest and forest edge. Forages in understory and mid-canopy. Sings sweet but erratic phrases that display a choppy, high-low pitch pattern. Calls with a sharp, slightly scratchy *chip-chewt!*

Golden-crowned Warbler (Reinita Coronigualda) *Basileuterus culicivorus*
5 in (13 cm). Note **black lateral crown stripe** and faint yellow superciliary; **narrow yellow-orange central crown stripe** is not always visible. From below, note all-yellow underparts and long, dark tail. Abundant on eastern slopes of Sierra Isabelia and Sierra Dariense; locally common in Saslaya NP; rare on Sierra Dipilto-Jalapa; above 2,600 ft (800 m). Found in cloud forest and adjacent secondary forest and forest edge. Mainly forages in understory; moves quickly, often flicking wings, and will join mixed-species flocks. Song is a clearly whistled *whi-wu widit-werdit-wee-DEE*; the second note is noticeably downslurred and the succeeding notes ascend in pitch. Call is a rapid, doubled-noted rattle.

Painted Redstart (Candelita Aliblanca) *Myioborus pictus*
5 in (13 cm). **Black, white, and red** coloration make it unmistakable. White wing patch and lower eye crescent further separate it from Slate-throated Redstart. Immature lacks the bright red breast and belly. From below, note long, white tail, black mottling on white undertail coverts, and gray vent and lower belly. Rare to uncommon in Northern Highlands, but locally common in Yalí NR and Sierra Dipilto-Jalapa; from 3,300 to 5,200 ft (1,000 to 1,600 m). Reaches southernmost distribution in Nicaragua. Found in highland pine forest and pine-oak forest and nearby cloud forest. While actively foraging on branches and tree trunks, frequently fans its cocked tail; sometimes joins mixed-species flocks. Whistles a variety of sweet songs that often include *see-su see-su* or *sweeta sweeta* phrases; when foraging, also makes a standalone *seeu*.

Slate-throated Redstart (Candelita Gorginegra) *Myioborus miniatus*
5 in (13 cm). **White-tipped outer rectrices** contrast with slate-gray upperparts; colorful breast and belly range from orange to red. **Red crown patch** (often not visible) and lack of white wing patch and lower eye crescent distinguish it from Painted Redstart. From below, note white-tipped outer rectrices on black tail. Locally uncommon on Sierra Dipilto-Jalapa, Tepesomoto-Pataste NR, and Saslaya NP; above 3,300 ft (1,000 m). Very rare and local on Casita Volcano; above 3,000 ft (900 m). Found in cloud forest and highland pine and pine-oak forest; joins mixed-species flocks and frequently fans tail. Sings a sharp *chi-chi-chi-chi-chiwá-chiwá-chiwá*. Call is a thin *chí*.

adult

Fan-tailed Warbler

immature

Rufous-capped Warbler

Golden-crowned Warbler

Painted Redstart

adult

Slate-throated Redstart

immature

Restricted to the New World, this family was formerly included in the family Emberizidae. These birds are characterized by stout bills, often conical, that are well suited for cracking seeds and mashing fruit, though they also sometimes catch insects. The family contains a number of colorful species, most exhibiting sexual dimorphism. Drab females are sometimes best identified by their proximity to the more readily identified male. Recent genetic studies resulted in the new inclusion of 3 tanager genera (*Piranga*, *Habia*, and *Chlorothraupis*), all previously within family Thraupidae. About half of the species in Nicaragua are NA migrants.

Hepatic Tanager (Tangara Rojiza)   *Piranga flava*

8 in (20 cm). On both sexes, **black bill** and **dusky ear patch** distinguish it from the respective sexes of Summer Tanager; also note **dark lore**. Male is dark red-orange (male Summer Tanager is bright red). Common in Northern Highlands; rare to uncommon in San Cristóbal-Casita NR; from 2,600 to 6,200 ft (800 to 1,900 m). Common in the Mosquitia. Associated with highland pine and pine-oak forest (and occasionally in adjacent cloud forest) and lowland pine savanna. In pairs, small groups, and mixed-species flocks, forages in mid-canopy and canopy for insects and fruit. Sings a rich, bubbly *woo-wit-witwit-woo-wí* (delivered in a mere second). Calls with a rich *chip, chip, chip…* (repeated at length) or a thinner, ascending *wee*.

Summer Tanager (Tangara Veranera)   *Piranga rubra*

7 in (18 cm). On both sexes, **pale bill** and lack of dusky ear patch distinguish it from the respective sexes of Hepatic Tanager; also note lack of dark lore. Male is **bright red** (male Hepatic Tanager is dark red-orange). Female potentially confused with female Scarlet Tanager, but is larger, has a longer bill, and lacks dusky wings and tail. Common winter resident (Sept to April). Found countrywide, at all elevations. Forages in canopy of most forest habitats. Found alone and in mixed-species flocks as it searches for fruit, wasps, bees, and other insects. Call is a descending, chatter-like *chi-chi-chi-chi-chup*, with a harsh and raspy quality.

Scarlet Tanager (Tangara Escarlata)   *Piranga olivacea*

6.5 in (17 cm). **Black wings and tail** in all plumages make male unmistakable. Female is the only female *Piranga* tanager with entirely dusky wings and tail (these contrast with olive-yellow underparts). Has a shorter bill than does Summer Tanager. Passage migrant; (mid-Sept to Nov and March to April), though some individuals can turn up slightly earlier or later. Common in Caribbean; uncommon in Northern Highlands; locally uncommon in Sierras Managua and Rivas Isthmus but rare in other parts of the Pacific; to 4,900 ft (1,500 m). Found in most forest habitats. Forages in canopy for fruit and insects, either in groups or mixed-species flocks of tanagers and warblers. Call is a rich *chip* or a double-noted *chip!-weeh*, with the second note soft and muffled.

Western Tanager (Tangara Cabecirroja)   *Piranga ludoviciana*

7 in (18 cm). On male, **orange-red head** is distinctive; brighter in boreal fall and spring, duller in winter. Both sexes show bold **wing bars**; upper wing bar is yellow, lower is white. Female is separated from female Flame-colored Tanager (p. 388) by lack of **streaking on dusky back**, and from female White-winged Tanager (p. 388) by lack of dark mask; also note paler bill. Winter resident (Oct to mid-April). Locally common in Sierras Managua, Los Pueblos Plateau, and other foothills in the Pacific (uncommon in Pacific lowlands); common in dry intermontane valleys of Northern Highlands; to 4,600 ft (1,400 m). Alone or in mixed-species flocks, forages for insects and fruit in mid-canopy and canopy of dry forest, woodland, scrub, second growth, and orchards. Usually moves within shaded vegetation and is not always easily detected, despite bright coloration. Call is a staccato chatter: *chi-chi-ché*.

male

Hepatic Tanager

female

adult female

immature male

Summer Tanager

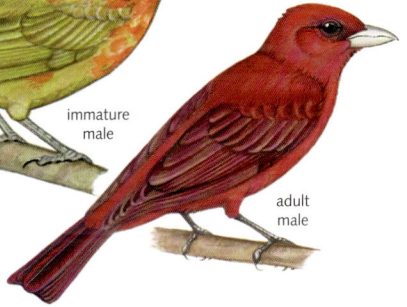

adult male

nonbreeding male

Scarlet Tanager

breeding male

nonbreeding male

adult female

breeding male

Western Tanager

Flame-colored Tanager (Tangara Dorsirrayada)   *Piranga bidentata*

7 in (18 cm). Combination of **streaked back** and conspicuous **white wing bars** distinguishes it from other *Piranga* tanagers. Also note **dusky ear patch** and **white-tipped tertials** and **outer rectrices**. Uncommon in Northern Highlands; from 3,300 to 5,900 ft (1,000 to 1,800 m). Prefers highland pine and pine-oak forest, but also frequents second growth and cloud forest edge. Alone, in pairs, or in mixed-species flocks, forages for insects and fruit in mid-canopy and canopy. Song is delivered slowly and consists of 2 or 3 burry whistles; each whistle notably fluctuates in pitch. Most frequent call is a curt, burry *ter-dék*.

White-winged Tanager (Tangara Aliblanca)   *Piranga leucoptera*

5.5 in (14 cm). Both sexes have broad **white wing bars on black wings**. On male note bold **black mask**. Female is similar to larger female Western Tanager (p. 386), but has **dusky lores**, brighter yellow underparts, and **bicolored bill**. Uncommon to locally common in Northern Highlands and rare in Saslaya NP; above 3,600 ft (1,100 m). Favors cloud forest and edge, highland pine and pine-oak forest, clearings with scattered trees, and tall second growth. Forages in pairs or small groups for fruit and insects; usually in canopy but also descends to mid-canopy; sometimes joins mixed-species flocks of tanagers and warblers. Whistles a short, high-pitched *swee-si-su*, with first note ascending in pitch and last note descending in pitch. Calls with an abrupt *twít* interspersed with a rising *sweet!*

Red-crowned Ant-Tanager (Tangara Hormiguera Coronirroja)   *Habia rubica*

7.5 in (19 cm). On male, note **red crown with black lateral border**, though sometimes hard to see; on female, crown is **bright yellow**. Both sexes lack dusky cheeks and lores, which rules out very similar Red-throated Ant-Tanager. Female distinguished from female Carmiol's Tanager by yellow crown, longer tail, and more yellowish coloration. Common in Northern Highlands; rare in Caribbean; generally rare in Pacific but locally common in foothills; to 5,000 ft (1,550 m). Favors cloud forest, dry forest, forest edge, adjacent tall second growth, thickets, and humid lowland forest edge. In pairs or small groups, searches for fruit and insects in the understory; joins mixed-species flocks and opportunistically attends army ant swarms. Produces rich, loudly whistled songs, including a repeated *fee-dúp fee-dúp fee-dúp…*, a slightly burry, repeated *tuer-tuer-tuer…*, and an ascending *fadá-fadá-fadá…*, with softening intensity at the end. Calls with a raspy, agitated, chattering *chuh-chuh-chuh…*.

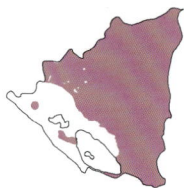

Red-throated Ant-Tanager   *Habia fuscicauda*
(Tangara Hormiguera Gorgirroja)

8 in (20 cm). Distinctive **dusky cheeks and lores** best distinguish it from very similar Red-crowned Ant-Tanager; also note longer bill. On male, red crown lacks a black border (as seen on male Red-crowned Ant-Tanager); also note **bright red throat**. On female, **bright yellow throat** contrasts with olive-brown plumage; further distinguished from smaller female Carmiol's Tanager by longer tail. Common in Caribbean; uncommon in Northern Highlands; locally rare in foothills of Sierras Managua and Volcán Mombacho NR; historical records exist from San Cristóbal-Casita NR (1961, 1985); to 5,000 ft (1,550 m). Found in humid lowland forest, cloud forest, slightly humid dry forest, adjacent forest patches, tall second growth, and thickets. In pairs or small groups, forages on fruit and insects in understory. Opportunistically attends ant swarms. Sings a rich, bubbly, musical series of notes that slide up and down in pitch and vary in length. Also makes a soft, nasal, chattering *queh-queh-queh…* (that seems to go on forever) and a deeper, harsher *EGHH* call.

Carmiol's Tanager (Tangara Olivácea)   *Chlorothraupis carmioli*

7 in (18 cm). No obvious field marks on uniform olive-green plumage; underparts are paler and throat is washed yellow (duller than on female Red-throated Ant-Tanager). Also note **robust body, stout bill,** and **short tail** (important to note when comparing to female ant-tanagers, which have a longer tail). Lacks bright yellow crown of female Red-crowned Ant-Tanager. Very similar to female Hepatic Tanager (p. 386), but there is no range overlap. Has a patchy distribution: Generally uncommon in the Caribbean, but abundant in Bosawas Biosphere Reserve and foothills of Indio Maíz BR and uncommon in Musún NR; to 3,600 ft (1,100 m). Favors stream edge in humid lowland forest, adjacent forest patches, and tall second growth. In noisy flocks of 10 to 20 individuals, travels from understory to mid-canopy in search of insects and fruit; joins mixed-species flocks (often with Tawny-crested Tanager, Buff-throated Foliage-Gleaner, and other furnarids). Song is a series of electric, tinkling, thin, sliding notes. Calls include a squeaky buzz (*zeáw*) and an arcing, mousy whistle (*sweeó*).

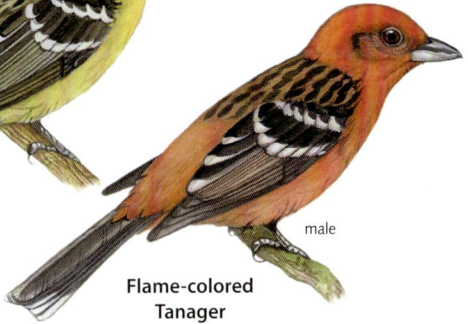

Flame-colored Tanager

female

male

White-winged Tanager

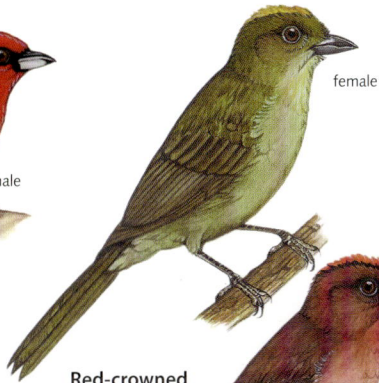

female

male

Red-crowned Ant-Tanager

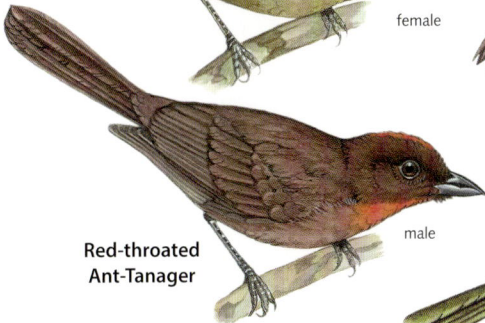

female

male

Red-throated Ant-Tanager

female

male

Carmiol's Tanager

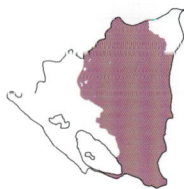

Black-faced Grosbeak (Piquigrueso Carinegro)   *Caryothraustes poliogaster*
7 in (18 cm). Combination of **black mask and throat** and **yellow head and breast** is diagnostic. Abundant in Saslaya NP and Indio Maiz BR but uncommon in other regions of the Caribbean; uncommon on eastern slopes of Northern Highlands (Peñas Blancas NR and Kilambé NR); to 3,900 ft (1,200 m). Dwells in humid lowland forest and edge, adjacent forest patches, forest light gaps with tall trees, and secondary forest. Moves in large flocks (up to 50 individuals), actively foraging for large insects, fruit, and sometimes nectar; generally forages in mid-canopy and canopy but also descends to fruiting trees on forest edge. Sometimes joined by tanagers and honeycreepers. Sweetly whistles a rich and clear *tu-wé-tu-tuwé-tu* (2 seconds in length). Calls with a high-pitched *twee-tu-tu-tu*, sometimes proceeded by a buzzy *bzzzt*.

Painted Bunting (Azulito Multicolor)   *Passerina ciris*
5 in (13 cm). Male is unmistakable. Female has **greenish upperparts, yellowish underparts,** and a **pale eye ring**. Female euphonias are smaller, shorter-tailed, and typically found higher in canopy. Winter resident (Nov to April). Common in Pacific lowlands and foothills; to 2,600 ft (800 m). Uncommon in Northern Highlands; to 4,600 ft (1,400 m). Rare in Guatuzo Plains. Favors woodland, dry forest, grassland, marshes, thickets, and forest edge. Feeds on grass seeds and insects; male stays hidden while foraging on the ground below vegetation, whereas female is more likely to be found in the open. Calls with a short, scratchy *chá* or an upslurred *wheá* whistle. **NT**

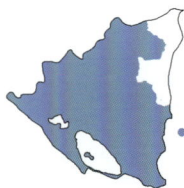

Rose-breasted Grosbeak   *Pheucticus ludovicianus*
(Piquigrueso Pechirrosado)
8 in (20 cm). Male is unmistakable. Female distinguished from female Black-headed Grosbeak (p. 420) by **entirely pale bill** (bicolored on Black-headed) and **extensive streaking on underparts**. Winter resident (mid-Sept to April). Common in Pacific; uncommon in Northern Highlands; locally uncommon in Bosawas Biosphere Reserve but rare to uncommon in other regions of the Caribbean. Found in slightly humid dry forest, open areas with scattered trees, woodland, thorn forest, humid lowland forest edge and forest light gaps, and cloud forest edge. Migrates in flocks; forages alone, on seeds, fruit, and insects, usually in canopy but sometimes descending to mid-canopy. Call is a screechy *chep!*

Dickcissel (Sabanero Común)   *Spiza americana*
6.5 in (16 cm). Both sexes have **rufous scapulars**. Male has bold **yellow superciliary and malar stripe** and **yellow wash on breast** (brighter in boreal spring). Female and immature show yellow of varying intensity. Winter resident (Sept to April). Abundant but erratic in Pacific lowlands and foothills. Uncommon in Northern Highlands; to 4,600 ft (1,400 m). Extremely large flocks move through as passage migrants (late Aug to early Oct and March to mid-April); during this time, common in the Pacific (east of Lake Nicaragua) and in southern Caribbean lowlands. Occurs in semi-open and open areas, grasslands, flooded fields, agricultural fields (grains), and shrubby fields. Feeds on seeds and insects on ground and in low vegetation. Calls are a thin *spit* and a short, buzzy *bzznt*.

Black-faced Grosbeak

female

Painted Bunting

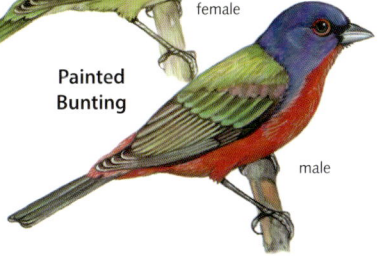

male

immature male

adult female

Rose-breasted Grosbeak

adult male

immature female

Dickcissel

breeding male

Blue Seedeater (Semillero Azulado) — *Amaurospiza concolor*

5 in (13 cm). Male resembles several other birds: bill is smaller and more conical than that of larger Blue-black Grosbeak; bill is stouter and shorter and body is more robust in comparison to Slaty Finch (p. 404); plumage is somewhat paler (sometimes appearing grayish in poor light) than that of the smaller Blue-black Grassquit (p. 406). Female is almost identical to female Blue Bunting, but note **shorter, dark gray bill** (not black) with **straighter culmen**. Locally very rare in Northern Highlands and Caribbean foothills and highlands (populations are irruptive and nomadic); generally above 1,300 ft (400 m), but some historical records reach as low as 700 ft (200 m). **Strictly associated with seeding bamboo** (*Chusquea* sp.) **in forested habitats**; habitat is often the key to identifying this species. Favors cloud forest, humid lowland forest and edges, and brushy woodland with bamboo. Solitary or in pairs, forages in bamboo thickets on bamboo sprouts, insects, and seeds. Whistles a high-pitched, 5-noted phrase: *feu-fí-fea-fí-tí* (1 second in length). Call is a short, sharp *chip!*

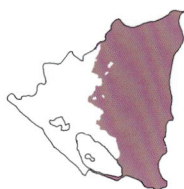

Blue-black Grosbeak (Piquigrueso Negriazulado) — *Cyanocompsa cyanoides*

7 in (18 cm). **Massive, black conical bill** distinguishes both sexes from Blue Seedeater and Blue Bunting. Female is similar to female Nicaraguan Seed-Finch (p. 406), but is notably larger and has a curved culmen. Common in Caribbean; to 3,000 ft (900 m). Accidental on eastern slopes of Northern Highlands; to 4,300 ft (1,300 m). Prefers humid lowland forest and edge, but also occurs in forest patches, forest light gaps, secondary forest, and tall second growth. Forages alone or in pairs for seeds and fruit in understory; often hidden in dense cover, so it is easily overlooked. Sings a descending series of slurred whistles, with the last note often fainter. Call is a squeaky *whit!*

Blue Bunting (Azulito Oscuro) — *Cyanocompsa parellina*

5 in (13 cm). Male is dark blue with **brighter blue head, scapular, and rump**; unmistakable in good light (appears black in poor light). Bill is smaller and more conical than that of larger Blue-black Grosbeak and on smaller Blue-black Grassquit. Female is almost identical to female Blue Seedeater, but note **longer, black bill** (not dark gray) with **more curved culmen**. Uncommon and local on Cosigüina Peninsula. Uncommon on Sierra Isabelia and Sierra Dariense of Northern Highlands. Reaches southernmost distribution in Nicaragua. Found in dry forest, cloud forest and edge, woodland, and thickets. Occurs alone or in pairs in understory. Delivers a moderately paced, 9-noted warble (2 seconds in length), with the last 2 notes noticeably dropping in pitch and intensity. Call is a nondescript *chip*.

Blue Grosbeak (Piquigrueso Azul) — *Passerina caerulea*

6.5 in (17 cm). **Broad rufous wing bars** distinguish both sexes from other birds with conical bills and similar coloration. Female is larger than female Indigo Bunting and has a more prominent bill. Common breeding resident in Pacific and in Northern Highlands. Winter residents (Oct to April) join local population; also rare to uncommon in Caribbean lowlands and foothills. Favors thorn forest, dry forest, scrub, agricultural areas with scattered trees, second growth, secondary forest, and semi-open areas with hedgerows; often perches on roadside wires. Alone or in pairs, feeds on large insects such as grasshoppers and crickets, and some seeds. Whistles a sweet warble with 2–3 short buzzy notes casually mixed in (2 seconds in length). Call is a metallic *teap!*

Indigo Bunting (Azulito Norteño) — *Passerina cyanea*

5 in (13 cm). Breeding male is unmistakable; nonbreeding male and female both have **blue scapular patch**, nonbreeding male also has **spotty blue patches**. Female distinguished from similar looking sparrows by unstreaked back; distinguished from smaller female Blue-black Grassquit (p. 406) by fainter streaking on breast. Winter resident (late Oct to April). Common to abundant in Pacific; to 3,000 ft (900 m). Common in Northern Highlands; to 4,600 ft (1,400 m). Uncommon in Caribbean lowlands and foothills (more numerous in foothills). Forages for seeds on ground and low vegetation in fields close to forest edge, hedgerows, grassland, low second growth, dry forest, and humid lowland forest edge. Often in large flocks during migration (smaller groups persist as winter residents). Moves tail side-to-side while perched. Calls with a thin *twit!* and a soft, buzzy, insect-like *zzzt*.

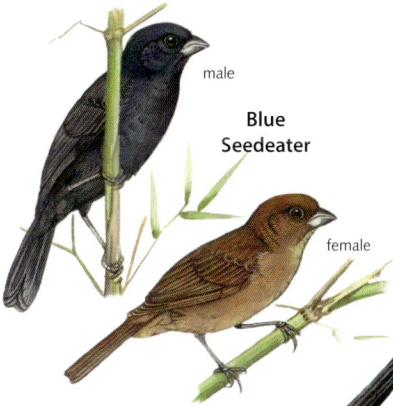

male

**Blue
Seedeater**

female

male

**Blue-black
Grosbeak**

female

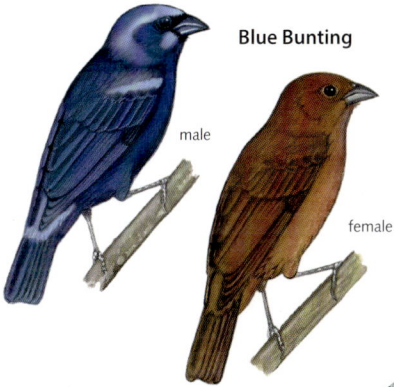

Blue Bunting

male

female

male

female

**Blue
Grosbeak**

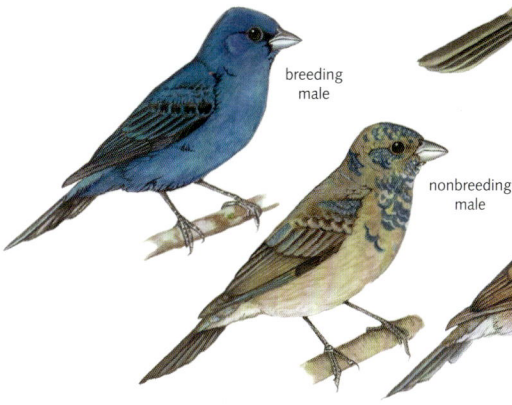

breeding
male

nonbreeding
male

female

**Indigo
Bunting**

The 4 members of this Neotropical family were recently split from the Thraupidae family. The Dusky-faced Tanager is the only species that occurs outside of South America.

Dusky-faced Tanager (Tangara Carinegra) *Mitrospingus cassinii*
7 in (18 cm). Combination of **pale iris** on **black face** and **yellow-olive crown** is diagnostic. Very rare in Indio Maíz BR, with 1 record close to San Juan River; to 1,300 ft (400 m). Reaches northernmost distribution in Nicaragua. Found in humid lowland forest edge and forest light gaps, adjacent second growth, and stream- and riverside vegetation. Noisy flocks roam in thickets and tangles of dense understory, and can also be found in the canopy of fruiting trees. Sings with a high-pitched, sharp warble and gives a harsh *chut*.

TANAGERS, SEEDEATERS, SALTATORS, and ALLIES Thraupidae

The taxonomy of this Neotropical family has gone through a recent shakeup based on genetic research. Species have been shifted among this family and 3 others: Passerellidae, Cardinalidae, and Mitrospingidae. As a result, birds whose common name is *tanager* can be found in all 3 families. Tanagers tend to be colorful birds with a variety of plumage patterns and body sizes. The most colorful representatives of the family famously join mixed-species flocks.

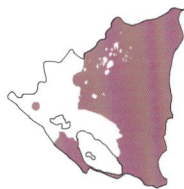

Gray-headed Tanager (Tangara Cabecigrís) *Eucometis penicillata*
6.5 in (17 cm). Combination of **gray head, olive-yellow upperparts**, and **bright yellow underparts** is distinctive. Has a patchy distribution; locally common from foothills of Sierras Managua and Los Pueblos Plateau to Mombacho Volcano and Rivas Isthmus; historical records suggest it could be present in San Cristóbal-Casita NR; to 2,600 ft (800 m). Uncommon in northern extent of Sierra Chontaleña, common on eastern slopes of Northern Highlands; to 4,400 ft (1,350 m). Common in Caribbean foothills (less numerous in lowlands); to 2,000 ft (600 m). Occurs in a variety of forest habitats, including both primary and secondary forests. Raises crest when excited. Song consists of a fast, jumbled series of squeaky, slurred, high-pitched notes. Calls include an abrupt *chet*.

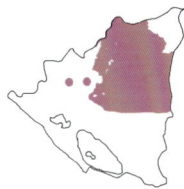

White-throated Shrike-Tanager (Tangara Piquiganchuda) *Lanio leucothorax*
8 in (20 cm). **Stout, hook-tipped bill** is distinctive. Color pattern on male resembles that of an oriole, but behavior and habitat differ. Female has brown upperparts and tawny underparts; larger size and bill structure set her apart from other brown female tanagers. Generally uncommon in northern Caribbean, but locally common in Bosawas Biosphere Reserve; to 3,300 ft (1,000 m). Historical records indicate that this species once occurred in the Northern Highlands (San Rafael del Norte, Feb and April 1909; Peñas Blancas NR, May 1909). Found in humid lowland forest with varied elevations. Sings a rapid-fire *TU-TU-TU…* that is repeated 10–15 times; series is introduced by a downward sliding note. Also constantly repeats an emphatic *which-DO*, with both notes descending in pitch.

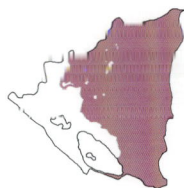

Passerini's Tanager (Tangara Rabirroja) *Ramphocelus passerinii*
6.5 in (17 cm). On male, **scarlet rump and undertail coverts** are unmistakable; on exposed perches, male often conspicuously displays his jutted rump. Female has **ochra-yellow underparts** and **silver-blue bill** and usually accompanies male. Abundant in Caribbean lowlands and common in Caribbean foothills; uncommon on Sierra Chontaleña; to 2,600 ft (800 m). Locally common on eastern slopes of Northern Highlands to 4,600 ft (1,400 m). Small flocks move among river- and roadside vegetation, scrub, second growth, woodland, coffee plantations, and forest edge. They forage for fruit and arthropods. Repeats evenly spaced, single or double-noted whistles, most high-pitched, and sometimes interspersed with nasal chatter: *chet-chet*.

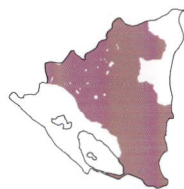

Crimson-collared Tanager (Tangara Nuquirroja) *Ramphocelus sanguinolentus*
7.5 in (19 cm). Adult is unmistakable; immature is duller. Locally common in Northern Highlands; to 4,900 ft (1,500 m). Uncommon in northern region of Sierra Chontaleña and uncommon to locally common in Caribbean lowlands and foothills; to 2,600 ft (800 m). Favors humid lowland forest edge, cloud forest edge, and tall old second growth. Alone or in pairs, forages in mid-canopy for fruit and arthropods; occasionally joins mixed-species flocks of other tanagers and saltators. Sings a moderately paced series of melodic, warbling high-pitched whistles. Call is a descending whistle: *sweeu*.

Dusky-faced
Tanager

Gray-headed
Tanager

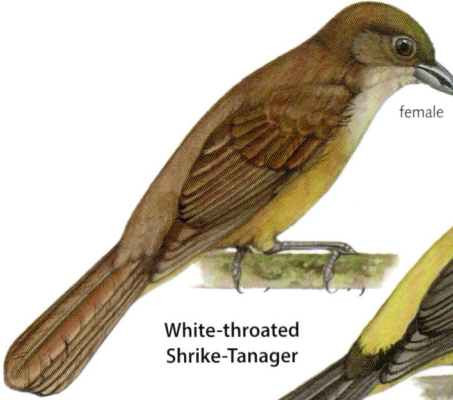

female

White-throated
Shrike-Tanager

male

female

Passerini's
Tanager

male

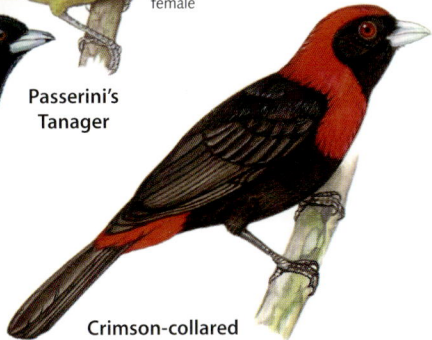

Crimson-collared
Tanager

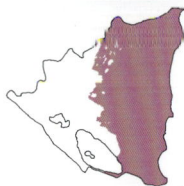

White-shouldered Tanager (Tangara Ribetiblanca)   *Tachyphonus luctuosus*
5.5 in (14 cm). On male, **white shoulder patch** is diagnostic; white shoulder patch on larger male White-lined Tanager is inconspicuous. Female has olive head and yellowish throat and underparts. Uncommon in Caribbean lowlands and foothills; to 2,300 ft (700 m). Accidental on eastern slopes of Northern Highlands. Inhabits humid lowland forest, river edge with forest, forest light gaps, forest edge with dense vegetation, secondary forest, and tall second growth. In pairs, actively moves from mid-canopy to canopy in search of fruit and insects; joins mixed-species flocks. Sings a short, high-pitched warble. Calls include a thin *zeet!* and a soft *chew*.

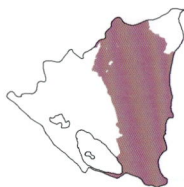

Tawny-crested Tanager (Tangara Coronidorada)   *Tachyphonus delatrii*
6 in (15 cm). On male **erectile orange-tawny crest** is distinctive (but not always easy to see). **Olive-brown** female is darker than any other brown tanager. Generally rare in Caribbean lowlands and foothills, but locally common in Saslaya NP and Indio Maíz BR; to 2,600 ft (800 m). Dwells in humid lowland forest, secondary forest, and forest patches. Flocks of 8 to 20 move noisily and rapidly through mid-canopy and canopy, occasionally being joined by Carmiol's Tanager and woodcreepers. Delivers a series of short notes that fluctuate in loudness and speed, and are often interspersed with short chatters.

White-lined Tanager (Tangara Forriblanca)   *Tachyphonus rufus*
7 in (17 cm). As male and female are nondescript, the best way to make an identification is to see them together. Male is larger than other black tanagers; the white wing patch is less conspicuous than that of the smaller male White-shouldered Tanager. Uncommon in far southern Caribbean lowlands and foothills and rare on southeast side of Lake Nicaragua (probably expanding range into deforested areas); to 1,300 ft (400 m). Reaches northernmost distribution in Nicaragua. Unlike other *Tachyphonus* tanagers, it prefers open habitats and avoids forest; found in woodland, scrub, second growth, secondary forest, and forest light gaps. In pairs, forages in understory for fruit, insects, and nectar; very approachable. Cheerful, melodious song is a slow-paced, repeated series: *chérp cher chur-i-whíp…*.

White-shouldered Tanager

female

male

Tawny-crested Tanager

female

male

White-lined Tanager

female

male

Blue-gray Tanager (Tangara Azulada) — *Thraupis episcopus*

6.5 in (17 cm). **Pale blue body** and **sky blue wings and tail** make this tanager unmistakable. In poor light or misty weather, can be confused with Yellow-winged Tanager and Palm Tanager. Common to locally abundant countrywide at all elevations. Prefers open areas with scattered trees, forest edge, secondary forest, second growth, scrub, and urban areas (parks and gardens). Usually in pairs, but also in small family groups after breeding; joins other frugivorous birds at fruiting trees. Delivers a very fast series of thin, downslurred, high-pitched notes (*tseeú*) and also calls with a more robust *tchew*.

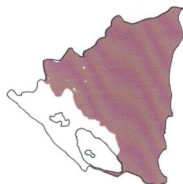

Yellow-winged Tanager (Tangara Aliamarilla) — *Thraupis abbas*

6.5 in (17 cm). **Bright yellow patch at base of primaries** is diagnostic. Also note **purplish-blue head and breast** and **black lore**; back shows black scaling. Beware of potential confusion with drabber Palm Tanager. Common in Northern Highlands; to 5,200 ft (1,600 m). Common in northern Caribbean lowlands and foothills and rare to uncommon in southern Caribbean lowlands. Occurs in humid lowland forest edge, cloud forest edge, pine and pine-oak forest, coffee plantations, secondary forest, gallery forest, second growth, open areas with scattered trees, and forest remnants. Alone or in flocks, searches in canopy for fruit and insects. Sings a rapid-fire, chatter-like *wijijiji...* that slightly descends in pitch at the end (1–2 seconds in length). Call is a sharply upslurred whistle: *seeAH*.

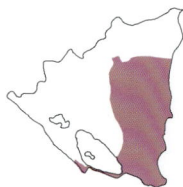

Palm Tanager (Tangara Palmera) — *Thraupis palmarum*

6 in (15 cm). Wing is bicolored, with **dull olive wing coverts** and **black flight feathers**. **Dull olive patch at base of primaries** helps to distinguish it from similarly patterned Yellow-winged Tanager (with bright yellow patch). Common in southern Caribbean, rare on Rivas Isthmus, and accidental in northern Caribbean lowlands and foothills; to 1,300 ft (400 m). Reaches northernmost distribution in Nicaragua (expanding north). Favors palm plantations, open areas with scattered trees, river edge with forest, secondary forest, gardens, and woodland. In pairs or small groups, moves in canopy, hanging upside down in palms as it gleans insects and picks fruit. Song is a fast series of reedy whistles and twitters, rising with the first note and with mostly downslurred notes following.

Plain-colored Tanager (Tangara Cenicienta) — *Tangara inornata*

5 in (12 cm). **Black lores** and **all-black wings** distinguish it from larger Palm Tanager. **Blue shoulder patch** is not always visible. Very rare along San Juan River and rivers in Indio Maíz BR; to 700 ft (200 m). Reaches northernmost distribution in Nicaragua. Found in humid lowland forest, secondary forest, and semi-open areas with scattered trees. Moves in pairs or small groups in canopy, foraging for fruit and arthropods; drops to lower levels to reach fruiting trees. Occasionally joins mixed-species flocks of other tanagers and honeycreepers. Song is a chattering series of sibilant notes that grow louder towards the end (2–3 seconds in length). Calls with a thin *swee*.

Blue-gray
Tanager

Yellow-winged
Tanager

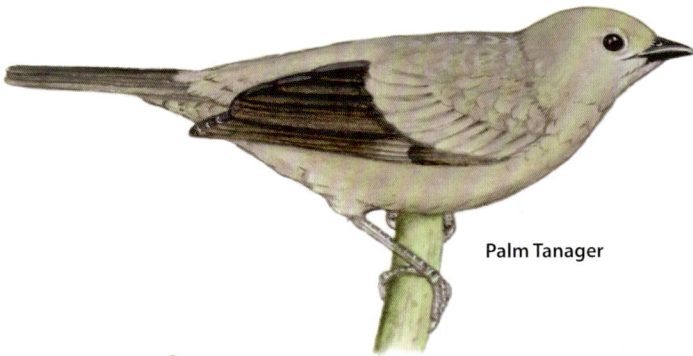

Palm Tanager

Plain-colored
Tanager

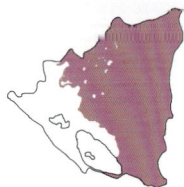

Golden-hooded Tanager (Tangara Capuchidorada)   *Tangara larvata*

5 in (13 cm). Unmistakable. Note **golden hood**, black lores, and turquoise on face. Also note **turquoise rump and flanks.** Immature is duller. Common on eastern slopes of Northern Highlands; to 4,600 ft (1,400 m). Common in Caribbean lowlands and foothills; less numerous on eastern slopes of Sierra Chonteleña and southeastern borders of Lake Nicaragua; to 2,600 ft (800 m). Found in humid lowland forest, cloud forest, forest edge, secondary forest, and tall second growth. Roams in pairs or small groups in canopy searching for fruit and insects; occasionally joins mixed-species flocks of other small frugivorous birds. Delivers an erratic series of *chit* notes that create a twittering, trilling sound.

Rufous-winged Tanager (Tangara Alirrufa)   *Tangara lavinia*

5.5 in (14 cm). On male, rufous wings are distinctive; **green from chin to upper breast** and **yellow nape and upper back** distinguish it from very similar male Bay-headed Tanager. On female and immature, typical tanager bill and traces of rufous on primaries and head distinguish them from female Green Honeycreeper (p. 402). Locally common on eastern slopes of Northern Highlands (absent from Sierra Dipilto-Jalapa); rare throughout Caribbean foothills (including eastern slopes of Sierra Chontaleña), but locally common in Saslaya NP; from 2,000 to 4,600 ft (600 to 1,400 m). Occurs in cloud forest, humid lowland forest, secondary forest, tall second growth, and open areas with scattered trees. In pairs or small groups, forages in canopy for fruit and insects, and descends to lower levels to reach fruiting trees; joins mixed tanager and honeycreeper flocks. Sings an erratic series of thin, sibilant notes; calls with an unassuming *tseet*.

Bay-headed Tanager (Tangara Cabecicastaña)   *Tangara gyrola*

5 in (13 cm). On male, combination of **turquoise-blue underparts** and **chestnut chin** distinguishes it from very similar male Rufous-winged Tanager; also note **narrow yellow nuchal collar** and lack of rufous on wings. Locally rare in Northern Highlands; above 3,900 ft (1,200 m). Locally rare in foothills of Saslaya NP; above 1,300 ft (400 m). Appears to make seasonal movements, as it is more likely to be seen in Northern Highlands during breeding season (Jan to July). Reaches northernmost distribution in Nicaragua. Alone or in pairs, roams canopy foraging for fruit and insects; joins mixed-species flocks. Distinct song consists of a sliding electronic note followed by a short twittering sound. Call is a weak *chit*.

Scarlet-thighed Dacnis (Mielero Negriazul)   *Dacnis venusta*

4.5 in (12 cm). Male is unmistakable; **red thighs** not always visible. On female, combination of **sky blue head** and **buffy underparts** is distinctive. Shorter, straight bill excludes honeycreepers. Locally rare in foothills of Saslaya NP; also locally rare in lowlands and foothills of Indio Maíz BR. Reaches northernmost distribution in Nicaragua. Occurs in humid lowland forest and edge, river edge with forest, and in forest light gaps with fruiting trees. Forages for fruit and seeds in canopy; also descends to mid-canopy to forage in vine tangles. Found in pairs or in mixed flocks of honeycreepers, tanagers, and warblers. Delivers a fast, erratic jumble of squeaky notes. Calls with a nasal *squee!* and a sibilant *rpeet*.

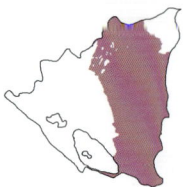

Blue Dacnis (Mielero Celeste)   *Dacnis cayana*

4.5 in (12 cm). Distinctive male is mostly **bright blue**, with **black wings, throat, and black back**. On female, bill (straight, fine-tipped, black), **bluish head**, and **dusky primaries** rule out larger female Green Honeycreeper (p. 402). Uncommon in Caribbean lowlands and foothills; to 2,300 ft (700 m). Prefers humid lowland forest and edge, river edge with forest, second growth, and open areas with scattered trees. Forages in canopy for fruit, nectar, and insects, but also descends to shrubbery in adjacent open areas; found in pairs or mixed flocks of honeycreepers, tanagers, and warblers. Calls with a thin *tseep* or *tsip* and a scratchy *chet!*

Golden-hooded Tanager

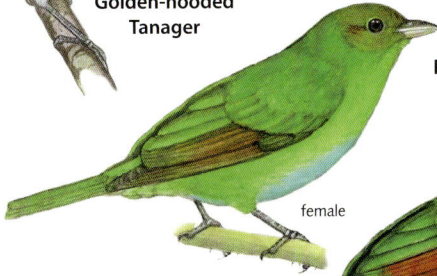

Rufous-winged Tanager

female

male

Bay-headed Tanager

female

male

Scarlet-thighed Dacnis

male

Blue Dacnis

female

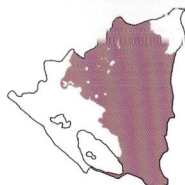

Green Honeycreeper (Mielero Verde) *Chlorophanes spiza*
5.5 in (14 cm). Male is unmistakable. On male and female note long yellow bill (brighter on male); it has a dark culmen and is slightly decurved. Female is distinguished from smaller female Blue Dacnis (p. 400) by bill color, darker legs, and green head (not bluish); female distinguished from female and immature Rufous-winged Tanager by bill color and shape and by lack of rufous on wings and head. Common in Caribbean lowlands and foothills; to 2,600 ft (800 m). Rare on eastern slopes of Northern Highlands; 4,200 ft (1,300 m). Favors humid lowland forest, open areas with scattered trees, secondary forest, and river edge with forest. Alone or in pairs, forages in canopy for fruit, nectar, and insects; joins mixed-species flocks (composed of honeycreepers, dacnis, and tanagers). Calls include a thin *seet* and a brief *chee*.

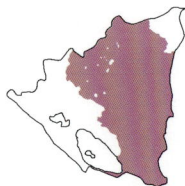

Shining Honeycreeper (Mielero Colicorto) *Cyanerpes lucidus*
4 in (11 cm). **Yellow legs** (brighter on male) are diagnostic. Male is further distinguished from male Red-legged Honeycreeper by **black throat** and **blue crown and back**. Female has buffy throat bordered by blue malar stripe and a **blue-streaked breast**. Locally common in Caribbean foothills, less numerous in lowlands; to 2,600 ft (800 m), but can go higher in Musún NR. Occasionally wanders up eastern slopes of Northern Highlands, where it is rare to uncommon; to 4,200 ft (1,300 m). Prefers humid lowland forest and edge, open areas with scattered trees, and tall second growth (especially terrain with varied elevation). In pairs or small groups, forages high in canopy on flowering trees; also forages in vine tangles, for fruit, insects, and nectar. Occasionally associates with mixed flocks of tanagers. Calls include an abrupt *twit!* and a thin, shrill *sweep!*

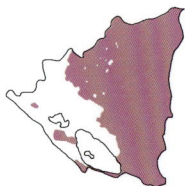

Red-legged Honeycreeper (Mielero Patirrojo) *Cyanerpes cyaneus*
4.5 in (12 cm). **Red legs** (brighter on male) are diagnostic. Breeding male is further distinguished from male Shining Honeycreeper by **black back** and **blue scapular stripes**; **turquoise crown** is prominent. In flight, male flashes **bright yellow wing linings**. Greenish female has **pale superciliary** and **indistinctly streaked breast**. Nonbreeding male resembles female, but with black wings. Common in Caribbean lowlands and foothills; locally uncommon in Pacific foothills, on San Cristóbal Volcano, from Sierras Managua to Mombacho Volcano, and on Rivas Isthmus; generally to 2,600 ft (800 m) but occasionally wanders into highlands, to 4,300 ft (1,300 m). Favors humid lowland forest and edge, forest light gaps, woodland, plantations, and slightly humid dry forest. In pairs, small groups, or mixed-species flocks, forages in canopy at flowering and fruiting trees for fruit, nectar, and small insects. Calls include a clear *tsee*, a nasal *speeh*, and a squeaky *sEEah*.

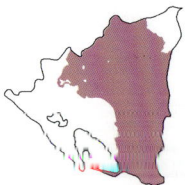

Bananaquit (Mielero Cejiblanco) *Coereba flaveola*
4 in (10 cm). Small, chunky, and warbler-like. Combination of **white superciliary on black head** and **yellow underparts and yellow rump** is distinctive. Also note **decurved, fine-tipped bill**. When only briefly glimpsed, note flashes of the small, white wing-patch. Immature is similar to adult but duller. Common throughout Caribbean and rare on eastern slopes of Northern Highlands; to 2,300 ft (1,300 m). Found in humid lowland forest, cloud forest, and adjacent open areas. In canopy, actively and acrobatically searches for flowers with nectar and small invertebrates; will drop to lower levels and into open areas when flowers are abundant. Sometimes joins mixed-species flocks. Sings a series of squeaky, pulsating notes (2 seconds in length), with an electronic, grating quality.

female

male

Green Honeycreeper

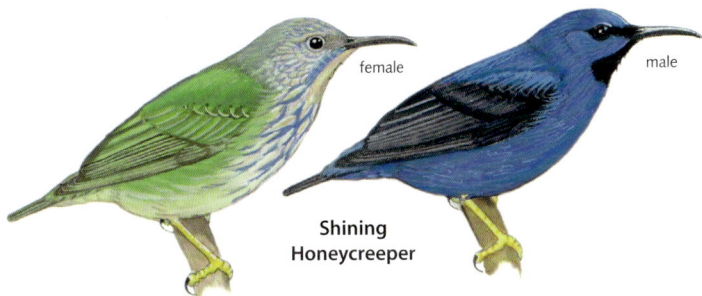

female

male

Shining Honeycreeper

male
underwing

female

Red-legged Honeycreeper

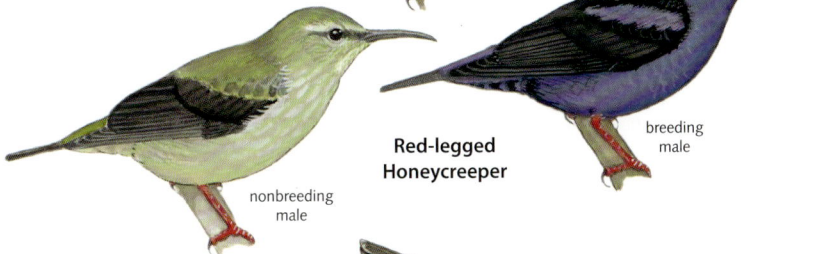

nonbreeding
male

breeding
male

Bananaquit

Cinnamon-bellied Flowerpiercer (Pinchaflor Canelo)   *Diglossa baritula*
4.5 in (11 cm). Hook-tipped upper mandible and **upturned lower mandible** are distinctive. Uncommon in a narrow band in the Northern Highlands, from Tepesomoto-Pataste NR to El Jaguar RSP; one record from Sierra Dariense (Santa Maria de Ostuma, April 1983); above 4,300 ft (1,300 m). Reaches southernmost distribution in Nicaragua. A nectar robbing specialist of cloud forest and highland pine and pine-oak forest and their edges; locates flowering shrubs in the understory or in adjacent gardens, probing for nectar or puncturing the flower base to steal it. Whistles a fast, 3-part song consisting of high-pitched, complex notes, with each segment lower in pitch than the last. Calls with a weak *tit*.

Slaty Finch (Pinzón Piquirrecto)   *Haplospiza rustica*
5 in (13 cm). Male is slate-gray. Female is olive-brown, with faint streaking on underparts. Both sexes have faint **rufous wing edgings**. Bill shape (longer and with a straight culmen) distinguishes it from Blue-black Grassquit (p. 406) and Blue Seadeater (p. 392). Rare and nomadic, with patchy distribution in Northern Highlands; above 4,600 ft (1,400 m). Also occurs on Mombacho Volcano. Found in cloud forest and forest edge. Appears during times of bamboo seed masts (genus *Chusquea*). Often forages in small flocks; picks seeds off the ground or from understory vegetation. Sings a distinctive *di-di-di-di-deeee-IP!*, consisting of 4 introductory notes, an insect-like trill, and finishing with a louder, abrupt note. Call is a weak, metallic *buzz*.

Yellow-faced Grassquit (Semillero Cariamarillo)   *Tiaris olivaceus*
4.5 in (11 cm). On male, **yellow superciliary and throat** are unmistakable. Female lacks most of the color of the male, but retains pale superciliary and throat; distinguished from similar-sized *Sporophila* females by olive-tinged upperparts. Common in Northern Highlands; on western slopes, from 3,900 ft (1,200 m) and above; found at all elevations on eastern slopes. Common in Caribbean lowlands and foothills. Locally rare on Sierras Managua. Found in open areas, highland pine and pine-oak forest, and lowland pine savanna. In small flocks, forages on the ground. Song is a sputtering, trilling noise (1 second in length). Calls with a *tip tip tip…*, given with varying speed.

Grassland Yellow-Finch (Pinzón Amarillo)   *Sicalis luteola*
4.5 in (11 cm). Only conical-billed passerine with predominately **yellow plumage** and **dark streaked crown, nape, and mantle**. On female, yellow is paler and brown upperparts are darker. Rare in Caribbean lowlands from the Mosquitia and south to Bluefields; to 300 ft (100 m). Very rare and local at Lake Apanás in Northern Highlands; at 3,300 ft (1,000 m). Found in pine savanna, grasslands, and other open areas with short vegetation. Picks and strips seeds while foraging on ground in short grasses. Gregarious and nomadic; foraging flocks can contain up to several hundred individuals; sometimes seen near slow burning savanna fires to capture fleeing insects. Song consists of a fast series of reedy, insect-like whistles. Calls from large flocks create a cacophony of soft, twittering notes.

Cinnamon-bellied Flowerpiercer

male

A Cinnamon-bellied Flowerpiercer piercing the base of a flower to seek out nectar.

female

male

Slaty Finch

female

male

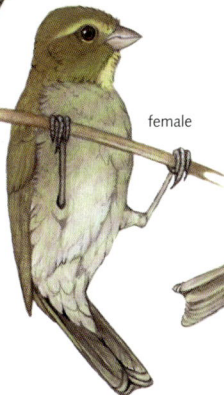

female

Yellow-faced Grassquit

Grassland Yellow-Finch

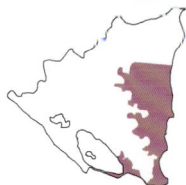

Nicaraguan Seed-Finch (Semillero Nicaragüense)   *Sporophila nuttingi*
6 in (15 cm). On both sexes, **massive bill** is diagnostic; only Thick-billed Seed-Finch has a bill that comes close. Male is uniform black and has a **flesh colored bill**; female is brown and has a black bill. Locally common in Caribbean lowlands, as far north as Prinzapolka River. Highly associated with marshes with tall grasses that are adjacent to humid lowland forest edge. Often found in pairs, but in the nonbreeding season will join mixed-species seedeater flocks to forage. Sings a long series of fast, sharp, squeaky, erratic whistles, interspersed with buzzy notes. Reaches northernmost distribution in Nicaragua. Endemic to Central American Caribbean slope EBA.

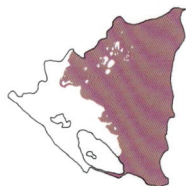

Thick-billed Seed-Finch (Semillero Piquigrueso)   *Sporophila funerea*
5 in (12 cm). **Stout, conical bill** is noticeably larger than bill on other *Sporophila* seedeaters (except for Nicaraguan Seed-Finch). Male has a **white patch at base of primaries** (only sometimes visible); distinguished from male Variable Seedeater by bill, which seems to form a continuous slope with the forehead; it also has a deeper base and a straighter culmen. Female very similar to female Nicaraguan Seed-Finch and distinguished mainly by smaller size. Also compare to females of Blue Seedeater, Blue Bunting, and Blue-black Grosbeak (p. 392). Common throughout Caribbean lowlands and foothills; to 2,000 ft (600 m). Rare on eastern slopes of Northern Highlands; to 4,600 ft (1,400 m). Frequents grasslands, marshes with tall grass, shrubby open areas, and forest edge; usually seen alone or in pairs. Song consists of a long series of sweet, erratic whistles, most of which are downslurred; series sometimes finishes with a sequence of soft twittering notes. Calls include a squeaky *wheá*, a descending *weeu*, and brief *swip!*

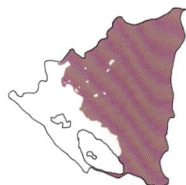

Variable Seedeater (Espiguero Variable)   *Sporophila corvina*
4.5 in (11 cm). Male has a white patch at base of primaries (sometimes not visible); curved culmen of bill distinguishes it from male Thick-billed Seed-Finch. On female, olive-brown upperparts and underparts are generally darker than those of similar female *Sporophila*. Common throughout Caribbean lowlands and foothills. Rare in most of Northern Highlands, but common on eastern slopes and northern region of Sierra Chontaleña; to 5,200 ft (1,600 m). Accidental in Pacific lowlands. Often found in mixed flocks of seedeaters, grassquits, and seed-finches. Forages in grasslands, gardens, marshes, river and forest edge, and other open areas. Whistles a fast-paced series of squeaky notes and occasional reedy trills. Calls include a sharp, downslurred *spíu* and a squeaky *wheú*.

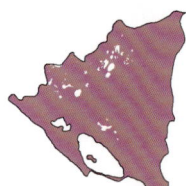

Blue-black Grassquit (Semillerito Negro)   *Volatinia jacarina*
4.5 in (11 cm). Male is a **uniform glossy blue** (appears black in poor light), with **black flight and tail feathers**; Blue Seedeater (p. 392) is somewhat paler. Female has heavy streaking on breast, belly, and flanks and lacks wing bars; unstreaked throat distinguishes her from larger female Slaty Finch (p. 404). Conical bill is shorter than that of larger Slaty Finch. Abundant throughout Pacific, uncommon throughout most of Northern Highlands, but abundant on southern sierras and northern extent of Sierra Chontaleña; to 4,600 ft (1,400 m). Common in Caribbean lowlands and foothills; to 2,000 ft (600 m). Occurs in all kinds of open places, including urban areas, where it is often found in small flocks foraging low in vegetation. Sings a twangy, insect-like *zeeee-ur* trill that finishes with an abrupt downslurred note. Calls with a buzzy *chu chu chu.*

female

Nicaraguan Seed-Finch

male

female

male

Thick-billed Seed-Finch

female

Variable Seedeater

male

Blue-black Grassquit

male

female

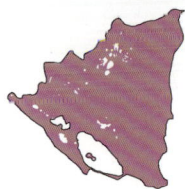

White-collared Seedeater (Espiguero Collarejo)   *Sporophila torqueola*
4.5 in (11 cm). On male, **black-and-white pattern** is unmistakable. On both sexes, note **two wing bars**; these distinguish female from other small-bodied female *Sporophila*. Common in lowlands and foothills; to 2,000 ft (600 m). Uncommon to common in Northern Highlands (no records from Sierra Dipilto-Jalapa); to 4,900 ft (1,500 m). Found in grasslands, agricultural areas, and marshes. Often forages in mixed-species flocks. Delivers a series of repeated, sweet whistles; repeats series 2–4 times, with pitch changing from series to series; sometimes ends with a buzzy trill. Calls include a sharp, descending *psEEUt*.

Yellow-bellied Seedeater (Espiguero Ventriamarillo)   *Sporophila nigricollis*
4 in (10 cm). On male, **combination of black, olive, and pale yellow** is distinctive. On female, note brown upperparts and yellowish underparts; paler than female Variable Seedeater (p. 406). Status unknown, with records from Datanlí-El Diablo NR and Guatuzos WR. Prefers grasslands, young second growth, and other open areas; appears to be nomadic during dry season, searching for water sources. Whistles a fast, high-pitched *we-we-we-WEAT-u-wu-uát* (1 second in length); gives variations of same, adding or subtracting a few notes.

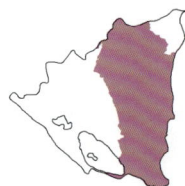

Slate-colored Seedeater (Espiguero Pizarroso)   *Sporophila schistacea*
4.5 in (11 cm). On male, combination of gray plumage and **yellow bill** is diagnostic; white markings less intense on younger males. On female, note brown upperparts; buffy underparts fade to white on lower belly; paler than Variable Seedeater (p. 406). Uncommon and local in Caribbean; populations are irruptive and nomadic. Large flocks of hundreds gather at seeding bamboo thickets, typically within riparian areas. Sings a high-pitched series of varied, metallic trills, sometimes finishing with a repeated *tup tup*.

Ruddy-breasted Seedeater (Espiguero Canelo)   *Sporophila minuta*
3.5 in (9 cm). On male, note **rufous underparts and rump** and gray upperparts; nonbreeding male often shows more subdued tones. Female is paler than female Variable Seedeater (p. 406), with buffy wing edgings on dusky wings; grayish-orange bill is distinctive. Common in Pacific lowlands, Sébaco Valley, Playitas-Moyúa-Tecomapa lagoons, and extreme southwestern Caribbean lowlands; accidental throughout rest of Caribbean lowlands. Prefers grasslands and marshes, but also found in thorn forest and scrub. Quite gregarious, this species sometimes forms flocks in large numbers, even flocking with other seedeaters. Song is a series of sweet whistles that rapidly change pitch and end with sharp grating notes. Calls with a rich *seeu*.

On a rural roadside barbed wire fence, sit (left to right), a male Variable Seedeater (p. 407) and female and male White-collared seedeaters.

female

male

White-collared Seedeater

male

Yellow-bellied Seedeater

female

female

Slate-colored Seedeater

male

Ruddy-breasted Seedeater

male

female

Grayish Saltator (Saltador Grisáceo)   *Saltator coerulescens*
8.5 in (22 cm). Only saltator with grayish upperparts and buffy underparts. Also note **short white superciliary** and **white throat bordered by black malar stripe**. Abundant in Pacific lowlands and foothills; common in other regions; to 4,600 ft (1,400 m). Favors dry forest, forest edge, scrub, woodland, wooded gardens, and open areas with scattered trees. Alone or in pairs, forages at mid-canopy for seeds, fruit, and insects. Sings from canopy. Song is a distinctive *juh-ju-hu-jeuWEE* whistle, with the final note rising in pitch. Also vocalizes with a bubbly, wrenlike chatter, including the occasional *zít* call.

Buff-throated Saltator (Saltador Enmedallado)   *Saltator maximus*
8.5 in (22 cm). **Grayish head** (not black) and **buffy throat** distinguish it from larger Black-headed Saltator. Common in Caribbean lowlands and foothills; to 2,000 ft (600 m). Common in Northern Highlands and locally uncommon in northern region of Sierra Chontaleña; to 5,300 ft (1,600 m). Locally uncommon in Pacific foothills (including Sierra Maribios and Sierras Managua). Favors humid lowland forest and cloud forest edge, second growth, shaded plantations, woodland, and slightly humid dry forest. Feeds in pairs, foraging mid-canopy to canopy on fruit, insects, flowers, and nectar. Whistles deliberate, repeated phrases, including a rapid *cheeríwee-cheeríwee*; sounds something like the song of a *Turdus* thrush. Calls include a thin *twít* and a shrill *seeet!*

Slate-colored Grosbeak (Piquigrueso Piquirrojo)   *Saltator grossus*
8 in (20 cm). On male, combination of **red-orange bill** and **white throat bordered by black** is distinctive. Female is duller. Locally uncommon in Saslaya NP and Indio Maíz BR; rare in other regions of Caribbean; to 3,300 ft (1,000 m). Two historical records in Peñas Blancas (June 1909) suggest it could be very rare on the eastern slopes of Northern Highlands, to 3,900 ft (1,200 m). Favors humid lowland forest and edge, river edge with forest, and adjacent forest patches. Alone or in pairs, forage for fruit and insects in mid-canopy and canopy, descending to understory at edges. Joins mixed-species flocks. Shy and skulking, it is more often detected by its various melodious songs, all of which contain 3–5 lilting and deliberate whistles. Calls include a metallic *spik!* and a jaylike *eahr*.

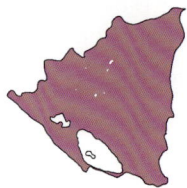

Black-headed Saltator (Saltador Cabecinegro)   *Saltator atriceps*
11 in (28 cm). **Darker head** and **white throat** distinguish it from smaller Buff-throated Saltator. Locally common in Pacific foothills and highlands (including San Cristóbal-Casita NR and from Sierras Managua to Mombacho Volcano) and rare in lowlands. Locally uncommon in Sierra Chontaleña, and common in Northern Highlands; to 5,300 ft (1,600 m). Common in Caribbean lowlands and foothills; to 2,600 ft (800 m). Found in a variety of forest habitats, shaded plantations, second growth, and scrub. Travels in noisy groups of 8 or more at all vegetation levels, feeding on fruit, insects, and flowers. Warbles a series of hoarse, burry notes, interspersed with shrill squeaks, and sometimes finishing with a raspy chuckle. Calls include a burry *cháu* and a shrill *cheat!*

Grayish
Saltator

Buff-throated
Saltator

Slate-colored
Grosbeak

Black-headed
Saltator

Hypothetical Species

Most of the following 43 species have yet to be recorded in Nicaragua but are likely to occur at some point in the future. Two species included have been reported as vagrants based on a single sighting, but more evidence is needed to make a definitive affirmation. Species are placed in the order in which they appear in the Checklist of North and Middle American Birds: seventh edition (updated through the fifty-eighth supplement), managed by the American Ornithological Society. Illustrations have been included when available.

Hooded Merganser   *Lophodytes cucullatus*
(Mergansa Capuchona)
18 in (46 cm). **Long, slender bill** (black on male; yellow with black upper mandible on female), **erectile crest**, pale iris, long tail, and slender body distinguish this merganser. On breeding male, white on the head looks like a fanned patch when the crest is erect; when the crest lies horizontally, the white color takes the form of a stripe. Female is brownish overall, with buffy crest that protrudes horizontally from the head. Nonbreeding male and immature resemble the female. Expected as an accidental winter resident; already recorded in Costa Rica. Prefers freshwater habitats, but is possible on brackish waters.

Eared Grebe   *Podiceps nigricollis*
(Zampullín Orejudo)
13 in (32 cm). In all plumages, note thin bill with a slightly upturned lower mandible, red iris (paler on immature), and **peaked crown**. In nonbreeding plumage, **white throat, neck, and belly** contrast with mostly black body. Golden plumes on the head make breeding adult unmistakable. Similar in size to Pied-billed Grebe (p. 44) but note much thinner bill and longer neck. Possible as an accidental winter resident (several records from El Salvador and 1 from Costa Rica). May be found on both fresh and brackish waters.

breeding

nonbreeding

Maroon-chested Ground-Dove   *Claravis mondetoura*
(Tortolita Serranera)
8.5 in (21 cm). On male, **maroon-purplish breast** and **white outer rectrices** distinguish it from male Blue Ground-Dove (p. 48). Female is similar to female Blue Ground-Dove, but note **white-tipped outer rectrices on rufous tail**, bill color, and pattern on wing coverts. Very rare breeding resident in Honduras and Costa Rica. Abundance and distribution is correlated with infrequent bamboo seeding; the birds seem to disappear for years when seeds are not available. Expected in Northern Highlands; above 2,300 ft (900 m). Prefers cloud forest with bamboo understory. Repeats a gentle and consistent *huwUUP huwUUP huwUUP...* (each note lasts 1 second).

male

female

White-tailed Nightjar   *Hydropsalis cayennensis*
(Pocoyo Coliblanco)
8.5 in (21 cm). Cinnamon nuchal collar and several rows of buffy spots on wing coverts are very similar to those found on Spot-tailed Nightjar (p. 68), but the ranges of the two birds would not overlap. Note white wing band on outermost primaries and white rectrices (female and immature lack white on tail). This bird is a breeding resident in Costa Rica, where it reaches its northernmost distribution in northwest Pacific region of the country. It could occur across the border in Nicaragua, on Rivas Isthmus. Prefers open areas; perches on the ground, where it roosts during the day; feeds at night by sallying for flying insects. Song is a staccato note immediately followed by a downslurred *tsi-seeu*.

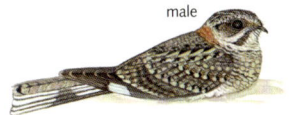

male

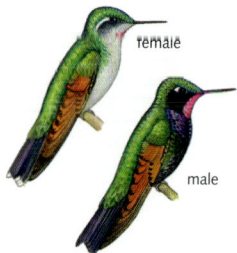

Garnet-throated Hummingbird
Lamprolaima rhami
(Colibrí Cuelligranate)

5 in (13 cm). Note **short, straight black bill** and **rufous wings** (rufous less extensive on female and immature). White postocular spot, pink-rose gorget, and purple breast make male unmistakable (though he might appear entirely dark in poor light). Female has white postocular stripe, dusky gray underparts, and white tipped outer rectrices. Immature resembles female. Recent records (2012) from Honduras very close to Tepesomoto-Pataste NR in the Northern Highlands suggest it could be found there. Inhabits cloud forest, highland pine and pine-oak forest and scrub; above 3,900 ft (1,200 m).

Berylline Hummingbird
Amazilia beryllina
(Amazilia Berilina)

4 in (10 cm). Very similar to Rufous-tailed Hummingbird (p. 90) but note **cinnamon rufous patch** on dusky gray wings, purplish-bronze uppertail coverts, and **undertail coverts with scaling that is pale buff and has cinnamon borders**. In Honduras, occurs adjacent to Apacunca GRR in Chinandega Dept. and close to Tepesomoto-Pataste NR and Dipilto Jalapa NR in the Northern Highlands. Recent sightings close to the border suggest it might occur in Nicaragua. Prefers highland pine and pine-oak forest, dry forest, and scrub; above 2,000 ft (600 m).

Black Rail
Laterallus jamaicensis
(Polluela Negra)

5.5 in (14 cm). A **very small, dark** rail with white speckling on upperparts. Possible local breeding resident throughout Caribbean lowlands. Known from the Mosquitia of Honduras and Guatuzo Plains of Costa Rica. Inhabits dense marshes and wet grasslands. Secretive; quickly dashes, mouse-like, within vegetation. Sings a quick, 3-noted phrase: *pee-pee-brrrr*; the first 2 notes are squeaky and high-pitched, and the third drops in pitch and has a burry quality. **NT**

Paint-billed Crake
Neocrex erythrops
(Polluela Piquirroja)

8 in (20 cm). Combination of bright, **bicolored bill** and **red legs** is diagnostic. Potential breeding resident in southern Caribbean lowlands; known from Guatuzo Plains in Costa Rica. Found in marshes, wet grasslands, and other habitats with standing water. Gives a variety of clucks and buzzy noises. Sings a series of grating notes: *chu-chu-chu . . .* .

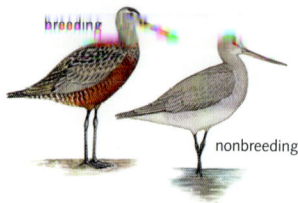

Hudsonian Godwit
Limosa haemastica
(Piquipando de Hudson)

15.5 in (39 cm). Note long, thick, slightly upturned, bicolored bill. In breeding plumage, underparts range from rufous (on male) to rufous barring (on female). In nonbreeding plumage, male and female are drab overall; orange on bill is limited to the base of the lower mandible. In flight, note black wing linings and a black tail with white uppertail coverts. Expected as a very rare NA passage migrant; records exist from the Pacific coast of El Salvador, Honduras, and Costa Rica. Probes for invertebrates on mudflats, salt ponds, flooded fields, and, occasionally, on sandy beaches.

Ruff
(Combatiente)

Calidris pugnax

M 11.5 in (29 cm); F 9.5 in (24 cm). Variation in size and plumage can make identification a challenge. All birds have a plump body and a relatively short bill; most have yellow legs. On breeding male, shaggy feathers around the base of the head make it unmistakable (though rarely observable); breeding female has a black bill and heavily marked upperparts and breast. In nonbreeding plumage, male is drab and shows a pale base around his bicolored bill; female is similar but has an almost entirely black bill. Immature resembles nonbreeding adult but has unmarked, buffy breast. In flight, note white U-shape on black tail. A Eurasian vagrant expected on both coasts; recorded on Pacific coast in Costa Rica. Prefers to forage in flooded fields and on mudflats.

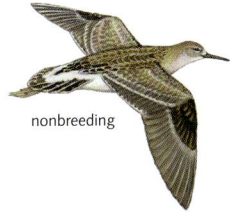

nonbreeding

Curlew Sandpiper
(Correlimos Zarapatín)

Calidris ferruginea

8 in (20 cm). In breeding plumage, face and underparts range from solid rufous (on male) to rufous with black barring (on female). In nonbreeding plumage, distinguished from Dunlin (p. 114) by more evidently decurved bill and by grayer upperparts. Black legs distinguish it from Stilt Sandpiper (p. 118). In flight, note large white patch on uppertail coverts. Expected as a vagrant on both coasts; there are records from the Pacific coasts of both Costa Rica and Panama. More likely in freshwater habitats, but also found on mudflats and sandy beaches; likely to join large flocks of peeps. **NT**

nonbreeding

South Polar Skua
(Salteador Polar)

Stercorarius maccormicki

21 in (53 cm); WS 52 in (132 cm). Large, bulky body, long wings, and short, wedge-shaped tail help distinguish it from the jaegers (p. 124). Both dark and pale morphs are possible, as are intermediates, which results in great variation in plumage. All morphs show a bold white patch at the base of the primaries on both upper- and underwings; also note black wing linings. Expected as a very rare pelagic migrant on Pacific and Caribbean pelagic and coastal waters.

dark morph

Black-legged Kittiwake
(Gaviota Patinegra)

Rissa tridactyla

17 in (43 cm). On mature adult, completely yellow bill is unique among the gulls. In breeding plumage, head is entirely white; in all other plumages, head shows a dusky ear patch. Immature and first year adult have a variable amount of black on the bill and a gray hindcrown; immature has a black nuchal collar. In flight, adults show black tips on outermost primaries; immature has a bold pattern on the upperwings created by black-tipped primaries, leading edge of wing, and covert feathers. Expected as an accidental winter resident on the Caribbean coast and pelagic waters; records exist in Costa Rica and Panama.

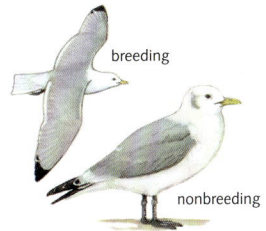

breeding

nonbreeding

Bonaparte's Gull
(Gaviota de Bonaparte)

Chroicocephalus philadelphia

13.5 in (34 cm). Has a smaller body and a more **petite bill** than do Laughing and Franklin's Gulls (p. 126). In nonbreeding plumage, black hood reduces to **dark ear coverts**. Immature has dusky crown and nape. In flight, note white outermost primaries with black tips. Expected as an accidental winter resident on both coasts; Pacific coast records exist in Costa Rica and Panama.

breeding

Black Noddy
Anous minutus

(Tiñosa Negra)

14 in (35 cm). In all plumages, the **white crown** contrasts with an all- (adult) or mostly (first year and immature) **black body**. Bill is relatively longer and thinner than that of Brown Noddy (p. 130). Expected as an accidental pelagic migrant on both the Pacific Ocean and the Caribbean Sea.

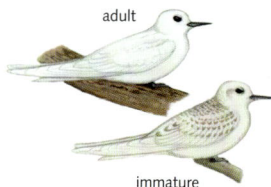

adult

immature

White Tern
Gygis alba

(Charrán Blanco)

12 in (30 cm). The only tern with **entirely white plumage**. Immature has brown mottling on upperparts and upperwings. Expected as pelagic migrant on pelagic waters of the Pacific; reported on Costa Rican waters not far south of Nicaragua's territory; also breeds on Cocos Island in Costa Rica.

Galapagos Petrel
Pterodroma phaeopygia

(Petrel de Galápagos)

16 in (41 cm); WS 39.5 in (100 cm). Upperparts are entirely dark except for **white forehead**. Dark hood extends below the cheeks and onto sides of the neck. Underparts are white; underwings are white with black on the primaries and leading and trailing edges. Black undertail is mostly concealed by white undertail coverts. Expected as a pelagic migrant on the Pacific Ocean. **CR**

Tahiti Petrel
Pterodroma rostrata

(Petrel de Tahiti)

15.5 in (39 cm); WS 41.5 in (105 cm). Upperparts are entirely dark; dark coloration on head extends to the upper breast and sharply contrasts with white underparts. Dark underwings and undertail also contrast with the white underparts. Wings are notably long and narrow. Expected as a pelagic migrant on the Pacific Ocean. **NT**

Cory's Shearwater
Calonectris diomedea

(Pardela de Cory)

19 in (48 cm); WS 46.5 in (118 cm). Largest of the shearwaters. Long, stout **yellow bill with a black tip** stands out on the gray head and neck. Has brownish-gray upperparts and upperwings, with black primaries. Underparts are entirely white and lead into black undertail; white underwings have black on the leading and trailing edges (black is more extensive on the primaries). Expected as a vagrant on coastal and pelagic waters of the Caribbean Sea.

Short-tailed Shearwater
Ardenna tenuirostris

(Pardela Colicorta)

16 in (40 cm); WS 38 in (97 cm). Dark overall; body looks compact. Dark underwings typically have pale bands on the wing linings. In flight, feet extend past the tail; head and neck projection is shorter than on Sooty Shearwater. Potential vagrant; previously recorded in Costa Rica.

Sooty Shearwater
Ardenna grisea

(Pardela Sombría)

17.5 in (45 cm); WS 41 in (104 cm). Dark overall. Dark underwings have whitish patches on the outer wing linings. In flight, feet sometimes extend past the tail; head and neck projection is longer than on Short-tailed Shearwater. Expected as a pelagic migrant on the Pacific Ocean; also possible on the Caribbean Sea. **NT**

Christmas Shearwater   *Puffinus nativitatis*
(Pardela de Navidad)
14 in (36 cm); WS 32.5 in (82 cm). **Entirely dark** with no pale or white coloration on the wing linings. Noticeably smaller than any other all-dark petrel or shearwater that might potentially occur on Nicaraguan waters. In flight, feet do not extend past the tail. Expected as a pelagic migrant on the Pacific Ocean.

Manx Shearwater   *Puffinus puffinus*
(Pardela Pichoneta)
13.5 in (34 cm); WS 32.5 in (82 cm). Dark upperparts are only interrupted by **white spur behind the ear coverts** and **white at the base of the bill**. Underparts are white and slightly extend onto the rump. Wings are narrower and tail is shorter than on Audubon's Shearwater; and face tends to be blacker. In flight, feet barely extend beyond the tail. Possible vagrant on the Caribbean Sea.

Audubon's Shearwater   *Puffinus lherminieri*
(Pardela de Audubon)
12 in (31 cm), WS 27.5 in (70 cm). Dark upperparts typically contrast sharply with the white underparts; white underparts sometimes slightly extend onto the rump. **White undertail coverts are black-tipped**, making the undertail look darker. Wings are broader and tail is longer than on Manx's Shearwater; and face tends to be whiter. In flight, feet typically do not extend beyond the tail. Expected as pelagic migrant on the Caribbean Sea.

Wilson's Storm-Petrel   *Oceanites oceanicus*
(Paíño de Wilson)
7 in (18 cm); WS 16 in (41 cm). Entirely dark except for broad white rump patch (extends onto lower flanks and undertail coverts) and pale upperwing band (does not reach leading edge of wing). In flight, trailing edge of the wing appears straight and the **feet extend beyond the tail**. Square tail sometimes appears to be slightly notched. Expected as pelagic migrant on Pacific coastal and pelagic waters; records exist from Costa Rica.

Band-rumped Storm-Petrel   *Oceanodroma castro*
(Paíño Rabifajeado)
7 in (18 cm); WS 18.5 in (47 cm). Entirely dark except for narrow white rump patch (extends onto lower flanks) and pale upperwing band (does not reach leading edge of the wing). In flight, feet do not extend beyond the tail. Square tail is slightly notched. Expected as a pelagic migrant on Pacific pelagic waters; records exist from Costa Rica.

Markham's Storm-Petrel   *Oceanodroma markhami*
(Paíño de Markham)
9 in (23 cm); WS 20 in (51 cm). Entirely dark except for broad, pale upperwing band (reaches leading edge of the wing); band is more extensive than band on Black Storm-Petrel. Tail is deeply forked. Expected as a pelagic migrant in Pacific pelagic waters; records exist from Costa Rica.

White-faced Ibis

Plegadis chihi

(Ibis Cariblanco)

23 in (58 cm). **Red iris and facial skin** (dull pink or grayish in nonbreeding plumage) and lack of bluish borders around facial skin rule out almost identical Glossy Ibis (p. 164); immature is indistinguishable from Glossy Ibis. Red iris is only observable up close. Expected as an accidental winter resident. Recently recorded in Honduras across the border from Apacunca GRR; one historical record exists from Costa Rica (late 1800s); and there is a recent record from Panama (March, 2016).

Short-eared Owl

Asio flammeus

(Búho Campestre)

15 in (38 cm). Short ear-tufts are barely visible. Heavily patterned plumage is spotted on upperparts and streaked on underparts. Male is typically paler than female. In flight, note short tail, long wings, and a buffy patch on the upperwing primaries. Two historical records from Costa Rica suggest it might occur as an accidental winter resident in Nicaragua. Prefers to hunt in grasslands, marshes, and agricultural fields. Flies slow and low over the ground to hunt for prey; flight behavior is similar to that of the Northern Harrier (p. 168).

Mountain Trogon

Trogon mexicanus

(Trogón Mexicano)

12 in (30 cm). On male, **broad, white tips on black undertail** rule out very similar male Collared and Elegant Trogons (p. 200). On female, brown band on lower breast distinguishes it from female Collared Trogon; lack of white ear patch further distinguishes it from female Elegant Trogon. Southernmost distribution occurs in the mountains at the Honduran border. Possibly occurs as a breeding resident on Sierra Dipilto-Jalapa and Tepesomoto-Pataste NR. Inhabits cloud forest and highland pine and pine-oak forest; above 3,900 ft (1,200 m). In succession, barks several subtle, slightly nasal *aow!* notes.

Streaked Xenops

Xenops rutilans

(Piquivuelto Listado)

4.5 in (12 cm). **White superciliary** and **white malar stripe** distinctly mark the head; and bill has **upturned lower mandible**. **Streaked underparts** distinguish it from very similar Plain Xenops (p. 250). Vagrant; a single record from Matagalpa Dept. (Esperanza Verde RSP, March 2004); at 4,000 ft (1,200 m). Observed with a mixed-species flock in tall secondary forest. Whistles a short series of piercing sharp notes immediately followed by a thin, arcing trill.

Blue-crowned Manakin

Lepidothrix coronata

(Saltarín Gorriazul)

3.5 in (9 cm). On male, bright **blue crown** is diagnostic. Female is brighter green than other female manakins. Vagrant; a single record of an adult male in Guatuzos WR (Jan 2001). Delivers a bubbly trill (usually 1 second in length).

Black-whiskered Vireo

Vireo altiloquus

(Vireo Bigotudo)

6 in (16 cm). **Dusky malar stripe** (not always visible) distinguishes it from all other vireos within its range. Further distinguished from Red-eyed Vireo and Yellow-green Vireo (p. 302) by browner back, heavier bill, and absence of black on border of white superciliary. Possible as a winter resident (expected from Sept to March) on the Corn Islands and Cayos Miskitos BR, and probably very rare on the Caribbean coast. Found in mangroves, swamps, and scrub.

Brown-chested Martin

Progne tapera

(Avión de Ríos)
7 in (18 cm). Broad, brown breast band on white throat is similar to that of Bank Swallow (p. 308). Larger size is key to distinguishing it; also note that white from throat does not extend to behind the ear coverts, as it does on the Bank Swallow. Expected as a vagrant from South America; several records exist in Costa Rica, one from a site directly across the border, in the Guatuzo Plains.

American Pipit

Anthus rubescens

(Bisbita Americana)
6.5 in (17 cm). Gray-brown upperparts are faintly streaked, underparts are buffy-whitish, and throat is streaked with black. Has a buffy superciliary and white malar stripe. Dark tail shows white outer rectrices. Expected as an accidental winter resident (Nov to March); known from the Pacific lowlands of El Salvador, Honduras, and Costa Rica. Found in open areas, often near aquatic habitats such as coastal marshes, river banks, ponds, mudflats, and flooded fields; these locations are generally not visited by migratory *Catharus* thrushes, with which it could be confused.

Clay-colored Sparrow

Spizella pallida

(Sabanero Pálido)
5.5 in (14 cm). In nonbreeding plumage, **brown rump**, buffy underparts, and **pale lore** distinguish it from nonbreeding Chipping Sparrow (p. 346). Also note gray nape, broad superciliary, and dark moustachial and lateral throat stripes. Expected as an accidental winter resident; there are several records from Costa Rica.

Lark Sparrow

Chondestes grammacus

(Sabanero Arlequín)
6.5 in (17 cm). On adult, **striking head pattern** (rufous, white, and black) and black spot on the white breast are unique. Immature lacks most of the color, but has the bold, white lower eye crescent, white malar stripe, and black breast spot. In flight, note white-tipped outer rectrices. Expected as an accidental winter resident in western regions of the country; already documented with multiple records in Costa Rica. Likely to be found in grasslands and semi-open areas.

Lincoln's Sparrow

Melospiza lincolnii

(Sabanero de Lincoln)
6 in (15 cm). Combination of **buffy malar stripe** and **heavy black streaking on buffy breast and flanks** is diagnostic. Also note broad, gray superciliary. Expected as an accidental winter resident. Regular winter resident distribution reaches southern Honduras; already documented with several records in Costa Rica. Most likely found foraging in grasslands and marshes.

Bullock's Oriole

Icterus bullockii

(Chichiltote de Bullock)
8 in (20 cm). **Black eye line** and extensive **white** on wing coverts (which forms **broad wing patches**) are diagnostic. Immature is very similar to immature Baltimore Oriole (p. 354) but paler. Expected as a winter resident. Several sightings in Guanacaste, Costa Rica, suggest it should occur in the Pacific lowlands of Nicaragua. Found in open woodland and dry forest edge. In canopy, gleans insects.

female

male

Orange-crowned Warbler   *Oreothlypis celata*
(Reinita Coroninaranja)
5 in (13 cm). Plumages range from very drab to yellow. In all plumages, note **fine-tipped bill**; thin, dark eye line; and broken eye ring. From below, also note **pale yellow undertail coverts** and **faintly streaked underparts**. Tail is relatively longer and primary projection is relatively shorter than on Tennessee Warbler (p. 366). Expected as an accidental winter resident; already documented with several records in Costa Rica. Likely to forage at all levels of vegetation for nectar, berry pulp, and insects.

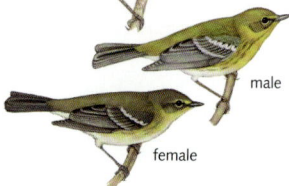
immature

Pine Warbler   *Setophaga pinus*
(Reinita del Pinar)
5.5 in (14 cm). In all plumages, note white wing bars, **ear coverts that slightly contrast with the sides of the neck**, broken eye ring, and streaking on flanks (variable amount); bill is relatively large for a warbler. Yellow on adult shows variable intensity, sometimes creating a yellow spectacle; immature is drab overall. From below, note white tail and undertail coverts and notched tail. Possible vagrant in the Caribbean, most likely during the boreal winter months. Two individuals were recorded in Costa Rica (1976). Often forages on branches; occasionally pumps tail.

male

female

Red-faced Warbler   *Cardellina rubrifrons*
(Reinita Carirroja)
5.5 in (14 cm). **Red head with black hood** (incomplete) is unmistakable. Expected as a very rare winter resident. Winter resident distribution reaches Sierra La Botija in Honduras. Nicaragua shares these mountains and also the bird's preferred habitats, highland pine and pine-oak forest.

male

Black-headed Grosbeak   *Pheucticus melanocephalus*
(Piquigrueso Cabecinegro)
8 in (20 cm). Male is unmistakable. On female, dark upper mandible, buffier underparts, and streaking that is mainly confined to the flanks distinguish it from female Rose-breasted Grosbeak (which has more densely streaked underparts), p. 390. Expected as an accidental winter resident; Nicaragua has a single unconfirmed record from Masaya Volcano NP (April 1990), but multiple records exist in Costa Rica. Possible at all elevations. Feeds on insects, spiders, fruit, and seeds at all levels of vegetation, even dropping to the ground to find seeds.

female

Checklist of the Birds of Nicaragua

Species are placed in the order in which they appear in the Checklist of North and Middle American Birds: seventh edition (updated through the fifty-eighth supplement), managed by the American Ornithological Society.

Tinamous (4)   Tinamidae

Great Tinamou	*Tinamus major*	
Little Tinamou	*Crypturellus soui*	
Thicket Tinamou	*Crypturellus cinnamomeus*	
Slaty-breasted Tinamou	*Crypturellus boucardi*	

Ducks (17)   Anatidae

Black-bellied Whistling-Duck	*Dendrocygna autumnalis*	
Fulvous Whistling-Duck	*Dendrocygna bicolor*	
Muscovy Duck	*Cairina moschata*	
Blue-winged Teal	*Spatula discors*	
Cinnamon Teal	*Spatula cyanoptera*	
Northern Shoveler	*Spatula clypeata*	
American Wigeon	*Mareca americana*	
Mallard	*Anas platyrhynchos*	
Northern Pintail	*Anas acuta*	
Green-winged Teal	*Anas crecca*	
Canvasback	*Aythya valisineria*	
Redhead	*Aythya americana*	
Ring-necked Duck	*Aythya collaris*	
Greater Scaup	*Aythya marila*	
Lesser Scaup	*Aythya affinis*	
Masked Duck	*Nomonyx dominicus*	
Ruddy Duck	*Oxyura jamaicensis*	

Chachalacas, Guans, & Curassows (6)   Cracidae

Plain Chachalaca	*Ortalis vetula*	
Gray-headed Chachalaca	*Ortalis cinereiceps*	
White-bellied Chachalaca	*Ortalis leucogastra*	
Crested Guan	*Penelope purpurascens*	
Highland Guan	*Penelopina nigra*	
Great Curassow	*Crax rubra*	

New World Quail (7)   Odontophoridae

Tawny-faced Quail	*Rhynchortyx cinctus*	
Buffy-crowned Wood-Partridge	*Dendrortyx leucophrys*	
Black-throated Bobwhite	*Colinus nigrogularis*	
Crested Bobwhite	*Colinus cristatus*	
Ocellated Quail	*Cyrtonyx ocellatus*	

| Black-eared Wood-Quail | Odontophorus melanotis | |
| Spotted Wood-Quail | Odontophorus guttatus | |

Grebes (2) Podicipedidae

| Least Grebe | Tachybaptus dominicus | |
| Pied-billed Grebe | Podilymbus podiceps | |

Pigeons & Doves (22) Columbidae

Rock Pigeon	Columba livia	
Pale-vented Pigeon	Patagioenas cayennensis	
Scaled Pigeon	Patagioenas speciosa	
White-crowned Pigeon	Patagioenas leucocephala	
Red-billed Pigeon	Patagioenas flavirostris	
Band-tailed Pigeon	Patagioenas fasciata	
Short-billed Pigeon	Patagioenas nigrirostris	
Eurasian Collared-Dove	Streptopelia decaocto	
Inca Dove	Columbina inca	
Common Ground-Dove	Columbina passerina	
Plain-breasted Ground-Dove	Columbina minuta	
Ruddy Ground-Dove	Columbina talpacoti	
Blue Ground-Dove	Claravis pretiosa	
Ruddy Quail-Dove	Geotrygon montana	
Violaceous Quail-Dove	Geotrygon violacea	
Olive-backed Quail-Dove	Leptotrygon veraguensis	
White-tipped Dove	Leptotila verreauxi	
Gray-chested Dove	Leptotila cassinii	
Gray-headed Dove	Leptotila plumbeiceps	
White-faced Quail-Dove	Zentrygon albifacies	
White-winged Dove	Zenaida asiatica	
Mourning Dove	Zenaida macroura	

Cuckoos (12) Cuculidae

Squirrel Cuckoo	Playa cayana	
Dark-billed Cuckoo	Coccyzus melacoryphus	
Yellow-billed Cuckoo	Coccyzus americanus	
Mangrove Cuckoo	Coccyzus minor	
Black-billed Cuckoo	Coccyzus erythropthalmus	
Striped Cuckoo	Tapera naevia	
Pheasant Cuckoo	Dromococcyx phasianellus	
Lesser Ground-Cuckoo	Morococcyx erythropygus	
Lesser Roadrunner	Geococcyx velox	
Rufous-vented Ground-Cuckoo	Neomorphus geoffroyi	

| Smooth-billed Ani | *Crotophaga ani* | |
| Groove-billed Ani | *Crotophaga sulcirostris* | |

Nightjars & Allies (12) — **Caprimulgidae**

Short-tailed Nighthawk	*Lurocalis semitorquatus*	
Lesser Nighthawk	*Chordeiles acutipennis*	
Common Nighthawk	*Chordeiles minor*	
Common Pauraque	*Nyctidromus albicollis*	
Ocellated Poorwill	*Nyctiphrynus ocellatus*	
Chuck-will's-widow	*Antrostomus carolinensis*	
Rufous Nightjar	*Antrostomus rufus*	
Tawny-collared Nightjar	*Antrostomus salvini*	
Buff-collared Nightjar	*Antrostomus ridgwayi*	
Eastern Whip-poor-will	*Antrostomus vociferus*	
Mexican Whip-poor-will	*Antrostomus arizonae*	
Spot-tailed Nightjar	*Hydropsalis maculicaudus*	

Potoos (3) — **Nyctibiidae**

Great Potoo	*Nyctibius grandis*	
Common Potoo	*Nyctibius griseus*	
Northern Potoo	*Nyctibius jamaicensis*	

Swifts (9) — **Apodidae**

Black Swift	*Cypseloides niger*	
White-chinned Swift	*Cypseloides cryptus*	
Chestnut-collared Swift	*Streptoprocne rutila*	
White-collared Swift	*Streptoprocne zonaris*	
Chimney Swift	*Chaetura pelagica*	
Vaux's Swift	*Chaetura vauxi*	
Gray-rumped Swift	*Chaetura cinereiventris*	
Lesser Swallow-tailed Swift	*Panyptila cayennensis*	
Great Swallow-tailed Swift	*Panyptila sanctihieronymi*	

Hummingbirds (37) — **Trochilidae**

White-necked Jacobin	*Florisuga mellivora*	
White-tipped Sicklebill	*Eutoxeres aquila*	
Bronzy Hermit	*Glaucis aeneus*	
Band-tailed Barbthroat	*Threnetes ruckeri*	
Long-billed Hermit	*Phaethornis longirostris*	
Stripe-throated Hermit	*Phaethornis striigularis*	
Brown Violetear	*Colibri delphinae*	
Mexican Violetear	*Colibri thalassinus*	
Purple-crowned Fairy	*Heliothryx barroti*	
Green-breasted Mango	*Anthracothorax prevostii*	

Black-throated Mango	Anthracothorax nigricollis	
Black-crested Coquette	Lophornis helenae	
Rivoli's Hummingbird	Eugenes fulgens	
Long-billed Starthroat	Heliomaster longirostris	
Plain-capped Starthroat	Heliomaster constantii	
Green-breasted Mountain-gem	Lampornis sybillae	
Purple-throated Mountain-gem	Lampornis calolaemus	
Sparkling-tailed Hummingbird	Tilmatura dupontii	
Ruby-throated Hummingbird	Archilochus colubris	
Canivet's Emerald	Chlorostilbon canivetii	
Violet-headed Hummingbird	Klais guimeti	
Emerald-chinned Hummingbird	Abeillia abeillei	
Scaly-breasted Hummingbird	Phaeochroa cuvierii	
Violet Sabrewing	Campylopterus hemileucurus	
Stripe-tailed Hummingbird	Eupherusa eximia	
Snowcap	Microchera albocoronata	
Bronze-tailed Plumeleteer	Chalybura urochrysia	
Crowned Woodnymph	Thalurania colombica	
White-bellied Emerald	Amazilia candida	
Blue-chested Hummingbird	Amazilia amabilis	
Azure-crowned Hummingbird	Amazilia cyanocephala	
Blue-tailed Hummingbird	Amazilia cyanura	
Steely-vented Hummingbird	Amazilia saucerottei	
Rufous-tailed Hummingbird	Amazilia tzacatl	
Cinnamon Hummingbird	Amazilia rutila	
Blue-throated Goldentail	Hylocharis eliciae	
White-eared Hummingbird	Hylocharis leucotis	

Rails, Crakes, & Gallinules (13)   Rallidae

Ruddy Crake	Laterallus ruber	
White-throated Crake	Laterallus albigularis	
Gray-breasted Crake	Laterallus exilis	
Mangrove Rail	Rallus longirostris	
Rufous-necked Wood-Rail	Aramides axillaris	
Russet-naped Wood-Rail	Aramides albiventris	
Uniform Crake	Amaurolimnas concolor	
Sora	Porzana carolina	
Yellow-breasted Crake	Hapalocrex flaviventer	
Spotted Rail	Pardirallus maculatus	
Purple Gallinule	Porphyrio martinicus	

| Common Gallinule | *Gallinula galeata* | |
| American Coot | *Fulica americana* | |

Finfoots (1) | **Heliornithidae** | |
| Sungrebe | *Heliornis fulica* | |

Limpkin (1) | **Aramidae** | |
| Limpkin | *Aramus guarauna* | |

Thick-knees (1) | **Burhinidae** | |
| Double-striped Thick-knee | *Burhinus bistriatus* | |

Stilts & Avocets (2) | **Recurvirostridae** | |
| Black-necked Stilt | *Himantopus mexicanus* | |
| American Avocet | *Recurvirostra americana* | |

Oystercatchers (1) | **Haematopodidae** | |
| American Oystercatcher | *Haematopus palliatus* | |

Plovers & Lapwings (10) | **Charadriidae** | |
Southern Lapwing	*Vanellus chilensis*	
Black-bellied Plover	*Pluvialis squatarola*	
American Golden-Plover	*Pluvialis dominica*	
Pacific Golden-Plover	*Pluvialis fulva*	
Collared Plover	*Charadrius collaris*	
Snowy Plover	*Charadrius nivosus*	
Wilson's Plover	*Charadrius wilsonia*	
Semipalmated Plover	*Charadrius semipalmatus*	
Piping Plover	*Charadrius melodus*	
Killdeer	*Charadrius vociferus*	

Jacanas (2) | **Jacanidae** | |
| Northern Jacana | *Jacana spinosa* | |
| Wattled Jacana | *Jacana jacana* | |

Sandpipers & Allies (29) | **Scolopacidae** | |
Upland Sandpiper	*Bartramia longicauda*	
Whimbrel	*Numenius phaeopus*	
Long-billed Curlew	*Numenius americanus*	
Marbled Godwit	*Limosa fedoa*	
Ruddy Turnstone	*Arenaria interpres*	
Red Knot	*Calidris canutus*	
Surfbird	*Calidris virgata*	
Stilt Sandpiper	*Calidris himantopus*	
Sanderling	*Calidris alba*	
Dunlin	*Calidris alpina*	

Baird's Sandpiper	*Calidris bairdii*	
Least Sandpiper	*Calidris minutilla*	
White-rumped Sandpiper	*Calidris fuscicollis*	
Buff-breasted Sandpiper	*Calidris subruficollis*	
Pectoral Sandpiper	*Calidris melanotos*	
Semipalmated Sandpiper	*Calidris pusilla*	
Western Sandpiper	*Calidris mauri*	
Short-billed Dowitcher	*Limnodromus griseus*	
Long-billed Dowitcher	*Limnodromus scolopaceus*	
Wilson's Snipe	*Gallinago delicata*	
Spotted Sandpiper	*Actitis macularius*	
Solitary Sandpiper	*Tringa solitaria*	
Wandering Tattler	*Tringa incana*	
Lesser Yellowlegs	*Tringa flavipes*	
Willet	*Tringa semipalmata*	
Greater Yellowlegs	*Tringa melanoleuca*	
Wilson's Phalarope	*Phalaropus tricolor*	
Red-necked Phalarope	*Phalaropus lobatus*	
Red Phalarope	*Phalaropus fulicarius*	

Jaegers (3) — **Stercorariidae**

Pomarine Jaeger	*Stercorarius pomarinus*	
Parasitic Jaeger	*Stercorarius parasiticus*	
Long-tailed Jaeger	*Stercorarius longicaudus*	

Gulls, Terns, & Skimmers (25) — **Laridae**

Swallow-tailed Gull	*Creagrus furcatus*	
Sabine's Gull	*Xema sabini*	
Laughing Gull	*Leucophaeus atricilla*	
Franklin's Gull	*Leucophaeus pipixcan*	
Ring-billed Gull	*Larus delawarensis*	
California Gull	*Larus californicus*	
Herring Gull	*Larus argentatus*	
Lesser Black-backed Gull	*Larus fuscus*	
Kelp Gull	*Larus dominicanus*	
Brown Noddy	*Anous stolidus*	
Sooty Tern	*Onychoprion fuscatus*	
Bridled Tern	*Onychoprion anaethetus*	
Least Tern	*Sternula antillarum*	
Large-billed Tern	*Phaetusa simplex*	
Gull-billed Tern	*Gelochelidon nilotica*	
Caspian Tern	*Hydroprogne caspia*	

Black Tern	Chlidonias niger	
Roseate Tern	Sterna dougallii	
Common Tern	Sterna hirundo	
Arctic Tern	Sterna paradisaea	
Forster's Tern	Sterna forsteri	
Royal Tern	Thalasseus maximus	
Sandwich Tern	Thalasseus sandvicensis	
Elegant Tern	Thalasseus elegans	
Black Skimmer	Rynchops niger	

Sunbittern (1) — **Eurypygidae**

| Sunbittern | Eurypyga helias | |

Tropicbirds (1) — **Phaethontidae**

| Red-billed Tropicbird | Phaethon aethereus | |

Shearwaters & Petrels (6) — **Procellariidae**

Black-capped Petrel	Pterodroma hasitata	
Parkinson's Petrel	Procellaria parkinsoni	
Wedge-tailed Shearwater	Ardenna pacifica	
Pink-footed Shearwater	Ardenna creatopus	
Galapagos Shearwater	Puffinus subalaris	
Black-vented Shearwater	Puffinus opisthomelas	

Storm-Petrels (4) — **Hydrobatidae**

Leach's Storm-Petrel	Oceanodroma leucorhoa	
Wedge-rumped Storm-Petrel	Oceanodroma tethys	
Black Storm-Petrel	Oceanodroma melania	
Least Storm-Petrel	Oceanodroma microsoma	

Storks (2) — **Ciconiidae**

| Jabiru | Jabiru mycteria | |
| Wood Stork | Mycteria americana | |

Frigatebirds (1) — **Fregatidae**

| Magnificent Frigatebird | Fregata magnificens | |

Boobies (5) — **Sulidae**

Masked Booby	Sula dactylatra	
Nazca Booby	Sula granti	
Blue-footed Booby	Sula nebouxii	
Brown Booby	Sula leucogaster	
Red-footed Booby	Sula sula	

Cormorants (1) — **Phalacrocoracidae**

| Neotropic Cormorant | Phalacrocorax brasilianus | |

Darters (1) — Anhingidae

Anhinga	*Anhinga anhinga*	

Pelicans (2) — Pelecanidae

American White Pelican	*Pelecanus erythrorhynchos*	
Brown Pelican	*Pelecanus occidentalis*	

Herons, Egrets, & Bitterns (19) — Ardeidae

Pinnated Bittern	*Botaurus pinnatus*	
American Bittern	*Botaurus lentiginosus*	
Least Bittern	*Ixobrychus exilis*	
Rufescent Tiger-Heron	*Tigrisoma lineatum*	
Fasciated Tiger-Heron	*Tigrisoma fasciatum*	
Bare-throated Tiger-Heron	*Tigrisoma mexicanum*	
Great Blue Heron	*Ardea herodias*	
Great Egret	*Ardea alba*	
Snowy Egret	*Egretta thula*	
Little Blue Heron	*Egretta caerulea*	
Tricolored Heron	*Egretta tricolor*	
Reddish Egret	*Egretta rufescens*	
Cattle Egret	*Bubulcus ibis*	
Green Heron	*Butorides virescens*	
Striated Heron	*Butorides striata*	
Agami Heron	*Agamia agami*	
Black-crowned Night-Heron	*Nycticorax nycticorax*	
Yellow-crowned Night-Heron	*Nyctanassa violacea*	
Boat-billed Heron	*Cochlearius cochlearius*	

Ibises & Spoonbills (4) — Threskiornithidae

White Ibis	*Eudocimus albus*	
Glossy Ibis	*Plegadis falcinellus*	
Green Ibis	*Mesembrinibis cayennensis*	
Roseate Spoonbill	*Platalea ajaja*	

New World Vultures (4) — Cathartidae

Black Vulture	*Coragyps atratus*	
Turkey Vulture	*Cathartes aura*	
Lesser Yellow-headed Vulture	*Cathartes burrovianus*	
King Vulture	*Sarcoramphus papa*	

Osprey (1) — Pandionidae

Osprey	*Pandion haliaetus*	

Hawks, Kites, & Eagles (37)	Accipitridae	
Gray-headed Kite	*Leptodon cayanensis*	
Hook-billed Kite	*Chondrohierax uncinatus*	
Swallow-tailed Kite	*Elanoides forficatus*	
Pearl Kite	*Gampsonyx swainsonii*	
White-tailed Kite	*Elanus leucurus*	
Snail Kite	*Rostrhamus sociabilis*	
Double-toothed Kite	*Harpagus bidentatus*	
Mississippi Kite	*Ictinia mississippiensis*	
Plumbeous Kite	*Ictinia plumbea*	
Black-collared Hawk	*Busarellus nigricollis*	
Northern Harrier	*Circus hudsonius*	
Tiny Hawk	*Accipiter superciliosus*	
Sharp-shinned Hawk	*Accipiter striatus*	
Cooper's Hawk	*Accipiter cooperii*	
Bicolored Hawk	*Accipiter bicolor*	
Crane Hawk	*Geranospiza caerulescens*	
Common Black Hawk	*Buteogallus anthracinus*	
Savanna Hawk	*Buteogallus meridionalis*	
Great Black Hawk	*Buteogallus urubitinga*	
Solitary Eagle	*Buteogallus solitarius*	
Barred Hawk	*Morphnarchus princeps*	
Roadside Hawk	*Rupornis magnirostris*	
Harris's Hawk	*Parabuteo unicinctus*	
White-tailed Hawk	*Geranoaetus albicaudatus*	
White Hawk	*Pseudastur albicollis*	
Semiplumbeous Hawk	*Leucopternis semiplumbeus*	
Gray Hawk	*Buteo plagiatus*	
Broad-winged Hawk	*Buteo platypterus*	
Short-tailed Hawk	*Buteo brachyurus*	
Swainson's Hawk	*Buteo swainsoni*	
Zone-tailed Hawk	*Buteo albonotatus*	
Red-tailed Hawk	*Buteo jamaicensis*	
Crested Eagle	*Morphnus guianensis*	
Harpy Eagle	*Harpia harpyja*	
Black Hawk-Eagle	*Spizaetus tyrannus*	
Ornate Hawk-Eagle	*Spizaetus ornatus*	
Black-and-white Hawk-Eagle	*Spizaetus melanoleucus*	

Barn Owls (1) **Tytonidae**

Barn Owl	Tyto alba	

Owls (14) **Strigidae**

Pacific Screech-Owl	Megascops cooperi	
Whiskered Screech-Owl	Megascops trichopsis	
Vermiculated Screech-Owl	Megascops guatemalae	
Crested Owl	Lophostrix cristata	
Spectacled Owl	Pulsatrix perspicillata	
Great Horned Owl	Bubo virginianus	
Northern Pygmy-Owl	Glaucidium gnoma	
Central American Pygmy-Owl	Glaucidium griseiceps	
Ferruginous Pygmy-Owl	Glaucidium brasilianum	
Burrowing Owl	Athene cunicularia	
Mottled Owl	Ciccaba virgata	
Black-and-white Owl	Ciccaba nigrolineata	
Stygian Owl	Asio stygius	
Striped Owl	Pseudoscops clamator	

Trogons (8) **Trogoninae**

Lattice-tailed Trogon	Trogon clathratus	
Slaty-tailed Trogon	Trogon massena	
Black-headed Trogon	Trogon melanocephalus	
Gartered Trogon	Trogon caligatus	
Black-throated Trogon	Trogon rufus	
Elegant Trogon	Trogon elegans	
Collared Trogon	Trogon collaris	
Resplendent Quetzal	Pharomachrus mocinno	

Motmots (6) **Momotidae**

Tody Motmot	Hylomanes momotula	
Lesson's Motmot	Momotus lessonii	
Rufous Motmot	Baryphthengus martii	
Keel-billed Motmot	Electron carinatum	
Broad-billed Motmot	Electron platyrhynchum	
Turquoise-browed Motmot	Eumomota superciliosa	

Kingfishers (6) **Alcedinidae**

Ringed Kingfisher	Megaceryle torquata	
Belted Kingfisher	Megaceryle alcyon	
Amazon Kingfisher	Chloroceryle amazona	
Green Kingfisher	Chloroceryle americana	
Green-and-rufous Kingfisher	Chloroceryle inda	
American Pygmy Kingfisher	Chloroceryle aenea	

Puffbirds & Nunbird (4)   Bucconidae

White-necked Puffbird	*Notharchus hyperrhynchus*	
Pied Puffbird	*Notharchus tectus*	
White-whiskered Puffbird	*Malacoptila panamensis*	
White-fronted Nunbird	*Monasa morphoeus*	

Jacamars (2)   Galbulidae

Rufous-tailed Jacamar	*Galbula ruficauda*	
Great Jacamar	*Jacamerops aureus*	

Toucans (5)   Ramphastidae

Northern Emerald-Toucanet	*Aulacorhynchus prasinus*	
Collared Aracari	*Pteroglossus torquatus*	
Yellow-eared Toucanet	*Selenidera spectabilis*	
Keel-billed Toucan	*Ramphastos sulfuratus*	
Yellow-throated Toucan	*Ramphastos ambiguus*	

Woodpeckers (16)   Picidae

Olivaceous Piculet	*Picumnus olivaceus*	
Acorn Woodpecker	*Melanerpes formicivorus*	
Black-cheeked Woodpecker	*Melanerpes pucherani*	
Hoffmann's Woodpecker	*Melanerpes hoffmannii*	
Golden-fronted Woodpecker	*Melanerpes aurifrons*	
Yellow-bellied Sapsucker	*Sphyrapicus varius*	
Ladder-backed Woodpecker	*Picoides scalaris*	
Smoky-brown Woodpecker	*Picoides fumigatus*	
Hairy Woodpecker	*Picoides villosus*	
Rufous-winged Woodpecker	*Piculus simplex*	
Golden-olive Woodpecker	*Colaptes rubiginosus*	
Northern Flicker	*Colaptes auratus*	
Cinnamon Woodpecker	*Celeus loricatus*	
Chestnut-colored Woodpecker	*Celeus castaneus*	
Lineated Woodpecker	*Dryocopus lineatus*	
Pale-billed Woodpecker	*Campephilus guatemalensis*	

Falcons & Caracaras (13)   Falconidae

Laughing Falcon	*Herpetotheres cachinnans*	
Barred Forest-Falcon	*Micrastur ruficollis*	
Slaty-backed Forest-Falcon	*Micrastur mirandollei*	
Collared Forest-Falcon	*Micrastur semitorquatus*	
Red-throated Caracara	*Ibycter americanus*	
Crested Caracara	*Caracara cheriway*	
Yellow-headed Caracara	*Milvago chimachima*	
American Kestrel	*Falco sparverius*	
Merlin	*Falco columbarius*	

Aplomado Falcon	*Falco femoralis*	
Bat Falcon	*Falco rufigularis*	
Orange-breasted Falcon	*Falco deiroleucus*	
Peregrine Falcon	*Falco peregrinus*	

Parrots (16) — Psittacidae

Olive-throated Parakeet	*Eupsittula nana*	
Orange-fronted Parakeet	*Eupsittula canicularis*	
Great Green Macaw	*Ara ambiguus*	
Scarlet Macaw	*Ara macao*	
Green Parakeet	*Psittacara holochlorus*	
Pacific Parakeet	*Psittacara strenuus*	
Crimson-fronted Parakeet	*Psittacara finschi*	
Barred Parakeet	*Bolborhynchus lineola*	
Orange-chinned Parakeet	*Brotogeris jugularis*	
Brown-hooded Parrot	*Pyrilia haematotis*	
Blue-headed Parrot	*Pionus menstruus*	
White-crowned Parrot	*Pionus senilis*	
White-fronted Parrot	*Amazona albifrons*	
Red-lored Parrot	*Amazona autumnalis*	
Mealy Parrot	*Amazona farinosa*	
Yellow-naped Parrot	*Amazona auropalliata*	

Antbirds (19) — Thamnophilidae

Fasciated Antshrike	*Cymbilaimus lineatus*	
Great Antshrike	*Taraba major*	
Barred Antshrike	*Thamnophilus doliatus*	
Black-crowned Antshrike	*Thamnophilus atrinucha*	
Russet Antshrike	*Thamnistes anabatinus*	
Plain Antvireo	*Dysithamnus mentalis*	
Streak-crowned Antvireo	*Dysithamnus striaticeps*	
White-flanked Antwren	*Myrmotherula axillaris*	
Slaty Antwren	*Myrmotherula schisticolor*	
Checker-throated Antwren	*Epinecrophylla fulviventris*	
Dot-winged Antwren	*Microrhopias quixensis*	
Dusky Antbird	*Cercomacroides tyrannina*	
Bare-crowned Antbird	*Gymnocichla nudiceps*	
Chestnut-backed Antbird	*Myrmeciza exsul*	
Zeledon's Antbird	*Myrmeciza zeledoni*	
Spotted Antbird	*Hylophylax naevioides*	
Wing-banded Antbird	*Myrmornis torquata*	
Bicolored Antbird	*Gymnopithys bicolor*	
Ocellated Antbird	*Phaenostictus mcleannani*	

Antpittas (3)	Grallariidae	
Scaled Antpitta	*Grallaria guatimalensis*	
Streak-chested Antpitta	*Hylopezus perspicillatus*	
Thicket Antpitta	*Hylopezus dives*	

Antthrushes (1)	Formicariidae	
Black-faced Antthrush	*Formicarius analis*	

Ovenbirds & Woodcreepers (24)	Furnariidae	
Tawny-throated Leaftosser	*Sclerurus mexicanus*	
Scaly-throated Leaftosser	*Sclerurus guatemalensis*	
Olivaceous Woodcreeper	*Sittasomus griseicapillus*	
Long-tailed Woodcreeper	*Deconychura longicauda*	
Ruddy Woodcreeper	*Dendrocincla homochroa*	
Tawny-winged Woodcreeper	*Dendrocincla anabatina*	
Plain-brown Woodcreeper	*Dendrocincla fuliginosa*	
Wedge-billed Woodcreeper	*Glyphorynchus spirurus*	
Northern Barred-Woodcreeper	*Dendrocolaptes sanctithomae*	
Black-banded Woodcreeper	*Dendrocolaptes picumnus*	
Strong-billed Woodcreeper	*Xiphocolaptes promeropirhynchus*	
Cocoa Woodcreeper	*Xiphorhynchus susurrans*	
Ivory-billed Woodcreeper	*Xiphorhynchus flavigaster*	
Black-striped Woodcreeper	*Xiphorhynchus lachrymosus*	
Spotted Woodcreeper	*Xiphorhynchus erythropygius*	
Brown-billed Scythebill	*Campylorhamphus pusillus*	
Streak-headed Woodcreeper	*Lepidocolaptes souleyetii*	
Spot-crowned Woodcreeper	*Lepidocolaptes affinis*	
Plain Xenops	*Xenops minutus*	
Scaly-throated Foliage-gleaner	*Anabacerthia variegaticeps*	
Ruddy Foliage-gleaner	*Clibanornis rubiginosus*	
Buff-throated Foliage-gleaner	*Automolus ochrolaemus*	
Striped Woodhaunter	*Automolus subulatus*	
Slaty Spinetail	*Synallaxis brachyura*	

Tyrant Flycatchers (66)	Tyrannidae	
Yellow-bellied Tyrannulet	*Ornithion semiflavum*	
Brown-capped Tyrannulet	*Ornithion brunneicapillus*	
Northern Beardless-Tyrannulet	*Camptostoma imberbe*	
Yellow Tyrannulet	*Capsiempis flaveola*	
Greenish Elaenia	*Myiopagis viridicata*	
Yellow-bellied Elaenia	*Elaenia flavogaster*	
Mountain Elaenia	*Elaenia frantzii*	
Ochre-bellied Flycatcher	*Mionectes oleagineus*	
Sepia-capped Flycatcher	*Leptopogon amaurocephalus*	
Paltry Tyrannulet	*Zimmerius vilissimus*	

Black-capped Pygmy-Tyrant	*Myiornis atricapillus*	
Scale-crested Pygmy-Tyrant	*Lophotriccus pileatus*	
Northern Bentbill	*Oncostoma cinereigulare*	
Slate-headed Tody-Flycatcher	*Poecilotriccus sylvia*	
Common Tody-Flycatcher	*Todirostrum cinereum*	
Black-headed Tody-Flycatcher	*Todirostrum nigriceps*	
Eye-ringed Flatbill	*Rhynchocyclus brevirostris*	
Yellow-olive Flycatcher	*Tolmomyias sulphurescens*	
Yellow-margined Flycatcher	*Tolmomyias assimilis*	
Stub-tailed Spadebill	*Platyrinchus cancrominus*	
Golden-crowned Spadebill	*Platyrinchus coronatus*	
Royal Flycatcher	*Onychorhynchus coronatus*	
Ruddy-tailed Flycatcher	*Terenotriccus erythrurus*	
Sulphur-rumped Flycatcher	*Myiobius sulphureipygius*	
Tawny-chested Flycatcher	*Aphanotriccus capitalis*	
Tufted Flycatcher	*Mitrephanes phaeocercus*	
Olive-sided Flycatcher	*Contopus cooperi*	
Greater Pewee	*Contopus pertinax*	
Western Wood-Pewee	*Contopus sordidulus*	
Eastern Wood-Pewee	*Contopus virens*	
Tropical Pewee	*Contopus cinereus*	
Yellow-bellied Flycatcher	*Empidonax flaviventris*	
Acadian Flycatcher	*Empidonax virescens*	
Alder Flycatcher	*Empidonax alnorum*	
Willow Flycatcher	*Empidonax traillii*	
White-throated Flycatcher	*Empidonax albigularis*	
Least Flycatcher	*Empidonax minimus*	
Hammond's Flycatcher	*Empidonax hammondii*	
Yellowish Flycatcher	*Empidonax flavescens*	
Buff-breasted Flycatcher	*Empidonax fulvifrons*	
Black Phoebe	*Sayornis nigricans*	
Vermilion Flycatcher	*Pyrocephalus rubinus*	
Long-tailed Tyrant	*Colonia colonus*	
Bright-rumped Attila	*Attila spadiceus*	
Rufous Mourner	*Rhytipterna holerythra*	
Dusky-capped Flycatcher	*Myiarchus tuberculifer*	
Ash-throated Flycatcher	*Myiarchus cinerascens*	
Nutting's Flycatcher	*Myiarchus nuttingi*	
Great Crested Flycatcher	*Myiarchus crinitus*	
Brown-crested Flycatcher	*Myiarchus tyrannulus*	
Great Kiskadee	*Pitangus sulphuratus*	
Boat-billed Flycatcher	*Megarynchus pitangua*	
Social Flycatcher	*Myiozetetes similis*	
Gray-capped Flycatcher	*Myiozetetes granadensis*	

White-ringed Flycatcher	*Conopias albovittatus*	
Streaked Flycatcher	*Myiodynastes maculatus*	
Sulphur-bellied Flycatcher	*Myiodynastes luteiventris*	
Piratic Flycatcher	*Legatus leucophaius*	
Tropical Kingbird	*Tyrannus melancholicus*	
Cassin's Kingbird	*Tyrannus vociferans*	
Western Kingbird	*Tyrannus verticalis*	
Eastern Kingbird	*Tyrannus tyrannus*	
Gray Kingbird	*Tyrannus dominicensis*	
Scissor-tailed Flycatcher	*Tyrannus forficatus*	
Fork-tailed Flycatcher	*Tyrannus savana*	
Gray-headed Piprites	*Piprites griseiceps*	

Tityras, Becards, & Allies (8) — Tityridae

Northern Schiffornis	*Schiffornis veraepacis*	
Speckled Mourner	*Laniocera rufescens*	
Masked Tityra	*Tityra semifasciata*	
Black-crowned Tityra	*Tityra inquisitor*	
Cinnamon Becard	*Pachyramphus cinnamomeus*	
White-winged Becard	*Pachyramphus polychopterus*	
Gray-collared Becard	*Pachyramphus major*	
Rose-throated Becard	*Pachyramphus aglaiae*	

Contingas (6) — Cotingidae

Purple-throated Fruitcrow	*Querula purpurata*	
Bare-necked Umbrellabird	*Cephalopterus glabricollis*	
Lovely Cotinga	*Cotinga amabilis*	
Rufous Piha	*Lipaugus unirufus*	
Three-wattled Bellbird	*Procnias tricarunculatus*	
Snowy Cotinga	*Carpodectes nitidus*	

Manakins (4) — Pipridae

Long-tailed Manakin	*Chiroxiphia linearis*	
White-ruffed Manakin	*Corapipo altera*	
White-collared Manakin	*Manacus candei*	
Red-capped Manakin	*Ceratopipra mentalis*	

Vireos (15) — Vireonidae

Rufous-browed Peppershrike	*Cyclarhis gujanensis*	
Green Shrike-Vireo	*Vireolanius pulchellus*	
Tawny-crowned Greenlet	*Tunchiornis ochraceiceps*	
Lesser Greenlet	*Pachysylvia decurtata*	
White-eyed Vireo	*Vireo griseus*	
Mangrove Vireo	*Vireo pallens*	

Bell's Vireo	Vireo bellii	
Yellow-throated Vireo	Vireo flavifrons	
Blue-headed Vireo	Vireo solitarius	
Plumbeous Vireo	Vireo plumbeus	
Philadelphia Vireo	Vireo philadelphicus	
Warbling Vireo	Vireo gilvus	
Brown-capped Vireo	Vireo leucophrys	
Red-eyed Vireo	Vireo olivaceus	
Yellow-green Vireo	Vireo flavoviridis	

Jays (7) Corvidae

White-throated Magpie-Jay	Calocitta formosa	
Brown Jay	Psilorhinus morio	
Green Jay	Cyanocorax yncas	
Bushy-crested Jay	Cyanocorax melanocyaneus	
Steller's Jay	Cyanocitta stelleri	
Unicolored Jay	Aphelocoma unicolor	
Common Raven	Corvus corax	

Swallows (12) Hirundinidae

Purple Martin	Progne subis	
Gray-breasted Martin	Progne chalybea	
Tree Swallow	Tachycineta bicolor	
Mangrove Swallow	Tachycineta albilinea	
Violet-green Swallow	Tachycineta thalassina	
Blue-and-white Swallow	Pygochelidon cyanoleuca	
Northern Rough-winged Swallow	Stelgidopteryx serripennis	
Southern Rough-winged Swallow	Stelgidopteryx ruficollis	
Bank Swallow	Riparia riparia	
Cliff Swallow	Petrochelidon pyrrhonota	
Cave Swallow	Petrochelidon fulva	
Barn Swallow	Hirundo rustica	

Treecreepers (1) Certhiidae

Brown Creeper	Certhia americana	

Wrens (19) Troglodytidae

Rock Wren	Salpinctes obsoletus	
Nightingale Wren	Microcerculus philomela	
House Wren	Troglodytes aedon	
Rufous-browed Wren	Troglodytes rufociliatus	
Sedge Wren	Cistothorus platensis	
Carolina Wren	Thryothorus ludovicianus	
Band-backed Wren	Campylorhynchus zonatus	
Rufous-naped Wren	Campylorhynchus rufinucha	

Black-throated Wren	Pheugopedius atrogularis	
Spot-breasted Wren	Pheugopedius maculipectus	
Rufous-and-white Wren	Thryophilus rufalbus	
Banded Wren	Thryophilus pleurostictus	
Stripe-breasted Wren	Cantorchilus thoracicus	
Cabanis's Wren	Cantorchilus modestus	
Canebrake Wren	Cantorchilus zeledoni	
Bay Wren	Cantorchilus nigricapillus	
White-breasted Wood-Wren	Henicorhina leucosticta	
Gray-breasted Wood-Wren	Henicorhina leucophrys	
Song Wren	Cyphorhinus phaeocephalus	

Gnatwrens & Gnatcatchers (5)   **Polioptilidae**

Tawny-faced Gnatwren	Microbates cinereiventris	
Long-billed Gnatwren	Ramphocaenus melanurus	
Blue-gray Gnatcatcher	Polioptila caerulea	
White-lored Gnatcatcher	Polioptila albiloris	
Tropical Gnatcatcher	Polioptila plumbea	

Dippers (1)   **Cinclidae**

| American Dipper | Cinclus mexicanus | |

Thrushes & Allies (14)   **Turdidae**

Eastern Bluebird	Sialia sialis	
Slate-colored Solitaire	Myadestes unicolor	
Orange-billed Nightingale-Thrush	Catharus aurantiirostris	
Ruddy-capped Nightingale-Thrush	Catharus frantzii	
Black-headed Nightingale-Thrush	Catharus mexicanus	
Spotted Nightingale-Thrush	Catharus dryas	
Veery	Catharus fuscescens	
Gray-cheeked Thrush	Catharus minimus	
Swainson's Thrush	Catharus ustulatus	
Wood Thrush	Hylocichla mustelina	
Black Thrush	Turdus infuscatus	
Mountain Thrush	Turdus plebejus	
Clay-colored Thrush	Turdus grayi	
White-throated Thrush	Turdus assimilis	

Mockingbirds (3)   **Mimidae**

Blue-and-white Mockingbird	Melanotis hypoleucus	
Gray Catbird	Dumetella carolinensis	
Tropical Mockingbird	Mimus gilvus	

Waxwings (1)   **Bombycillidae**

| Cedar Waxwing | Bombycilla cedrorum | |

Olive Warbler (1)	Peucedramidae	
Olive Warbler	*Peucedramus taeniatus*	

Munias (1)	Estrildidae	
Tricolored Munia	*Lonchura malacca*	

Old World Sparrows (1)	Passeridae	
House Sparrow	*Passer domesticus*	

Finches, Euphonias, & Chlorophonias (11)	Fringillidae	
Blue-crowned Chlorophonia	*Chlorophonia occipitalis*	
Golden-browed Chlorophonia	*Chlorophonia callophrys*	
Scrub Euphonia	*Euphonia affinis*	
Yellow-crowned Euphonia	*Euphonia luteicapilla*	
Yellow-throated Euphonia	*Euphonia hirundinacea*	
Elegant Euphonia	*Euphonia elegantissima*	
Olive-backed Euphonia	*Euphonia gouldi*	
White-vented Euphonia	*Euphonia minuta*	
Red Crossbill	*Loxia curvirostra*	
Black-headed Siskin	*Spinus notatus*	
Lesser Goldfinch	*Spinus psaltria*	

New World Sparrows (14)	Passerellidae	
Orange-billed Sparrow	*Arremon aurantiirostris*	
Chestnut-capped Brushfinch	*Arremon brunneinucha*	
Olive Sparrow	*Arremonops rufivirgatus*	
Black-striped Sparrow	*Arremonops conirostris*	
White-naped Brushfinch	*Atlapetes albinucha*	
Rusty Sparrow	*Aimophila rufescens*	
White-eared Ground-Sparrow	*Melozone leucotis*	
Stripe-headed Sparrow	*Peucaea ruficauda*	
Botteri's Sparrow	*Peucaea botterii*	
Chipping Sparrow	*Spizella passerina*	
Savannah Sparrow	*Passerculus sandwichensis*	
Grasshopper Sparrow	*Ammodramus savannarum*	
Rufous-collared Sparrow	*Zonotrichia capensis*	
Common Chlorospingus	*Chlorospingus flavopectus*	

Yellow-breasted Chat (1)	Icteriidae	
Yellow-breasted Chat	*Icteria virens*	

Blackbirds, Orioles, & Oropendolas (23)	Icteridae	
Yellow-headed Blackbird	*Xanthocephalus xanthocephalus*	
Bobolink	*Dolichonyx oryzivorus*	

Eastern Meadowlark	*Sturnella magna*	
Red-breasted Blackbird	*Leistes militaris*	
Yellow-billed Cacique	*Amblycercus holosericeus*	
Chestnut-headed Oropendola	*Psarocolius wagleri*	
Montezuma Oropendola	*Psarocolius montezuma*	
Scarlet-rumped Cacique	*Cacicus uropygialis*	
Black-vented Oriole	*Icterus wagleri*	
Black-cowled Oriole	*Icterus prosthemelas*	
Orchard Oriole	*Icterus spurius*	
Yellow-backed Oriole	*Icterus chrysater*	
Yellow-tailed Oriole	*Icterus mesomelas*	
Streak-backed Oriole	*Icterus pustulatus*	
Spot-breasted Oriole	*Icterus pectoralis*	
Altamira Oriole	*Icterus gularis*	
Baltimore Oriole	*Icterus galbula*	
Red-winged Blackbird	*Agelaius phoeniceus*	
Bronzed Cowbird	*Molothrus aeneus*	
Giant Cowbird	*Molothrus oryzivorus*	
Melodious Blackbird	*Dives dives*	
Great-tailed Grackle	*Quiscalus mexicanus*	
Nicaraguan Grackle	*Quiscalus nicaraguensis*	

New World Warblers (48)	**Parulidae**	
Ovenbird	*Seiurus aurocapilla*	
Worm-eating Warbler	*Helmitheros vermivorum*	
Louisiana Waterthrush	*Parkesia motacilla*	
Northern Waterthrush	*Parkesia noveboracensis*	
Golden-winged Warbler	*Vermivora chrysoptera*	
Blue-winged Warbler	*Vermivora cyanoptera*	
Black-and-white Warbler	*Mniotilta varia*	
Prothonotary Warbler	*Protonotaria citrea*	
Crescent-chested Warbler	*Oreothlypis superciliosa*	
Tennessee Warbler	*Oreothlypis peregrina*	
Nashville Warbler	*Oreothlypis ruficapilla*	
Connecticut Warbler	*Oporornis agilis*	
Gray-crowned Yellowthroat	*Geothlypis poliocephala*	
MacGillivray's Warbler	*Geothlypis tolmiei*	
Mourning Warbler	*Geothlypis philadelphia*	
Kentucky Warbler	*Geothlypis formosa*	
Olive-crowned Yellowthroat	*Geothlypis semiflava*	
Common Yellowthroat	*Geothlypis trichas*	
Hooded Warbler	*Setophaga citrina*	
American Redstart	*Setophaga ruticilla*	
Cape May Warbler	*Setophaga tigrina*	
Cerulean Warbler	*Setophaga cerulea*	
Northern Parula	*Setophaga americana*	

Tropical Parula	Setophaga pitiayumi	
Magnolia Warbler	Setophaga magnolia	
Bay-breasted Warbler	Setophaga castanea	
Blackburnian Warbler	Setophaga fusca	
Yellow Warbler	Setophaga petechia	
Chestnut-sided Warbler	Setophaga pensylvanica	
Blackpoll Warbler	Setophaga striata	
Black-throated Blue Warbler	Setophaga caerulescens	
Palm Warbler	Setophaga palmarum	
Yellow-rumped Warbler	Setophaga coronata	
Yellow-throated Warbler	Setophaga dominica	
Prairie Warbler	Setophaga discolor	
Grace's Warbler	Setophaga graciae	
Townsend's Warbler	Setophaga townsendi	
Hermit Warbler	Setophaga occidentalis	
Golden-cheeked Warbler	Setophaga chrysoparia	
Black-throated Green Warbler	Setophaga virens	
Buff-rumped Warbler	Myiothlypis fulvicauda	
Fan-tailed Warbler	Basileuterus lachrymosus	
Rufous-capped Warbler	Basileuterus rufifrons	
Golden-crowned Warbler	Basileuterus culicivorus	
Canada Warbler	Cardellina canadensis	
Wilson's Warbler	Cardellina pusilla	
Painted Redstart	Myioborus pictus	
Slate-throated Redstart	Myioborus miniatus	

Mitrospingid Tanagers (1) **Mitrospingidae**

| Dusky-faced Tanager | Mitrospingus cassinii | |

Grosbeaks, Buntings, & Allies (18) **Cardinalidae**

Hepatic Tanager	Piranga flava	
Summer Tanager	Piranga rubra	
Scarlet Tanager	Piranga olivacea	
Western Tanager	Piranga ludoviciana	
Flame-colored Tanager	Piranga bidentata	
White-winged Tanager	Piranga leucoptera	
Red-crowned Ant-Tanager	Habia rubica	
Red-throated Ant-Tanager	Habia fuscicauda	
Carmiol's Tanager	Chlorothraupis carmioli	
Black-faced Grosbeak	Caryothraustes poliogaster	
Rose-breasted Grosbeak	Pheucticus ludovicianus	
Blue Seedeater	Amaurospiza concolor	
Blue-black Grosbeak	Cyanocompsa cyanoides	

Blue Bunting	Cyanocompsa parellina	
Blue Grosbeak	Passerina caerulea	
Indigo Bunting	Passerina cyanea	
Painted Bunting	Passerina ciris	
Dickcissel	Spiza americana	

Tanagers, Seedeaters, Saltators, & Allies (36)	**Thraupidae**	
Blue-gray Tanager	Thraupis episcopus	
Yellow-winged Tanager	Thraupis abbas	
Palm Tanager	Thraupis palmarum	
Golden-hooded Tanager	Tangara larvata	
Plain-colored Tanager	Tangara inornata	
Rufous-winged Tanager	Tangara lavinia	
Bay-headed Tanager	Tangara gyrola	
Grassland Yellow-Finch	Sicalis luteola	
Slaty Finch	Haplospiza rustica	
Cinnamon-bellied Flowerpiercer	Diglossa baritula	
Green Honeycreeper	Chlorophanes spiza	
Blue-black Grassquit	Volatinia jacarina	
Gray-headed Tanager	Eucometis penicillata	
White-shouldered Tanager	Tachyphonus luctuosus	
Tawny-crested Tanager	Tachyphonus delatrii	
White-lined Tanager	Tachyphonus rufus	
White-throated Shrike-Tanager	Lanio leucothorax	
Crimson-collared Tanager	Ramphocelus sanguinolentus	
Passerini's Tanager	Ramphocelus passerinii	
Shining Honeycreeper	Cyanerpes lucidus	
Red-legged Honeycreeper	Cyanerpes cyaneus	
Scarlet-thighed Dacnis	Dacnis venusta	
Blue Dacnis	Dacnis cayana	
Bananaquit	Coereba flaveola	
Yellow-faced Grassquit	Tiaris olivaceus	
Thick-billed Seed-Finch	Sporophila funerea	
Nicaraguan Seed-Finch	Sporophila nuttingi	
Variable Seedeater	Sporophila corvina	
Slate-colored Seedeater	Sporophila schistacea	
White-collared Seedeater	Sporophila torqueola	
Yellow-bellied Seedeater	Sporophila nigricollis	
Ruddy-breasted Seedeater	Sporophila minuta	
Black-headed Saltator	Saltator atriceps	
Buff-throated Saltator	Saltator maximus	
Slate-colored Grosbeak	Saltator grossus	
Grayish Saltator	Saltator coerulescens	

Bibliography

American Ornithologists' Union. 1998. Checklist of North and Middle American Birds. 7th edition and supplements.

BirdLife International. 2017. Endemic Bird Areas. Available from http://www.birdlife.org/datazone/eba. Accessed: 1 September, 2017.

Chantler, P. & G. Driessens. 1995. Swift: A Guide to the Swifts and Treeswifts of the World. Sussex.

Chavarría - Duriaux, L. & A. C. Vallely. 2017. Preliminary list of the birds of Parque Nacional Saslaya, Reserva de Biosfera Bosawás, Nicaragua. Cotinga 39: 63–77.

Chavarría, L. & G. Duriaux. 2013. Estado del Hormiguero Alifranjeado *Myrmornis torquata* en Nicaragua. Cotinga 35: 71–75.

Chavarría, L. & R. L. Batchelder. 2012. Seven new records for Nicaragua and range extensions for two additional species. Cotinga 34: 28–32.

Chesser, R. T. et al. 2017. Fifty-eighth supplement to the American Ornithological Society's Check-list of North American Birds. The Auk 134:751-773.

Clark, W. S. & B. K. Wheeler. 1995. Field identification of Common and Great Black-Hawks in Mexico and Central America. Birding, February 1995:33-37.

Clark, W. S. & B. K. Wheeler. 2001. A Field Guide to the Hawks of North America, Volume 35. Houghton Mifflin Harcourt.

Clark, W. S., H. L. Jones, C. D. Benesh & N. J. Schmitt. 2006. Field identification of the Solitary Eagle. Birding, November/December 2006:66-74.

Cleere, N. 2010. Nightjars, Potoos, Frogmouths, Oilbird and Owlet-nightjars of the World. Princeton University Press. Princeton, New Jersey.

Crossley, R., J. Liguori & B. Sullivan. 2013. The Crossley ID Guide: Raptors. Princeton University Press. Princeton, New Jersey.

Dunn, J. L. & K. L. Garret. 1997. A field guide to Warblers of North America. Houghton Mifflin Harcourt. USA.

Dunn, J. L. & J. Alderfer. 2006. National Geographic Field Guide to the Birds of North America. National Geographic, Washington, D. C.

eBird. 2017. eBird: An online database of bird distribution and abundance. eBird, Ithaca, New York. Available from http://www.ebird.org. Accessed: 1 September, 2017.

Fagan, J. & O. Komar. 2016. Petereson Field Guide to Birds of Northern Central America. Houghton Mifflin Harcourt, New York.

Ferguson-Lees, J. & D. A. Christie. 2001. Raptors of the World. Princeton University Press. Princeton, New Jersey.

Gallardo, R. J. 2014. Guide to the Birds of Honduras. Mountain Gem Tours. Honduras.

Garrigues, R & R. Dean. 2014. The Birds of Costa Rica: a field guide, 2nd ed. Cornell University Press, Ithaca, New York.

Howell, S. N. G & A. Whittaker. 1995. Field identification of Orange-breasted and Bat Falcons. Cotinga 4:36-43.

Howell, S. N. G. & S. Webb. 1994. Field Identification of Myiarchus flycatchers in Mexico. Cotinga 2:20-25.Howell, T. R. 1971. An ecological study of the birds of the lowland savanna and adjacent rainforest in northeastern Nicaragua. The Living Bird, tenth annual: 185-242.

Howell, T. 1965. New subspecies of birds from the lowland pine savanna of northwestern Nicaragua. Auk 82:438-464.

Howell, T. R. 1964. Birds collected in Nicaragua by Bernardo Ponsol. Condor 66: 151-158.

Howell, S. N. G. 2012. Petrels, Albatrosses, and Storm-Petrels of North America: A Photographic Guide. Princeton University Press. Princeton, New Jersey.

Howell, S. N. G. & J. Dunn. 2007. Gulls of the Americas. Houghton Mifflin Company. New York, New York.

Howell, S. N. G. & S. Webb. 1995. A Guide to the Birds or Mexico and Northern Central America. Oxford University Press. New York.

Hubert, W. 1932. Birds Collected in Northeastern Nicaragua. Proceedings of the Academy of Natural Science of Philadelphia. Vol. LXXXIV: 245-249.

Incer, J. 1995. Geografía Dinámica de Nicaragua. Editorial Hispamer. Managua. Nicaragua.

IUCN. 2017. The IUCN Red List of Threatened Species. Available from http://www.iucnredlist.org. Checked: 1 September, 2017.

Knapp, D. & T. Will. 2008. Birdsounds of Nicaragua. CD-mp3.

Larsson, H. & K. M. Olsen. 1995. Terns of Europe and North America. Christopher Helm Publishers. London.

Lee, C., A. Birch & T. L. Eubanks. 2008. Field identification of Western and Eastern Wood-Pewees. Birding, September/October 2008:34-40.

Macaulay Library. 2017. Macaulay Library at the Cornell Lab of Ornithology. Ithaca, New York. Available from http://www.macaulaylibrary.org. Accessed: 1 September, 2017.

Martínez-S., J. 2007. Lista Patrón de las Aves de Nicaragua. Alianza para las Areas Silvestres, Nicaragua.

Martínez-S., J. C., L. Chavarría-D. & F. J. Muñoz. 2014. A Guide to the Birds of Nicaragua: una Guía de Aves. Westarp Verlagsservicegesellschaft mbH. Germany.

Martínez-S., J. C. & T. Will, Eds. 2010. Thomas R. Howells Check-list of the Birds of Nicaragua as of 1993. Ornithological Monographs, No. 68.

Mikkola, H. 2012. Owls of the World: A Photographic Guide, 2nd edition. Firefly Books. Buffalo, New York.

Miller W. DeW. & L. Griscom. Ca. 1929. Distribution of Bird-Life in Nicaragua. Unpublished manuscript deposited in American Museum of Natural History, New York.

Neotropical Birds. 2017. Neotropical Birds: Cornell Lab of Ornithology. Available from http://neotropicalbirds.cornell.edu. Accessed: 1 September, 2017.

O'Brien, M., R. Crossley & K. Karlson. 2006. The Shorebird Guide. Houghton Mifflin Company. New York, New York.

Onley, D. & P. Scofield. 2007. Albatrosses, Petrels, & Shearwaters of the World. Princeton University Press. Princeton, New Jersey.

Ridgely, R. S. & J. A. Gwynne. 1989. A Guide to the Birds of Panama. Princeton University Press. London, UK.

Ripley, S. D. 1977. Rails of the World: A Monograph of the Family Rallidae. David R. Godine. Boston, Massachusetts.

Rowland, F. 2009. Identifying Empidonax flycatchers: the ratio approach. Birding, March 2009:30-38.

Sibley, D. A. 2000. The Sibley Guide to Birds. A. A. Knopf. New York.

Stephenson, T. & S. Whittle. 2013. The Warbler Guide. Princeton University Press. Princeton, New Jersey.

Stiles F. G. & A. F. Skutch. 1989. A Guide to the Birds of Costa Rica. Cornell University Press. New York.

Valley, A. C. & L. Chavarría-Duriaux. 2014. Notes on the birds of Parque Nacional Saslaya, Reserva de Biosfera Bosawás, Nicaragua. BOC 134: 23-29.

Xeno-Canto. 2017. Xeno-Canto: Sharing birdsounds from around the world. Available from http://xeno-canto.org. Accessed: 1 September, 2017.

About the Authors and Illustrator

Nicaraguan ornithologist **Liliana Chavarría-Duriaux** has dedicated the past 20 years to avian research and conservation. She is co-founder of El Jaguar Reserve, where she serves as ornithological research director. A certified bander, she runs two banding stations within an IBP (Institute for Bird Populations) monitoring program. Liliana has participated in numerous monitoring studies of Neotropical migrants in conjunction with IBP, Audubon NC, and the American Bird Conservancy. She annually leads expeditions to document the distribution and diversity of Nicaraguan avifauna and serves as the Nicaraguan regional eBird reviewer, two activities that have contributed greatly to the range maps in this book. Previous publications include A Guide to the Birds of Nicaragua-Una Guía de Aves (ALAS). It probably goes without saying that Liliana is an avid birder.

David C. Hille was born and raised in the Pacific Northwest of the United States. While developing an interest in ornithology as an undergraduate biology student at Northwest Nazarene University, he visited Central America for the first time in 2002 and was propelled down the path of Neotropical ornithology and conservation. His 15 years of work in Central America include most recently the study of the population ecology and conservation of parrots in Nicaragua. Before that, he and his wife Sarah were field station managers of the Quetzal Education Research Center in Costa Rica. He is currently finishing a Doctorate of Philosophy in the Ecology and Evolutionary Biology program at the University of Oklahoma. Every year, David does research and teaching in Nicaragua and Costa Rica, where he is often joined by his wife Sarah, daughter Adele, and son Henry.

Robert Dean has been studying and illustrating birds for more than twenty years. Born and raised in London, England, he was a musician and recording artist for eighteen years, living for extended periods in the United States and Australia before finally settling in Costa Rica in the early 1990s, where he re-discovered his childhood passion for art and wildlife. Dean has illustrated field guides for the birds of Costa Rica, Panama, and Aruba, Bonaire, and Curaçao (all published by Cornell University Press). He illustrated a guide to the birds of northern Central America, as well as foldouts for Rainforest Publications on the birds of Costa Rica, Nicaragua, Panama, Belize, Guatemala, Mexico, Colombia, Trinidad and Tobago, the Galapagos Islands, and Peru.

Species Index

Birds of Nicaragua / 453

Index of Spanish Common Names

Notes

Notes

Notes

Notes

Notes

Notes

Adult Raptors and Vultures in Flight

Included here are only the raptors and vultures you are likely to see in flight.

Turkey Vulture, p. 167

Black Vulture, p. 167

King Vulture, p. 167

Zone-tailed Hawk, p. 185

Short-tailed Hawk , p. 183
dark morph

Red-tailed Hawk, p. 185
dark morph *calurus* ssp.

Common Black Hawk, p. 179

Great Black Hawk, p. 179

Solitary Eagle, p. 179

Crested Caracara, p. 229

Crane Hawk, p. 185

Osprey, p. 169

Gray-headed Kite, p. 171

Black Hawk-Eagle, p. 189

Ornate Hawk-Eagle, p. 189

Plumbeous Kite, p. 173

Mississippi Kite, p. 173

Swallow-tailed Kite, p. 173

American Kestrel, p. 231

Merlin, p. 231

Peregrine Falcon, p. 231

Illustrations are not to scale.

Swainson's Hawk, p. 185
pale morph

Swainson's Hawk, p. 185
dark morph

Black-collared Hawk, p. 175

Harris's Hawk, p. 175

Savanna Hawk, p. 175

Sharp-shinned Hawk, p. 177
velox ssp.

Cooper's Hawk , p. 177

Hook-billed Kite, p. 171
female

Red-tailed Hawk, p. 185
calurus ssp.

Red-tailed Hawk, p. 185
kemsiesi ssp.

Double-toothed Kite, p. 171

Broad-winged Hawk, p. 183

Barred Hawk, p. 181

White Hawk, p. 181

White-tailed Hawk, p. 183

Short-tailed Hawk, p. 183
pale morph

Gray Hawk, p. 183

Sharp-shinned Hawk, p. 177
chinogaster ssp.

Black-and-white Hawk-Eagle,
p. 189

Quick-Find Index

The purpose of the quick-find index is to help you find the section of this field guide in which a given bird species is described. The page numbers indicate the first occurrence of a group in the field guide and any other occurrences in another family or in the Hypothetical Species section.